Transport and Chemical Transformation of Pollutants in the Troposphere

Series editors: Peter Borrell, Patricia M. Borrell, Tomislav Cvitaš, Kerry Kelly and Wolfgang Seiler

Springer-Verlag Berlin Heidelberg GmbH

Transport and Chemical Transformation
of Pollutants in the Troposphere

Volume 3

Chemical Processes
in Atmospheric Oxidation

Laboratory Studies of Chemistry
Related to Tropospheric Ozone

Georges Le Bras (Orléans)
Editor and Coordinator

LACTOZ Steering Group

Georges Le Bras	Orléans
Karl-Heinz Becker	Wuppertal
R. Anthony Cox	Cambridge
Geert K. Moortgat	Mainz
Howard W. Sidebottom	Dublin
Reinhard Zellner	Essen

Additional Contributors

Ian Barnes	Wuppertal
Richard P. Wayne	Oxford

Springer

Editor
Dr. GEORGES LE BRAS
CNRS-LCSR
1-c Av. de la Recherche Scientifique
F-45071 Orléans Cedex 2
France

With 43 Figures and 48 Tables

ISBN 978-3-642-63902-9

Library of Congress Cataloging-in-Publication Data. Chemical processes in atmospheric oxidation: laboratory studies of chemistry related to tropospheric ozone / Georges le Bras, editor and coordinator. p. cm. – (Transport and chemical transformation of pollutants in the troposphere; v. 3) Includes index.
ISBN 978-3-642-63902-9   ISBN 978-3-642-59216-4 (eBook)
DOI 10.1007/978-3-642-59216-4
1. Atmospheric ozone. 2. Troposphere. 3. Atmospheric chemistry. 4. Photochemical oxidants. I. Le Bras, Georges. II. Series. QC879.7.C53 1997, 551.5'11–dc20, 96-41132 CIP

© Springer-Verlag Berlin Heidelberg 1997
Softcover reprint of the hardcover 1st edition 1997
Originally published by Springer-Verlag Berlin Heidelberg New York in 1997
The use of general descriptive names, registered names, trademarks, etc. in this publication does not imply, even in the absence of a specific statement, that such names are exempt from the relevant protective laws and regulations and therefore free for general use.

Cover Design: Struve & Partner, Heidelberg

Coverpicture from Georges Le Bras (LACTOZ)

SPIN 10514750   30/3136-5 4 3 2 1 0 – Printed on acid-free paper

Transport and Chemical Transformation of Pollutants in the Troposphere

Series editors: Peter Borrell, Patricia M. Borrell, Tomislav Cvitaš, Kerry Kelly
and Wolfgang Seiler

Foreword by the Series Editors

EUROTRAC is the European co-ordinated research project, within the EUREKA initiative, studying the transport and chemical transformation of pollutants in the troposphere. The project has achieved a remarkable scientific success since its start in 1988, contributing substantially both to the scientific progress in this field and to the improvement of the scientific basis for environmental management in Europe. EUROTRAC, which at its peak comprised some 250 research groups organised into 14 subprojects, brought together international groups of scientists to work on problems directly related to the transport and chemical transformation of trace substances in the troposphere. In doing so, it helped to harness the resources of the participating countries to gain a better understanding of the trans-boundary, interdisciplinary environmental problems which beset us in Europe.

The scientific results of EUROTRAC are summarised in this report which consists of ten volumes.

Volume 1 provides a general overview of the scientific results, prepared by the Scientific Steering Committee (SSC) and the International Scientific Secretariat (ISS) of EUROTRAC, together with brief summaries of the work of the fourteen individual subprojects prepared by the respective subproject coordinators.

Volumes 2 to 9 comprise detailed overviews of the subproject achievements, each prepared by the respective subproject coordinator and steering group, together with summaries of the work of the participating research groups prepared by the principal investigators. Each volume also includes a full list of the scientific publications from the subproject.

The final volume, 10, is the complete report of the Application Project, which was set up in 1993 to assimilate the scientific results from EUROTRAC and present them in a condensed form so that they are suitable for use by those responsible for environmental planning and management in Europe. It illustrates how a scientific project such as EUROTRAC can contribute practically to providing the scientific consensus necessary for the development of a coherent atmospheric environmental policy for Europe.

A multi-volume work such as this has many contributors and we, as general editors, would like to express our thanks to all of them: to the subproject co-ordinators who have borne the brunt of the scientific co-ordination and who have contributed so much to the success of the project and the quality of this report; to the principal investigators who have carried out so much high-quality scientific work; to the members of the International Executive Committee (IEC) and the SSC for their enthusiastic encouragement and support of EUROTRAC; to the participating governments in EUROTRAC, and in particular the German Government (BMBF) for funding, not only the research, but also the ISS publication activities; and finally to Mr. Christian Witschell and his colleagues at Springer Verlag for providing the opportunity to publish the results in a way which will bring them to the notice of a large audience.

Peter Borrell (Scientific Secretary, ISS)   EUROTRAC ISS
Patricia May Borrell   Fraunhofer Institute (IFU)
Tomislav Cvitaš   Garmisch-Partenkirchen
Kerry Kelly
Wolfgang Seiler (Director, ISS)

Preface by the LACTOZ steering group

The LACTOZ project has been co-ordinated successively by R. Anthony Cox, Karl-Heinz Becker and Georges Le Bras. This report has been prepared by the Steering Committee of LACTOZ with contributions from I. Barnes and R.P. Wayne. The support of the European Commission for co-ordination of the project and the personal involvement of Heinz Ott and Giovanni Angeletti are greatfully acknowledged.

In Chapter 2 of the report describing the scientific results, when the work of an author, whose individual contribution can be found in a later chapter of this book, is mentioned, the appropriate chapter and section numbers have been provided. However when an author's name appears with a year it refers to a EUROTRAC final report.

Table of Contents

Transport and Chemical Transformation of
Pollutants in the Troposphere, **volume 3**.

Chemical Processes in Atmospheric Oxidation

Authors' names and addresses

Dr. Ian Barnes
Physikalische Chemie
Bergische Universität Wuppertal
FB 9
Gaußstr. 20
D-42119 Wuppertal

Dr. D. Bauer
Max Planck Institut für Chemie
Abt. Luftchemie
Postfach 3060
D-55020 Mainz

Dr. Karl-Heinz Becker
Bergische Universität Wuppertal
Physikalische Chemie, Fachbereich 9
Gaußstr. 20
D-42119 Wuppertal

Dr. Thomas Behmann
Universität Bremen
Fachbereich Physik
Institut für Fernerkundung
Postfach 33 04 40
D-28334 Bremen

Dr. H.-J. Benkelberg
Max-Planck-Institut für Chemie
Abt. Biogeochemie
Postfach 3060
D-55020 Mainz

Dr. A. Bierbach
Physikalische Chemie
Bergische Universität Wuppertal
FB 9
Gaußstr. 20
D-42119 Wuppertal

Dr. B. Bigand
Université des Sciences et Techniques de
Lille-Flandres-Artois
Lab. de Cinétique et Chimie de
la Combustion (LC3)
Bâtiment C 11
F-59655 Villeneuve D'Ascq Cedex

Dr. Peter Biggs
Oxford University
Physical Chemistry Laboratory
South Parks Road
GB-Oxford OX1 3QZ

Dr. Birger Bohn
Fraunhofer-Institut
I T A
Physikalische Chemie
Nikolai-Fuchs-Str.1
D-30625 Hannover

Dr. Werner Boullart
University of Leuven
Lab. of Analytical and Inorganic chemistry
Celestijnenlaan 200F
B-3001 Heverlee-Leuven

Dr. C. Bourbon
Université des Sciences et Techniques de
Lille-Flandres-Artois
Lab. de Cinétique et Chimie de
la Combustion (LC3)
Bâtiment C 11
F-59655 Villeneuve D'Ascq Cedex

Dr. Andrew Boyd
Laboratoire de Photophysique et
Photochimie Moleculaire
Université de Bordeaux I
F-33405 Talence Cedex

Dr. I. Bridier
Laboratoire de Photophysique et
Photochimie Moleculaire
Université de Bordeaux I
F-33405 Talence Cedex

Dr. Klaus J. Brockmann
Bergische Universität Wuppertal
Physikalische Chemie, Fachbereich 9
Gaußstr. 20
D-42119 Wuppertal

Dr. J.P. Burrows
Universität Bremen
Fachbereich Physik
Institut für Fernerkundung
Postfach 33 04 40
D-28334 Bremen

Dr. N. Butkovskaya
Centre de Recherches sur la Chimie de la
Combination et des Hautes Temperatures
C.N.R.S.
1-c Avenue de la Recherche Scientifique
F-45045 Orléans Cedex

Prof. Dr. K. Bächmann
Technische Hochschule Darmstadt
Fachbereich Chemie
Institut für Anorganische Chemie
Petersenstraße 18
D-64287 Darmstadt

Dr. Carlos E. Canosa-Mas
Oxford University
Physical Chemistry Laboratory
South Parks Road
GB-Oxford OX1 3QZ

Dr. F. Capellani
Joint Research Center
ISPRA Establishment
I-21020 Ispra (Varese)

Dr. F. Caralp
Laboratoire de Photophysique et
Photochimie Moleculaire
Université de Bordeaux I
F-33405 Talence Cedex

Dr. V. Catoire
Laboratoire de Photophysique et
Photochimie Moleculaire
Université de Bordeaux I
F-33405 Talence Cedex

Dr. A. Chakir
Laboratoire de Chimie Physique
Structure Moleculaire et Spectroscopic
Faculté des Sciences
Moulin de la Housse
B.P. 347
F-51062 Reims Cedex

Dr. B. Coquart
Laboratoire de Chimie Physique
Structure Moleculaire et Spectroscopic
Faculté des Sciences
Moulin de la Housse
B.P. 347
F-51062 Reims Cedex

Dr. R. Anthony Cox
Centre for Atmospheric Science
Department of Chemistry
University of Cambridge
Lensfield Road
GB-Cambridge CB2 1EP

Dr. John Crowley
MPI für Chemie
Abt. Luftchemie
Saarstraße 23
D-55122 Mainz

Dr. V. Daele
Centre de Recherches sur la Chimie de la
Combination et des Hautes Temperatures
C.N.R.S.
1-c Avenue de la Recherche Scientifique
F-45045 Orléans Cedex

Dr. D. Daumont
Laboratoire de Chimie Physique
Structure Moleculaire et Spectroscopic
Faculté des Sciences
Moulin de la Housse
B.P. 347
F-51062 Reims Cedex

Dr. P. Devolder
Université des Sciences et Techniques de
Lille-Flandres-Artois
Lab. de Cinétique et Chimie de
la Combustion (LC3)
Bâtiment C 11
F-59655 Villeneuve D'Ascq Cedex

Dr. J. Eberhard
EAWAG
CH-8600 Dübendorf-Zurich

Dr. L. Elmaimouni
Université des Sciences et Techniques de
Lille-Flandres-Artois
Lab. de Cinétique et Chimie de
la Combustion (LC3)
Bâtiment C 11
F-59655 Villeneuve D'Ascq Cedex

Dr. Frederick F. Fenter
Ecole Polytechnique Fédérale de
        Lausanne (EPFL)
Lab. de Pollution atmosphérique et sol
CH-1015 Lausanne

Dr. C. Fittschen
Université des Sciences et Techniques de
Lille-Flandres-Artois
Lab. de Cinétique et Chimie de
la Combustion (LC3)
Bâtiment C 11
F-59655 Villeneuve D'Ascq Cedex

Dr. A. Goumri
Université des Sciences et Techniques de
Lille-Flandres-Artois
Lab. de Cinétique et Chimie de
la Combustion (LC3)
Bâtiment C 11
F-59655 Villeneuve D'Ascq Cedex

Dr. S. Gäb
Fachbereich 9/Analytik
BUGH Wuppertal
Gaußstraße 20
D-42097 Wuppertal

Dr. M. Hallqvist
University of Göteborg
Department of Inorganic Chemistry
S-412 96 Göteborg

Dr. M. Hartmann
Technische Hochschule Darmstadt
Fachbereich Chemie
Institut für Anorganische Chemie
Petersenstraße 18
D-64287 Darmstadt

Dr. J. Hauptmann
Technische Hochschule Darmstadt
Fachbereich Chemie
Institut für Anorganische Chemie
Petersenstraße 18
D-64287 Darmstadt

Dr. G.D. Hayman
National Environmental Technology Centre
AEA Technology
E5, Culham
GB-Abingdon, Oxon OX14 3DB

Dr. A.E. Heard
University of Leeds
School of Chemistry
GB-Leeds LS2 9JT

Dr. Adolphe Heiss
CNRS - URA 879
Laboratoire de Mécanique Physique
Université Paris 6
2 Place de la Gare de Ceinture
F-78210 Saint-Cyr l'Ecole

Dr. Frank Helleis
Max Planck Institut für Chemie
Abt. Luftchemie
Postfach 3060
D-55020 Mainz

Dr. Jens Hjorth
CEC Joint Research Centre
Environmental Institute
Bldg. 27b
I-21027 Ispra (VA)

Dr. Jan van Hoeymissen
Katholiske Universiteit Leuven
Department Scheikunde
Celestijnenlaan 200 F
B-3001 Heverlee-Leuven

Dr. Axel Hoffmann
Institut f. Physikalische Chemie, FB 8
Universität - GH - Essen
Universitätsstraße
D-45117 Essen

Dr. Osamu Horie
Max-Planck-Institut für Chemie
Abtlg. Luftchemie
Saarstr. 23
D-55122 Mainz

Dr. B. Jean
Laboratoire de Chimie Physique
Structure Moleculaire et Spectroscopic
Faculté des Sciences
Moulin de la Housse
B.P. 347
F-51062 Reims Cedex

Dr. M.E. Jenkin
National Environmental Technology
Centre
AEA Technology
E5, Culham
GB-Abingdon, Oxon OX14 3DB

Dr. Alain Jenouvrier
Laboratoire de Chimie Physique
Structure Moleculaire et Spectroscopic
Faculté des Sciences
Moulin de la Housse
B.P. 347
F-51062 Reims Cedex

Dr. Niels R. Jensen
JRC. Env. Institute TP272
I-Ispra(Varese)

Dr. D. Johnstone
Centre de Recherches sur la Chimie de la
Combination et des Hautes Temperatures
C.N.R.S.
1-c Avenue de la Recherche Scientifique
F-45045 Orléans Cedex

Dr. R. Karlsson
University of Göteborg
Department of Inorganic Chemistry
S-412 96 Göteborg

Prof. J. Alistair Kerr
EAWAG
CH-8600 Dübendorf-Zurich

Dr. F. Kirchner
Bergische Universität Wuppertal
Physikalische Chemie, Fachbereich 9
Gaußstr. 20
D-42119 Wuppertal

Mr. Björn Klotz
Physikalische Chemie
Bergische Universität Wuppertal
FB 9
Gaußstr. 20
D-42119 Wuppertal

Dr. R. Knispel
Fraunhofer-Institut
I T A
Physikalische Chemie
Nikolai-Fuchs-Str.1
D-30625 Hannover

Dr. S. Koch
Max Planck Institut für Chemie
Abt. Luftchemie
Postfach 3060
D-55020 Mainz

Dr. Rainald Koch
Fraunhofer-Institut
I T A
Physikalische Chemie
Nikolai-Fuchs-Str.1
D-30625 Hannover

Dr. P. Komnick
Ruhr-Universität Bochum
Fakultät für Chemie
Lehrstuhl für Organische Chemie II
Gebäude NC
Universitätsstraße 150
D-44780 Bochum

Dr. B. Kusserow
Technische Hochschule Darmstadt
Fachbereich Chemie
Institut für Anorganische Chemie
Petersenstraße 18
D-64287 Darmstadt

Dr. I.T. Lancar
Centre de Recherches sur la Chimie de la
Combination et des Hautes Temperatures
C.N.R.S.
1-c Avenue de la Recherche Scientifique
F-45045 Orléans Cedex

Dr. S. Langer
University of Göteborg
Department of Inorganic Chemistry
S-412 96 Göteborg

Dr. I. Langhans
University of Leuven
Lab. of Analytical and
  Inorganic chemistry
Celestijnenlaan 200F
B-3001 Heverlee-Leuven

Dr. A. Lastätter-Weißenmayer
Universität Bremen
Fachbereich Physik
Institut für Fernerkundung
Postfach 33 04 40
D-28334 Bremen

Dr. G. Laverdet
Centre de Recherche sur la Chimie de la
Combustion et des Hautes Temperatures
C.N.R.S.-CRCHHT
1-c Avenue de la Recherche Scientifique
F-45045 Orléans Cedex 2

Dr. Georges Le Bras
CNRS - LCSR
1-c Avenue de la Recherche Scientifique
F-45071 Orléans Cedex 2

Dr. Robert Lesclaux
Laboratoire de Photophysique et
Photochimie Moleculaire
Université de Bordeaux I
F-33405 Talence Cedex

Dr. H.G. Libuda
Physikalische Chemie
Bergische Universität Wuppertal
FB 9
Gaußstr. 20
D-42119 Wuppertal

Dr. P.D. Lightfoot
Laboratoire de Photophysique et
Photochimie Moleculaire
Université de Bordeaux I
F-33405 Talence Cedex

Dr. S. Limbach
Max Planck Institut für Chemie
Abt. Luftchemie
Postfach 3060
D-55020 Mainz

Dr. E. Ljungström
University of Göteborg
Department of Inorganic Chemistry
S-412 96 Göteborg

Dr. J. Malicet
Université Reims
Faculté des Sciences
Lab. de Chemie Physique
Moulin de la Housse
B.P. 347
F-51062 Reims Cedex

Dr. Wolfgang Malms
Universität Essen
Institut für Physikalische Chemie
Universitätsstraße 5-7
D-45117 Essen

Dr. D. Maric
Universität Bremen
Fachbereich Physik
Institut für Fernerkundung
Postfach 33 04 40
D-28334 Bremen

Ms. Andrea Mayer-Figge
Physikalische Chemie
Bergische Universität Wuppertal
FB 9
Gaußstr. 20
D-42119 Wuppertal

Dr. A. Mellouki
Centre de Recherche sur la Chimie de la
Combustion et des Hautes Temperatures
C.N.R.S.-CRCHHT
1-c Avenue de la Recherche Scientifique
F-45045 Orléans Cedex 2

Dr. M.F. Merienne
Université Reims
Faculté des Sciences
Lab. de Chemie Physique
Moulin de la Housse
B.P. 347
F-51062 Reims Cedex

Dr. Geert K. Moortgat
Max Planck Institut für Chemie
Abt. Luftchemie
Postfach 3060
D-55020 Mainz

Dr. T.P. Murrells
National Environmental Technology
 Centre
AEA Technology
E5, Culham
GB-Abingdon, Oxon OX14 3DB

Dr. C. Müller
EAWAG
CH-8600 Dübendorf-Zurich

Dr. S. Mönninghoff
Physikalische Chemie
Bergische Universität Wuppertal
FB 9
Gaußstr. 20
D-42119 Wuppertal

Dr. V. Mörs
Universität Essen
Institut für Physikalische Chemie
Universitätsstraße 5-7
D-45117 Essen

Mr. Peter Neeb
Max Planck Institut für Chemie
Abt. Luftchemie
Postfach 3060
D-55020 Mainz

Dr. Ole John Nielsen
Ford Forschungszentrum Aachen
Dennewartstraße 25
D-52068 Aachen

Dr. B. Nozière
Laboratoire de Photophysique et
Photochimie Moleculaire
Université de Bordeaux I
F-33405 Talence Cedex

Prof. Jozef Peeters
University of Leuven
Lab. of Analytical and Inorganic chemistry
Celestijnenlaan 200F
B-3001 Heverlee-Leuven

Dr. Dieter Perner
Max-Planck-Institut für Chemie
Abteilung Luftchemie
Postfach 3060
D-55020 Mainz

Prof. Michael Pilling
University of Leeds
School of Chemistry
GB-Leeds LS2 9JT

Dr. J. Polzer
Technische Hochschule Darmstadt
Fachbereich Chemie
Institut für Anorganische Chemie
Petersenstraße 18
D-64287 Darmstadt

Dr. Gilles Poulet
Centre de Recherches sur la Chimie de la
Combination et des Hautes Temperatures
C.N.R.S.
1-c Avenue de la Recherche Scientifique
F-45045 Orléans Cedex

Mr. Veerle Pultau
University of Leuven
Lab. of Analytical and Inorganic chemistry
Celestijnenlaan 200F
B-3001 Heverlee-Leuven

Dr. W. Raber
Max Planck Institut für Chemie
Abt. Luftchemie
Postfach 3060
D-55020 Mainz

Dr. A. Ray
Centre de Recherches sur la Chimie de la
Combination et des Hautes Temperatures
C.N.R.S.
1-c Avenue de la Recherche Scientifique
F-45045 Orléans Cedex

Dr. Giambattista Restelli
c/o Prof. F. Cariatintre
Inst.of Inorganic & Analytical Chemistry
University of Milano
Via Venezian 21
I-20133 Milano

Dr. D.M. Rowley
Laboratoire de Photophysique et
Photochimie Moleculaire
Université de Bordeaux I
F-33405 Talence Cedex

Dr. L. Ruppert
Physikalische Chemie
Bergische Universität Wuppertal
FB 9
Gaußstr. 20
D-42119 Wuppertal

Dr. Krikor Sahetchian
CNRS - URA 879
Laboratoire de Mécanique Physique
Université Paris 6
2 Place de la Gare de Ceinture
F-78210 Saint-Cyr l'Ecole

Prof. Wolfram Sander
Ruhr-Universität Bochum
Fakultät für Chemie
Lehrstuhl für Organische Chemie II
Gebäude NC
Universitätsstraße 150
D-44780 Bochum

Mr. Frank Sauer
Max-Planck-Institut für Chemie
Abtlg. Luftchemie
Saarstr. 23
55122 Mainz

Dr. S.M. Saunders
University of Leeds
School of Chemistry
GB-Leeds LS2 9JT

Dr. Jean-Pierre Sawerysyn
Laboratoire de Cinétique et
Chemie de la Combustion
Université des Sciences et Technologies
de Lille Flandres Artois
Bâtiment C11
F-59655 Villeneuve d'Ascq Cedex

Prof. Dr. R.N. Schindler
Universität Kiel
Institut für Physikalische Chemie
Olshausenstr. 40
D-24098 Kiel

Dr. W. Schneider
Max Planck Institut für Chemie
Abt. Luftchemie
Postfach 3060
D-55020 Mainz

Mr. Christian Schäfer
Max-Planck-Institut für Chemie
Abtlg. Luftchemie
Saarstr. 23
D-55122 Mainz

Dr. J. Sehested
Riso National Laboratory
Chemical Department
Postbox 49
DK-4000 Roskilde

Dr. M. Semadeni
EAWAG
CH-8600 Dübendorf-Zurich
Schweiz

Dr. R. Seuwen
Max-Planck-Institut für Chemie
Abt. Biogeochemie
Postfach 3060
D-55020 Mainz

Dr. Howard W. Sidebottom
University College Dublin
Department of Chemistry
Belfield
IRL-Dublin 4

Dr. M. Siese
Fraunhofer-Institut
I T A
Physikalische Chemie
Nikolai-Fuchs-Str.1
D-30625 Hannover

Dr. F. Simon
Max Planck Institut für Chemie
Abt. Luftchemie
Postfach 3060
D-55020 Mainz

Dr. Henrik Skov
National Environmental
Research Institute
Division of Emissions and
Air Pollution
Frederiksborgvej 399
DK-4000 Roskilde

Dr. P.I. Smurthwaite
University of Leeds
School of Chemistry
GB-Leeds LS2 9JT

Dr. D.W. Stocker
EAWAG
CH-8600 Dübendorf-Zurich

Dr. W. Thomas
Physikalische Chemie
Bergische Universität Wuppertal
FB 9
Gaußstr. 20
D-42119 Wuppertal

Dr. Jack Tracy
Department of Chemistry
Dublin Institute of Technology
IRL-Dublin

Dr. M. Träubel
Ruhr-Universität Bochum
Fakultät für Chemie
Lehrstuhl für Organische Chemie II
Gebäude NC
Universitätsstraße 150
D-44780 Bochum

Dr. W.V. Turner
Fachbereich 9/Analytik
BUGH Wuppertal
Gaußstraße 20
D-42097 Wuppertal

Dr. S. Téton
Centre de Recherches sur la Chimie de la
Combination et des Hautes Temperatures
C.N.R.S.
1-c Avenue de la Recherche Scientifique
F-45045 Orléans Cedex

Dr. I. Vassali
Centre de Recherches sur la Chimie de la
Combination et des Hautes Temperatures
C.N.R.S.
1-c Avenue de la Recherche Scientifique
F-45045 Orléans Cedex

Dr. J. Vertommen
University of Leuven
Lab. of Analytical and
  Inorganic chemistry
Celestijnenlaan 200F
B-3001 Heverlee-Leuven

Dr. B. Veyret
Université de Bordeaux I
Laboratoire de Photophysique et
  Photochemie Moléculaire
F-33405 Talence Cedex

Mr. Eric Villenave
Laboratoire Photophysique Photochimie
Moléculaire
Université Bordeaux I - CNRS URA 348
351 Cours de la Libération
F-33405 Talence Cedex

Prof. Dr. Peter Warneck
Max-Planck-Institut für Chemie
Abt. Biogeochemie
Postfach 3060
D-55020 Mainz

Dr. Richard P. Wayne
Oxford University
Physical Chemistry Laboratory
South Parks Road
GB-Oxford OX1 3QZ

Dr. Michael Weißenmayer
Universität Bremen
Fachbereich Physik
Institut für Fernerkundung
Postfach 33 04 40
D-28334 Bremen

Ms. Evelyn Wiesen
Bergische Universität GH Wuppertal
Gaußstraße 20
D-42119 Wuppertal

Dr. K. Wirtz
Bergische Universität Wuppertal
Physikalische Chemie, Fachbereich 9
Gaußstr. 20
D-42119 Wuppertal

Dr. F. Witte
Fraunhofer-Institut
I T A
Physikalische Chemie
Nikolai-Fuchs-Str.1
D-30625 Hannover

Ms. Silke Wolff
Fachbereich 9/Analytik
BUGH Wuppertal
Gaußstraße 20
D-42097 Wuppertal

Dr. I. Wängberg
University of Göteborg
Department of Inorganic Chemistry
S-412 96 Göteborg

Dr. Friedhelm Zabel
Bergische Universität Wuppertal
GH Wuppertal/FB 9
Gaußstr. 20
D-42097 Wuppertal

Dr. C. Zahn
Max Planck Institut für Chemie
Abt. Luftchemie
Postfach 3060
D-55020 Mainz

Dr. Reinhard Zellner
Universität Essen
Institut für Physikalische Chemie
Universitätsstraße 5-7
D-45117 Essen

Prof. Dr. Cornelius Zetzsch
Fraunhofer-Institut
I T A
Physikalische Chemie
Nikolai-Fuchs-Str.1
D-30625 Hannover

Chapter 1

1.1 Summary

The EUROTRAC project has addressed the problem of transport and transformation of photo-oxidants in north-western Europe, including global-, regional- and local-scale effects occurring in the atmospheric boundary layer and free troposphere. Tropospheric ozone is also important for the climate because O_3 is an active greenhouse gas. LACTOZ (LAboratory studies of Chemistry related to Tropospheric Ozone), one of two EUROTRAC laboratory subprojects, was begun with the aim of providing, through laboratory investigations of the kinetics and mechanisms of relevant gas-phase reactions, a quantitative description of the chemical production and loss of ozone from its precursor molecules – nitrogen oxides and volatile organic compounds – in the atmospheric boundary layer and free troposphere.

The results of the project have greatly improved our knowledge about the following identified problems corresponding to the specific objectives defined at the start of LACTOZ:

-   the kinetics and mechanisms for the oxidative breakdown of simple organic compounds present in the free troposphere;
-   the kinetics and mechanisms of the reactions involved in the transport and interconversion of NO_x into and within the free troposphere;
-   the production and loss of odd hydrogen radicals required for accurate calculation of tropospheric free radical concentrations;
-   the determination of the rates and mechanisms for OH attack on more complex organic compounds, including aromatic, oxygenated and nitrate compounds;
-   the determination of the rate constants and branching ratios for reactions of peroxy and alkoxy radicals, derived from higher-molecular-weight organic compounds including isoprene, naturally emitted;
-   the elucidation of NO_3 chemistry with a view to modelling the night-time removal of NO_x and VOCs from the atmosphere and the impact of the overall oxidation of VOCs;

-   the investigation of reactions acting as sources and sinks of HO_x and RO_2 radicals in the boundary layer in support of anticipated field measurements of free radical concentrations.

The results of LACTOZ have provided an extended kinetic data base for the following classes of reactions: reactions of OH with VOCs, reactions of NO_3 with VOCs and peroxy radicals, reactions of O_3 with alkenes, reactions of peroxy radicals (self reactions, reaction with HO_2, other RO_2, NO, NO_2), reactions of alkoxy radicals (reactions with O_2, decomposition, isomerisation), thermal decomposition of peroxynitrates. Photolysis parameters (absorption cross-section, quantum yields) have been refined or obtained for the first time for species which photolyse in the troposphere. Significantly new mechanistic information has also been obtained for the oxidation of aromatic compounds and biogenic compounds (especially isoprene). These different data allow the rates of the processes involved to be modelled, especially the ozone production from the oxidation of hydrocarbons. The data from LACTOZ are summarised in the tables given in this report and have been used in evaluations of chemical data for atmospheric chemistry conducted by international evaluation groups of NASA and IUPAC.

The data base has been used to establish structure-reactivity relationships and to provide quantitative rate coefficients and reliable reaction pathways for reactions for which no experimental data are available or have proved to be experimentally inaccessible. Such a strategy was adopted in view of the very large number of VOCs involved in the generation of tropospheric ozone. These structure-reactivity relationships are discussed in this report and in other publications.

As an application of LACTOZ, a concerted effort has been made during the last phase of LACTOZ to incorporate the data obtained into models describing and predicting ozone and photo-oxidant formation. As an example, the input data of the EMEP mechanism have been updated and extended by LACTOZ.

1.2   Original Objectives and LACTOZ Achievements

1.2.1   Aims and objectives

The aims of LACTOZ have been to provide, through laboratory investigation of the kinetics and mechanisms of relevant gas-phase reactions, a quantitative description of the chemical production and loss of ozone from its precursor molecules, nitrogen oxides and volatile organic compounds, in the regions of the troposphere.

The first area of focus was on the reactions involved in the photochemically initiated oxidative degradation of CO and simpler hydrocarbons (CH_4, C_2 - C_5 alkanes, C_2 and C_3 unsaturated hydrocarbons, *etc.*) and the chemistry of NO_x, relevant for ozone production in the global troposphere. The objective was to provide rate coefficients and reaction mechanisms for a complete model description of the chemistry of NO_x and low molecular weight VOCs in the free troposphere and in air parcels transporting ozone precursors away from source regions.

The second area of focus was placed on the oxidative degradation of more complex volatile organic compounds (higher alkanes and alkenes; benzene, toluene and higher aromatic compounds; isoprene and other biogenic compounds; and oxygenated VOCs) which are important for ozone generation in the atmospheric boundary layer, closer to the sources of precursors. The chemistry of VOC oxidation initiated by the nitrate radical and by the reaction with ozone was also investigated, in addition to the photochemical oxidation initiated by OH. The reactions involved in the change in nitrogen speciation resulting from the formation of organic nitrates by the heavier VOCs were also studied.

Within the context of the above project foci, the following specific objectives were defined for LACTOZ

- to define quantitatively the kinetics and mechanisms for the elementary reactions involved in ozone production in the free troposphere, in particular:
  - the oxidative breakdown of simple organic compounds present in the free troposphere (CH_4, HCHO, C_2-C_5 alkanes, *etc.*);
  - the transport and interconversion of NO_x into and within the free troposphere;
  - the production and loss of odd hydrogen radicals required for accurate calculation of tropospheric free radical concentrations;

- to provide kinetic and mechanistic data necessary for the formulation of models describing the production of ozone in the polluted boundary layer including:
  - the determination of the rates and mechanisms for OH attack on more complex organic compounds including oxygenated species and nitrates;
  - the determination of the rate constants and branching ratios for reactions of peroxy and alkoxy radicals, derived from higher molecular weight organic compounds including natural hydrocarbons, isoprene and terpenes;
  - the elucidation of night-time NO_3 chemistry with a view to modelling the removal of NO_x and VOCs from the atmosphere and the impact on the overall oxidation of VOCs;
  - the investigation of reactions acting as sources and sinks of HO_x (OH, HO_2) and RO_2 radicals in the boundary layer in support of anticipated field measurements of free radical concentrations.

In view of the very large number of VOCs involved in the generation of tropospheric ozone it was recognised that it will not be feasible to investigate the elementary steps in the oxidation of every VOC. Therefore, a strategy based on the establishment of structure-reactivity relationships was adopted to provide quantitative rate coefficients and reliable reaction pathways for the diverse reactions of the many individual species.

The sequence of chemical reactions leading to oxidative degradation and ozone formation in the atmosphere is often determined by the relative rates of competing reaction pathways at several critical points in the degradation. The protocol adopted was therefore to establish these critical points and emphasis was then placed on provision of data allowing the relative importance of these competing processes to be quantitatively defined for a range of VOCs. These critical points were found to lie for example in the reactions of RO_2 and RO radicals.

1.2.2 Achievements

The high level of LACTOZ achievements can be best illustrated by comparing the results of the project as summarised in Chapter 2 of this report with the specific objectives, as defined in Section 1.2.1.

1.2.2a The kinetics and mechanisms for the oxidative breakdown of simple organic compounds present in the free troposphere

Hydroxyl radical attack on VOCs is the primary source of RO_2 radicals in the troposphere, and consequently the rates of the OH + VOC reactions have a strong influence on the local rate of O_3 formation (Fig. 1). Moreover, the mechanism of the OH reaction determines the structure of the peroxy radical formed, and hence, provides information on its subsequent degradation pathway.

The rates and mechanisms for OH attack on simple organic compounds have been established for sometime, and any uncertainty in their atmospheric lifetimes largely resides in the value of the concentration of the OH radical in the atmosphere. In LACTOZ, effort has been concentrated on oxidation mechanisms, particularly on the reactions of RO_2 and RO radicals, for understanding the degradation of small organic compounds.

The rate of ozone generation in a particular region of the troposphere is determined by the perturbation of the local photostationary state of NO_x due to oxidation of NO to NO_2, primarily by peroxy radicals, RO_2, in the reaction:

$$RO_2 + NO \quad \rightarrow \quad RO + NO_2 \qquad (R = H, \text{organic radical})$$

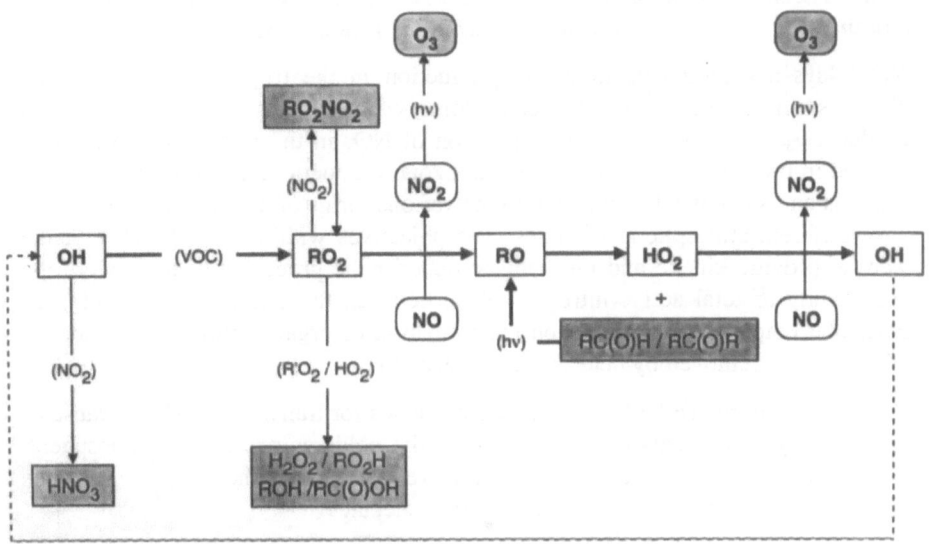

Fig. 1: VOC oxidation mechanism during daytime.

The importance of this reaction depends on the rates of competing reactions of RO_2, in particular, reaction with NO_2 to form peroxynitrates, with HO_2 to form hydroperoxides and the reaction with other RO_2, forming a variety of products:

$$RO_2 + NO_2 + M \rightarrow RO_2NO_2 + M$$
$$RO_2 + HO_2 \rightarrow ROOH + O_2$$
$$RO_2 + RO_2 \rightarrow products$$

In the sunlit free troposphere, competition between $RO_2 + HO_2$ and $RO_2 + NO$ is the key to the effective local ozone production, and the work in LACTOZ has led to the establishment of the rate parameters for these two reactions for a variety of organic peroxy radicals. The kinetic data base created is sufficiently broad to enable structure-reactivity criteria to be applied to peroxy radicals for which there is as yet no experimental data. The change in the rate coefficients at 298 K for these two competing processes as a function of the organic group tends to show opposite trends. It is therefore important to take reactivity changes into account when the propensity for ozone formation is assessed for a given type of VOC under atmospheric regimes in which the NO and HO_2 reactions are competing.

A complete quantitative mechanism for the atmospheric oxidation of C_1 and C_2 hydrocarbons was assembled and published as an interim product of LACTOZ.

1.2.2b The kinetics and mechanisms of the reactions involved in the transport and interconversion of NOₓ into and within the free troposphere

NO_x plays a central role in ozone production in the troposphere through the photodissociation of NO_2 to produce O atoms, which subsequently react with O_2 to produce O_3. It follows that the distribution of NO_x in the troposphere has to be defined in order to assess photochemical ozone production. The transport of NO_x depends on its chemical conversion to the various forms of nitrogen species, which have different atmospheric lifetimes. The objectives within this LACTOZ theme were to provide kinetic and mechanistic data for the processes which govern the partitioning of total active nitrogen (NO_y) between the different forms (Fig. 2). Particular emphasis was placed on the formation of organic nitrogen compounds which are not removed by heterogeneous reactions.

Peroxyacetylnitrate (PAN) is an important species for transport of NO_y because of its relatively low reactivity, particularly in the cold regions of the troposphere. Work in LACTOZ provided the first definitive rate data for the formation from its precursor species and decomposition of this molecule:

$$CH_3C(O)O_2 + NO_2 + M \quad \leftrightarrow \quad CH_3C(O)O_2NO_2 + M$$

Data have also been obtained for various other peroxyacylnitrate species formed from a number of different VOCs. Structure-reactivity relationships for the thermal stability of PAN type species have been established which provide the needed information for assessing the transport of NO_y in this form, and also the relationships required in order to estimate the relative rates of the reactions of RO_2 with NO (leading to ozone production) and NO_2 (leading to the PAN reservoirs) have been established. Although the peroxyacylnitrates are the most stable, other peroxynitrates could also be significant reservoirs for NO_x in the coldest regions of the troposphere.

Alkyl nitrates, formed in a secondary channel of the reaction of RO_2 with NO, are another "stable" form of NO_y. LACTOZ has provided a comprehensive data set for the formation of nitrates, mainly from the higher molecular weight VOCs. In addition, the rates of degradation of organic nitrates, by photolysis and by their reaction with OH radicals, can be determined from LACTOZ results, although some uncertainties remain in the production and loss of alkyl nitrates.

The project has provided the data necessary for a much more robust estimation of the partitioning of NO_y between the active NO_x forms and the reservoirs, both in the source regions and the remote troposphere.

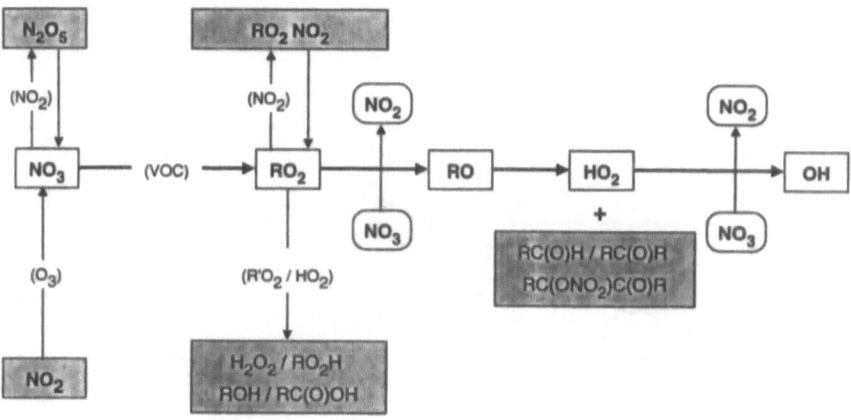

Fig. 2: NO$_y$ reaction system.

1.2.2c The production and loss of odd hydrogen radicals required for accurate calculation of tropospheric free radical concentrations

Atmospheric photolysis provides the predominant source of free radicals in the atmosphere. Photolysis of NO$_2$ produces O atoms which form ozone; ozone photolysis in the near UV produces O(^{1}D) which reacts with H$_2$O to produce OH radicals. A number of organic species absorb UV light and dissociate to yield organic peroxy and HO$_2$ radicals in the presence of O$_2$.

One of the objectives of LACTOZ has been to provide absorption cross-sections and quantum yields so that photolysis rates can be calculated under atmospheric conditions. Emphasis has been on (a) refinement of data for the simple, well-characterised molecules (O$_3$, HCHO) involved in HO$_x$ radical production, and (b) determination of data for photolysis of organic compounds formed as products in the degradation of alkanes, alkenes and selected multifunctional VOCs, mainly carbonyl compounds and nitrates, which produce HO$_x$ indirectly.

The reaction of ozone with alkenes produces OH and other radicals. Work in LACTOZ has provided quantitative data for the yield of OH for a variety of alkenes, including biogenic hydrocarbons. Yields of around 50 % were obtained for the reaction of ozone with typical alkenes, making this a source of atmospheric free radicals, which is especially significant at night-time when photolytic sources are absent.

The elucidation of the detailed chemistry of NO$_3$ radicals was a primary objective of LACTOZ since it was recognised that NO$_3$ initiated degradation of VOC could be significant for VOC oxidation and nitrogen speciation. NO$_3$ is formed from reaction of NO$_2$ with O$_3$ and is a key oxidising agent during night-time when photochemically produced radicals, primarily OH, are much less abundant. When NO$_3$ reacts with VOCs, RO$_2$ and ultimately HO$_x$ radicals are formed (Fig. 3).

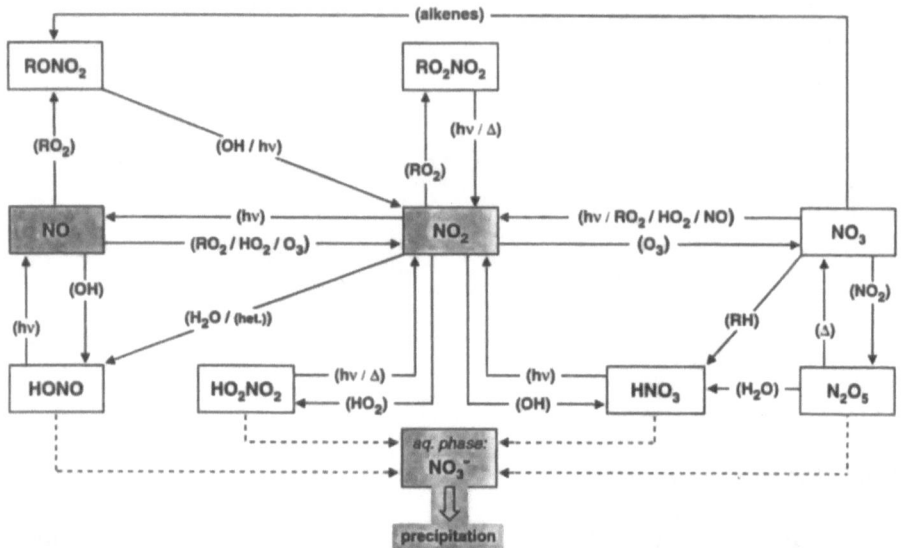

Fig. 3: VOC oxidation by NO_3 during night-time.

Early in the LACTOZ project a significant discovery was made of the mechanism by which oxidation chains involving NO_3 might be propagated through reactions such as:

$$NO_3 + RO_2 \quad \rightarrow \quad RO + NO_2 + O_2 \ (R = H \text{ or organic radical})$$

When R = H, OH radicals are generated. Kinetic and mechanistic data necessary for the description of the night-time radical chemistry initiated by NO_3 has been now established.

1.2.2d The determination of the rates and mechanisms for OH attack on more complex organic compounds, including oxygenated and nitrate compounds

An objective of LACTOZ has been to provide accurate rate coefficient data for OH attack on the more complex, higher molecular weight VOCs. As a result of the studies there is now a much improved data set for aromatic compounds, alkenes, including isoprene and other biogenic hydrocarbons. Kinetic data have also been obtained for some oxygenated compounds and nitrates for the first time.

Understanding of the mechanisms of oxidative degradation following OH attack has improved substantially, in particular for the aromatic compounds and for higher alkanes (C_5 and C_6). These results provide a firmer basis for the formulation of models for assessing ozone production from these important primary pollutants. Work in LACTOZ has established for the first time a validated model for isoprene oxidation under both high and low NO_x conditions. This provides a basis for assessing ozone formation resulting from emissions of this important biogenic hydrocarbon.

Although ozone-alkene reactions were not specified in the original objectives of LACTOZ they constitute a loss process for ozone and are important for degradation of unsaturated hydrocarbons, in particular alkenes with multiple double bonds and complex structures, such as are found in the biogenic hydrocarbons. Emphasis in LACTOZ has been on the rates and mechanisms under atmospheric conditions. These studies led to downward revision of the rate constants of the O_3 alkene reactions, due to complications arising in many earlier investigations from secondary reactions of radicals produced in the primary step. Some important hitherto unknown aspects of ozone reactions, such as the formation of peroxides and their dependence on the water vapour concentrations, have been discovered in LACTOZ.

1.2.2e The determination of the rate constants and branching ratios for reactions of peroxy and alkoxy radicals

The initial attack of radicals on VOCs gives rise to the formation of RO_2 radicals which can undergo a variety of reactions. These reactions may lead to propagation of the radical chain, *e.g.* formation of RO, or termination of the chain leading to radical loss and formation of stable products, *e.g.* hydroperoxides, peroxynitrates and nitrates. Propagation is necessary for ozone production, and thus these reactions are central to the scientific aim of LACTOZ: quantification of tropospheric ozone budget.

Outstanding progress has been made in RO_2 chemistry, both in terms of reactivity and reaction mechanisms. The data have been compiled and reviewed [1], and further progress in defining structure-reactivity aspects for more complex VOCs is documented in this Final Report. For example, the mechanism of isoprene oxidation involving six different RO_2 radicals, for low NO_x and high NO_x conditions, has recently been validated against observations.

The general mechanism for the atmospheric degradation of VOCs follows a pathway in which the initially produced RO_2 radical is transformed into an alkoxy radical, RO. The fate of the alkoxy radicals determines the nature of the first stable products of oxidation, the degree of fragmentation of the original carbon chain and the potential for photo-oxidant formation.

Alkoxy radicals have several potential reaction channels which compete under atmospheric conditions. The study of these reactions is difficult due to their complexity and the absence of suitable techniques for direct monitoring of the kinetics. The relative importance of these channels has been established accurately for several individual RO radicals and also for groups of RO radicals of similar structural type in the LACTOZ project. This aids the definition of the ozone forming potential for alkanes. A particularly important achievement is the characterisation of the isomerisation reactions of larger alkoxy radicals, which has been a significant uncertainty in modelling photochemical ozone formation in the polluted boundary layer.

Studies of the oxidation of aromatic compounds have established that the reaction of the initial adduct, formed by attack by OH, reacts with O_2 to produce HO_2 on a short timescale. This observation, together with the fast reaction rate for the OH reaction with substituted aromatic compounds, accounts for their efficient ozone production.

1.2.2f The elucidation of night-time NO_3 chemistry with a view to modelling the removal of NO_x and VOCs from the atmosphere and the impact on the overall oxidation of VOCs

Quantification of the NO_3 initiated degradation of VOC at night-time has required the establishment of rate coefficients for a wide range of VOCs present in the lower atmosphere. A substantial body of data for the reaction rates of NO_3 with VOCs has been obtained in LACTOZ, covering both man-made and naturally produced organic compounds. The mechanisms and products have also been investigated. As a result, it is now possible to make quantitative estimates of the rate of night-time HNO_3 production from NO_x via the homogeneous pathway as opposed to the heterogeneous route via N_2O_5. Night-time production of organic nitrates, which are reservoirs for NO_x, can also be estimated. The atmospheric chemistry of the NO_3 radical has been reviewed within LACTOZ [2].

1.2.2g The investigation of reactions acting as sources and sinks of HO_x and RO_2 radicals in the boundary layer in support of anticipated field measurements of free radical concentrations

At the beginning of LACTOZ there was a well developed theory of the 'fast photochemistry' governing tropospheric free radical concentrations. However, the theory had not been validated by field measurements. Successful development of instruments for field measurements of OH and RO_2 radicals have led to the anticipated capability to observe atmospheric free radical chemistry and validate the models describing their production and loss, and the related production rates of ozone. A number of field campaigns for the study of radical chemistry have been conducted in Europe. Work in LACTOZ has made a substantial contribution to the data base for gas-phase reactions which control OH, HO_2 and related radical concentrations in the daytime troposphere; this information is needed to interpret the results of these experiments.

The issue of night-time radical chemistry initiated by NO_3 was not prominent at the start of LACTOZ. Although O_3 is not generated in this process, NO_3 chemistry is now considered to be of importance for the modification of the precursors for ozone formation, for example at high latitudes in wintertime, when anthropogenic pollutants can accumulate. Observations of night-time oxidation chemistry, including NO_3 and RO_2 radicals in the boundary layer, are also now feasible. LACTOZ data will enable the interpretation of these experiments.

1.3   Application of LACTOZ Results

We consider here some applications of work performed within LACTOZ and discuss ways in which the information gained has been disseminated.

Laboratory data on the kinetics, reaction mechanisms and pathways of elementary reactions occurring in the atmosphere form the basis for models describing and predicting photo-oxidant formation. Experience in the incorporation of the chemical information into the models by the Chemical Mechanism Working Group (CMWG) has shown that critical evaluation of currently used chemistry is required. In this respect, a successful exercise was the evaluation and improvement carried out by LACTOZ of the chemistry contained within the EMEP MSC-W model of photo-oxidants.

One of the objectives of the modelling procedures is to predict the extent to which specific VOCs contribute to ozone formation in the atmosphere. Kinetic information on the initiation step (attack by OH, NO_3, O_3) is required, and a great deal of new data of this kind has been acquired within LACTOZ. However, work within LACTOZ has also confirmed that the efficiency of ozone generation is not determined by the efficiency of the initial step alone (even in a single VOC system). In particular, the conversion of NO to NO_2 is required for ozone to be formed, and the length of the radical chain thus has a direct bearing on ozone generation. A great deal of new information on this aspect of VOC oxidation has been provided by LACTOZ. The VOC/NO_x ratio in turn influences the nature and extent of the chain propagation and termination reactions. Thus, detailed understanding of all the steps in the mechanism (including photochemical processes) is necessary in defining useful measures of ozone generation in mixtures of VOC representative of real air masses.

A consequence of these considerations is that it is not possible to give a straightforward definition of an ozone forming potential of a given VOC. One index of ozone formation potential that has found some favour is the Photo-oxidant Creation Potential (POCP). This index is derived from a model calculation that includes detailed chemistry, but it has to be interpreted with caution because it applies to a single model domain. Another measure, which has the virtue of being objectively calculable from laboratory data, is the NOCON factor, defined as the total number of NO molecules converted to NO_2 for the first stage of the oxidation of each VOC molecule. The parameters required for the determination of the NOCON factor are the branching ratios for chain termination in the interaction of RO_2 with NO and for reaction with O_2, compared with decomposition and isomerisation, of the corresponding RO radicals.

It has become increasingly apparent that omission of biogenic species from tropospheric models leads to incorrect evaluation of photo-oxidant chemistry, particularly in rural areas. LACTOZ has provided substantial new data, largely on isoprene, which is an important component of the biogenic emissions. However, there are many other species, including the terpenes, which have not yet been subject to detailed investigation because of considerable experimental difficulties. The aim of future work should be to describe the impact of the major biogenic components both on the formation and the destruction of ozone in the atmosphere for meteorological conditions that apply in Europe.

In 1992, a complete mechanism for the oxidation of C_1 and C_2 hydrocarbons was formulated by the LACTOZ Steering Committee. The intended application is in global models for studies of methane oxidation and the tropospheric ozone budget. These studies have provided useful input to the Intergovernmental Panel on Climate Change (IPCC) assessment of the atmospheric chemistry of greenhouse gases.

Much of the work within LACTOZ has been published in the open literature and has also been recorded in the Annual Reports to EUROTRAC. Workshops have been held annually, and the proceedings published as reports by the European Commission. In addition, groups of workers within LACTOZ have written two extensive reviews summarising knowledge on the physics and chemistry of two of the key radicals and their involvement in atmospheric chemistry. These reviews discuss (i) the nitrate radical and (ii) peroxy radicals, RO_2, and were both published by *Atmospheric Environment* and by the European Commission. Similarly, reviews are currently in preparation on (i) the chemistry, and especially the oxidation pathways and mechanisms, of aromatic compounds, and (ii) chemical mechanisms for use in models and the methods for their experimental evaluation.

Chapter 2

Scientific Results

G. Le Bras and the LACTOZ Steering Group

Laboratoire de Combustion et Systèmes Réactifs-CNRS, F-45071 Orléans - cedex 2, France

2.1 Reaction of OH radicals

2.1.1 Introduction

Reactions with the hydroxyl radical provide the major removal pathway for most VOCs in the global troposphere. As a consequence, the rate constants for OH + VOC reactions determine the timescales, and hence the spatial distribution of VOCs. The timescales of VOC loss by reaction with OH are also identical with the timescales of formation of secondary pollutants and photochemical oxidants, including ozone. The data base for OH reaction rate coefficients is generally well defined and has been the subject of regular evaluations and updates including contributions from LACTOZ. The specific objectives of LACTOZ contributions are:

- to define and extend the current kinetic data base for compounds relevant to the formation of ozone, in particular for compounds which have only recently been identified as oxidation products of primary pollutants;
- to determine the product distribution for OH initiated oxidation reactions and to derive structure-reactivity relationships for OH attack at different sites of the parent VOC;
- to develop techniques for the estimation of photochemical ozone formation potentials of VOCs and for the reduction of oxidation mechanisms for incorporation into tropospheric chemical-dynamic models.

Of these objectives, only the first is related to overall OH radical reaction rate studies. The latter two rely largely on the full complexity of the oxidation chain including reactions of RO_2 and RO radicals as well as their interactions with NO_x.

2.1.2 Results

A number of new kinetic studies of reactions of OH radicals with VOCs have been performed in LACTOZ. The results are summarised in Tables 1a and 1b. Pilling has studied the rate coefficients for reactions of OH with C_4 alcohols and esters using DF/LIF and LP/LIF techniques. Becker has used a relative rate technique to determine the rate coefficients for reactions of OH with products originating from the oxidation of isoprene, 2- and 3-methyl-3-butene-1,2-diol and 1,2-epoxy-3-methyl-3-butene. Similar experiments have also been performed by this group on doubly unsaturated 1,6 and 1,4-dicarbonyl compounds, carbonyl nitrates and dinitrates. These compounds are suspected to be the primary ring-opening products in the oxidation of aromatic compounds and from the NO_3 initiated oxidation of alkenes, respectively. The rate coefficients for the reactions of OH with a series of alkenes and with methyl glyoxal, a product from the oxidation of isoprene, have been determined using a DF/RF technique (Devolder). Rate data for the reactions of OH radicals with various ethers have been obtained using LP-LIF (Le Bras/Poulet) and relative rate techniques (Kerr). Surprisingly, all the ethers show negative temperature dependence. Schindler has measured rate coefficients for reactions of OH with a number of oxiranes. These compounds are secondary pollutants which arise from the NO_3 and O_3 initiated oxidation of alkenes. Kinetic studies for reactions of OH with benzene and its substituted analogues (toluene, phenol, benzaldehyde, cresols) have been performed as a function of temperature using a relative rate technique in an irradiated flow reactor over the pressure range 130–1000 mbar (Kerr). Nielsen has measured rate coefficients for reactions of OH with a number of n-nitroalkanes, n-alkyl nitrates and nitrites using a pulse radiolysis technique. These studies have been complemented by Sidebottom using a relative rate technique. An important mechanistic finding arising from this work is that for the compounds CH_3NO_2, CH_3ONO_2 and CH_3ONO both abstraction and addition channels appear to be important.

Table 1a: Rate constants for reactions of OH radicals with VOCs determined at 298 K.

VOC	k (cm^3 s^{-1})	Method [a]	Reference [b]
Alkanes			
methane	5.4×10^{-15} [c]	LP/LIF	Pilling (92)
	6.3×10^{-15}	DF/LIF	Le Bras (93)
propane	1.1×10^{-12}	DF/LIF	Le Bras (93)
cyclohexane	6.7×10^{-12}	DF/LIF	Pilling (92)
1,2-dichloroethane	2.0×10^{-12}	LP/LIF	Pilling (93)
Alkenes			
isobutene	5.7×10^{-11}	LP/RF	Devolder (92)
1,2-butadiene	1.9×10^{-11}	DF/LIF	Pilling (92)
isoprene	9.7×10^{-11}	FP/RF	Zetzsch (93)
	9.7×10^{-11}	LP/RF	Devolder (92)
cyclohexene	6.6×10^{-11}	LP/RF	Devolder (92)
trans-1,2-dichloroethene	1.9×10^{-12}	DF/LIF	Pilling (93)
Alcohols			
n-butanol	8.4×10^{-12}	LP/LIF	Pilling (92)
i-butanol	9.0×10^{-12}	LP/LIF	Pilling (92)
t-butanol	8.1×10^{-13}	DF/LIF	Pilling (92)
	1.1×10^{-12}	LP/LIF	Le Bras (94)
i-amylalcohol	1.3×10^{-11}	DF/LIF	Pilling (92)
2-methyl-3-butene-1,2-diol	8.2×10^{-11}	RR/FTIR	Becker (93)
3-methyl-3-butene-1,2-diol	1.3×10^{-10}	RR/FTIR	Becker (93)
3-hexene-3,4-diol-2,5-dione	2.7×10^{-10}	RR/FTIR	Becker (94)
Ethers			
di-methyl ether	2.8×10^{-12}	LP/LIF	Le Bras (94)
di-ethyl ether	1.3×10^{-11}	LP/LIF	Le Bras (93)
	1.3×10^{-11}	RR/GC	Kerr (92)
di-*n*-propyl ether	2.2×10^{-11}	LP/LIF	Le Bras (94)
di-*i*-propyl ether	1.0×10^{-11}	LP/LIF	Le Bras (93)
methyl-*n*-butyl ether	1.5×10^{-11}	RR/GC	Kerr (92)
methyl-*t*-butyl ether	3.1×10^{-12}	LP/LIF	Le Bras (94)
methyl-*i*-pentyl ether	6.3×10^{-12}	LP/LIF	Le Bras (94)
methyl-*t*-amyl ether	5.0×10^{-12}	LP/LIF	Pilling (92)
ethyl-*n*-butyl ether	2.3×10^{-11}	RR/GC	Kerr (92)
	2.1×10^{-11}	LP/LIF	Le Bras (94)
ethyl-*t*-butyl ether	8.8×10^{-12}	LP/LIF	Le Bras (94)
di-*n*-butyl ether	2.7×10^{-11}	LP/LIF	Le Bras (93)
	3.0×10^{-11}	RR/GC	Kerr (92)
di-n-pentyl ether	3.4×10^{-11}	RR/GC	Kerr (92)
Carbonyls			
methyl-ethyl ketone	1.2×10^{-12}	RR/GC	Moortgat (90)
methyl-vinyl ketone	1.9×10^{-11}	RR/GC	Moortgat (90)
methyl glyoxal	7.7×10^{-12}	LP/RF	Devolder (93)
	1.7×10^{-11}	RR/GC	Moortgat (90)
methacrolein	3.3×10^{-11}	RR/GC	Moortgat (90)
cyclohexanone	4.3×10^{-12}	LP/LIF	Pilling (93)

Table 1a: continued

VOC	k (cm^3 s^{-1})	Method [a]	Reference [b]
Carbonyls (unsaturated)			
cis-butenedial	5.2×10^{-11}	RR/FTIR	Becker (92)
trans-butenedial	2.4×10^{-11}	RR/FTIR	Becker (92)
cis/trans-4-oxo-2-pentenal	5.5×10^{-11}	RR/FTIR	Becker (92)
cis-3-hexene-2,5-dione	6.9×10^{-11}	RR/FTIR	Becker (93)
trans-3-hexene-2,5-dione	4.0×10^{-11}	RR/FTIR	Becker (93)
E,Z-hexa-2,4-dienedial	1.1×10^{-10}	RR/FTIR	Becker (93)
E,E-hexa-2,4-dienedial	8.8×10^{-11}	RR/FTIR	Becker (93)
E,E-2-methyl-hexa-2,4-dienedial	1.2×10^{-10}	RR/FTIR	Becker (93)
maleic anhydride	1.5×10^{-12}	RR/FTIR	Becker (92)
3H-furan-2-one	4.4×10^{-11}	RR/FTIR	Becker (92)
5-methyl-3H-furan-2-one	6.9×10^{-11}	RR/FTIR	Becker (92)
Esters			
n-propyl acetate	2.9×10^{-12}	LP/LIF	Pilling (93)
vinylacetate	2.5×10^{-11}	DF/LIF	Pilling (93)
methyl methacrylate	2.6×10^{-11}	DF/LIF	Pilling (92)
Oxiranes			
ethylene oxide	1.1×10^{-13}		Schindler (92)
1,2-epoxipropane	5.1×10^{-13}		Schindler (92)
1,2-epoxibutane	3.1×10^{-12}		Schindler (92)
1,2-epoxihexane	1.2×10^{-11}		Schindler (92)
tetramethyloxirane	1.4×10^{-12}		Schindler (92)
butadienmonoxide	1.2×10^{-11}		Schindler (92)
2-methyl-2-vinyl-oxirane	2.6×10^{-11}		Schindler (92)
1,2-epoxi-5-hexene	5.4×10^{-11}		Schindler (92)
2-propenyl-oxirane	5.8×10^{-11}	RR/FTIR	Becker (94)
Nitroalkanes			
nitromethane	1.6×10^{-13}	PR/UV	Nielsen (94)
nitroethane	1.5×10^{-13}	PR/UV	Nielsen (94)
nitro-*n*-propane	3.4×10^{-13}	PR/UV	Nielsen (94)
nitro-*n*-butane	1.5×10^{-12}	PR/UV	Nielsen (94)
nitro-*n*-pentane	3.3×10^{-12}	PR/UV	Nielsen (94)
Nitrates			
methyl nitrate	3.3×10^{-13}	PR/UV	Nielsen (91)
ethyl nitrate	5.3×10^{-13}	PR/UV	Nielsen (91)
n-propyl nitrate	8.0×10^{-13}	PR/UV	Nielsen (91)
n-butyl nitrate	1.7×10^{-12}	PR/UV	Nielsen (91)
n-pentyl nitrate	3.1×10^{-12}	PR/UV	Nielsen (91)
1,2-propandiol dinitrate	$< 3 \times 10^{-13}$	RR/FTIR	Becker (90)
1,2-butandiol dinitrate	1.7×10^{-12}	RR/FTIR	Becker (90)
2,3-butandiol dinitrate	1.1×10^{-12}	PR/FTIR	Becker (90)
a-nitrooxy acetone	$< 4.3 \times 10^{-13}$	RR/FTIR	Becker (90)
1-nitrooxy-butanone-2	9.1×10^{-13}	RR/FTIR	Becker (90)
3-nitooxy-butanone-2	1.3×10^{-12}	RR/FTIR	Becker (90)
cis-1,4-dinitrooxy butene-2	1.5×10^{-11}	RR/FTIR	Becker (90)
3,4-dinitrooxy butene-1	1.0×10^{-11}	RR/FTIR	Becker (90)

Table 1a: continued

VOC	k (cm^3 s^{-1})	Method [a]	Reference [b]
Nitrites			
methylnitrite	2.6×10^{-13}	PR/UV	Nielsen (90)
ethylnitrite	7.0×10^{-13}	PR/UV	Nielsen (90)
n-propyl nitrite	1.2×10^{-12}	PR/UV	Nielsen (90)
n-butyl nitrite	2.7×10^{-12}	PR/UV	Nielsen (90)
n-pentyl nitrite	4.2×10^{-12}	PR/UV	Nielsen (90)
Aromatics			
benzene	1.2×10^{-12}	RR/GC	Kerr (93)
	$k_\infty = 1.0 \times 10^{-12}$	DF/RF	Devolder (94)
	$k_0 = 1.7 \times 10^{-29}$ cm^6/s	DF/RF	Devolder (94)
toluene	6.1×10^{-12}	RR/GC	Kerr (93)
	$k_\infty = 6.0 \times 10^{-12}$	DF/RF	Devolder (94)
	$k_0 = 4.0 \times 10^{-28}$ cm^6/s	DF/RF	Devolder (94)
	4.7×10^{-12} (N$_2$)	RR/FTIR	Becker (94)
	6.6×10^{-12} (air)	RR/FTIR	Becker (94)
p-xylene	1.1×10^{-11} (N$_2$)	RR/FTIR	Becker (94)
	1.5×10^{-11} (air)	RR/FTIR	Becker (94)
phenol	2.9×10^{-11}	RR/GC	Kerr (93)
benzaldehyde	1.2×10^{-11}	RR/GC	Kerr (93)
o-cresol	4.9×10^{-11}	RR/GC	Kerr (93)
m-cresol	5.3×10^{-11}	RR/GC	Kerr (93)
p-cresol	5.9×10^{-11}	RR/GC	Kerr (93)
naphtalene	2.3×10^{-11}	FP/RF	Zetzsch (93)

[a] $T = 292$ K

[b] These references are to EUROTRAC Annual Reports. The year in parentheses is the year of the report not the publishing year.

[c]
LP/LIF:	laser photolysis/laser induced fluorescence
DF/LIF:	discharge flow/laser induced fluorescence
FP/RF:	flash photolysis/resonance fluorescence
LP/RF:	laser photolysis/resonance fluorescence
RR/GC:	relative rate technique/gas chromatography
PR/UV:	pulse radiolysis/UV adsorption
RR/FTIR:	relative rate technique/Fourier transform IR absorption

Table 1b: Rate constants for reactions of OH radicals with VOCs. Results from temperature dependence studies.

VOC	A (cm³ s⁻¹)	E_a/R (K)	Reference*
Alkanes			
methane	2.6×10^{-12}	1765	Le Bras (93)
Alkenes			
isobutene	3.6×10^{-12}	−840	Devolder (92)
isoprene	7.2×10^{-12}	−783	Devolder (92)
cyclohexene	1.4×10^{-11}	−490	Devolder (92)
Ethers			
di-methyl ether	6.4×10^{-12}	234	Le Bras (94)
di-ethyl ether	5.2×10^{-12}	−262	Kerr (92)
	6.6×10^{-12}	−208	Le Bras (93)
di-*n*-propyl ether	1.1×10^{-11}	−210	Le Bras (94)
di-*i*-propyl ether	4.1×10^{-12}	−274	Le Bras (93)
di-*n*-butyl ether	5.5×10^{-12}	−502	Kerr (92)
	3.8×10^{-12}	−599	Le Bras (93)
di-*n*-pentyl ether	8.5×10^{-12}	−417	Kerr (92)
methyl-*n*-butyl ether	5.4×10^{-12}	−309	Kerr (92)
methyl-*t*-butyl ether	5.0×10^{-12}	133	Le Bras (94)
methyl-*i*-pentyl ether	4.7×10^{-12}	−82	Le Bras (94)
ethyl-*n*-butyl ether	6.6×10^{-12}	−362	Le Bras (94)
	7.3×10^{-12}	−335	Kerr (92)
ethyl-*t*-butyl ether	4.4×10^{-12}	−210	Le Bras (94)
Aromatics			
benzene	2.6×10^{-12}	−231	Kerr (94)
toluene	7.9×10^{-13}	−614	Kerr (94)
phenol	3.7×10^{-13}	−1270	Kerr (94)
benzaldehyde	5.3×10^{-12}	−243	Kerr (94)
o-cresol	9.8×10^{-13}	−1170	Kerr (94)
m-cresol	5.2×10^{-12}	−686	Kerr (94)
p-cresol	2.2×10^{-12}	−943	Kerr (94)
naphtalene	4.7×10^{-12}	−475	Zetzsch (93)

* These references are to EUROTRAC Annual Reports. The year in parentheses is the year of the report not the publishing year.

Kinetic data for reactions of OH with VOCs determined in LACTOZ show good agreement with literature data. However, the LACTOZ data base also contains OH rate coefficients for compounds which have not previously been investigated.

Based on an analysis of the overall rate coefficients for the reactions of OH radicals with alkenes, Peeters has developed a structure-reactivity relationship for prediction of the relative yields of OH addition to specific sites in the parent alkenes. For mono-alkenes and non-conjugated poly-alkenes the site-specific rate coefficients depend on the stability of the ensuing hydroxyalkyl radical, which can be either a primary, secondary or tertiary radical. For conjugated dienes, the structure-reactivity analysis requires additional consideration of the resonance stabilisation energies of the adduct radicals. The quality of the predictions have been tested for a large number of alkenes and found in most cases to be in agreement with experiment within 10 %.

Whereas kinetic measurements of OH reactions are usually designed to monitor the loss of OH, measurements of the kinetic behaviour of OH in oxidation studies can also be utilised to extract mechanistic information. Zellner has applied time-resolved laser absorption studies of OH to monitor the formation of HO_2 radicals (converted into OH in the presence of NO) in the Cl atom initiated oxidation of a large number of alkanes, alkenes and aromatic compounds. It has been demonstrated that by this technique the total yield of hydroxyperoxy radicals and, by implication, the number of NO molecules converted into NO_2, can be monitored. Zetzsch has applied highly sensitive OH measurements in oxidation studies of aromatic compounds to extract both the thermal stability of OH aromatic adducts and the extent of radical recycling occurring in the loss reactions of these adducts in the presence of O_2 and NO.

Mechanistic studies, aimed at direct determination of product distributions in the photo-oxidation of VOCs under simulated tropospheric conditions, have been performed (Warneck, Becker, Kerr). From the product distributions observed in the steady-state photo-oxidation of a number of C_5-C_7 alkanes and alkenes, branching ratios for site specific H atom abstraction and OH addition, respectively, have been determined (Warneck). For 2,3-dimethyl butane the ratio of tertiary to primary H atom abstraction was found to be 1.0:0.20. The dominance of abstraction from the tertiary site was also confirmed for 2-methyl butane (i-pentane). In addition, reactions of OH with asymmetric alkenes show a higher propensity for addition to a terminal position at room temperature. For instance, the fractions of OH addition to the 1-position of 1-butene, 2-methyl-1-butene and 3-methyl-1-butene have been determined as 0.74, 0.89 and 0.76, respectively (Warneck). Warneck has also investigated the product distribution in the oxidation of toluene under NO_x free conditions. The major finding was that 56 % of the total products could be assigned to ring-retaining mechanisms, whereas 25 % resulted from ring cleavage (see Section 4.7). Becker has performed product investigations for the oxidation of isoprene and aromatic compounds (toluene and p-xylene). For isoprene the identification of diols such as 2- and 3-methyl-3-butene-1,2-diol were taken as evidence for the mechanism shown in Fig. 11. In contrast to Warneck the experiments of Becker were carried out under conditions closer to those of the atmosphere. Under these conditions, ring-retaining (phenol-type) products are negligible. The OH initiated photo-oxidation of n-hexane has been studied using a collapsible Teflon smog chamber (Kerr). The primary products quantified include 3-hexylnitrate, 2-hexylnitrate, n-butylnitrate, 3-hexanone, 2-hexanone, formaldehyde and acet-aldehyde. From these product distributions valuable information on the branching ratios for reactions of the various alkoxy radicals produced in the system were obtained.

2.2 Reactions of NO$_3$ radicals

2.2.1 Introduction

Evidence for the presence of the nitrate radical during night-time in the troposphere has led to considerable effort in establishing its role in atmospheric chemistry. Two main types of reactions of NO$_3$ with volatile organic compounds have been identified, hydrogen atom abstraction and addition to unsaturated systems:

$$NO_3 + RH \rightarrow HNO_3 + R$$
$$NO_3 + R^1R^2C = CR^3R^4 \rightarrow R^1R^2C - CR^3R^4ONO_2$$

Abstraction reactions of NO$_3$ are generally slow, although reaction with dimethyl sulfide and oxygenated aromatics are facile and probably dominate the atmospheric removal of these species. Since the reaction rate of NO$_3$ with alkenes is often higher than that for the reaction with O$_3$, production of NO$_3$ radicals from O$_3$ leads to a faster effective oxidation rate by O$_3$. The major products identified from reactions of NO$_3$ with alkenes are carbonyl nitrates and unsubstituted carbonyl compounds. Thus, the reactions of NO$_3$ radicals play an important role in the conversion of NO$_x$ into both the permanent sink species, HNO$_3$, and into temporary reservoir species such as carbonyl nitrates which can lead to long-range transport of odd nitrogen. Nitric acid results from both hydrogen abstraction from saturated organic compounds or from hydrolysis of N$_2$O$_5$ formed from the reaction of NO$_3$ with NO$_2$.

It has recently been shown that the reaction of NO$_3$ with organic compounds initiates a chain reaction and can lead to a night-time source of OH radical [3]:

$$NO_3 + VOC \rightarrow R$$
$$R + O_2 + M \rightarrow RO_2 + M$$
$$RO_2 + NO_3 \rightarrow RO + NO_2 + O_2$$
$$RO + O_2 \rightarrow products + HO_2$$
$$HO_2 + NO_3 \rightarrow OH + NO_2 + O_2$$
$$OH + VOC \rightarrow R$$

The importance of this oxidation sequence depends on the availability of NO$_3$ to propagate the chain via reaction with alkylperoxy and HO$_2$ radicals.

2.2.2   Reactions of NO$_3$ with organic molecules

Abstraction reactions of NO$_3$ radicals

Rate constants for the reaction of NO$_3$ radicals with a number of alkanes have been determined using conventional discharge flow-visible absorption methods (Ljungström, Wayne) and by a stopped-flow technique (Wayne) (Table 2). Rate constants, determined as a function of temperature, for the reaction of NO$_3$ with ethane, n-butane and 2-methyl propane show that, as expected, the activation energy increases in the order primary > secondary > tertiary. Product studies have not been reported for the reaction of NO$_3$ with alkanes although the available rate data are consistent with a hydrogen atom abstraction process. Alkyl radical formation has also been used to explain the secondary reactions observed in discharge-flow investigations of NO$_3$ alkane reactions (Wayne). The reactivity of NO$_3$ is low and hence loss of alkanes in the troposphere due to reaction with NO$_3$ is minor compared to their removal by the daytime attack of hydroxyl radicals.

The reactivity of various oxygenated organic compounds with respect to reactions with NO$_3$ radicals has been reported (Ljungström, Schindler, Wayne) (Table 2). Many of these compounds have been used as industrial solvents and fuel additives, and are formed in the troposphere as degradation products from a variety of VOCs. Wayne has determined rate constants for the reactions of NO$_3$ with CH$_3$CHO and CH$_3$COCH$_3$ using discharge-flow and stopped-flow methods, respectively. The low activation energy for the reaction with CH$_3$CHO indicates a process involving hydrogen atom abstraction from the aldehyde group. The reactivity of CH$_3$COCH$_3$ is similar to that observed for ethane, suggesting that the carbonyl group has little influence on the CH$_3$ group reactivity. Kinetic and mechanistic studies on the reactions of NO$_3$ radicals with a series of alcohols, ethers and esters have been described (Ljungström). Rate constants were measured as a function of temperature at low pressures with a discharge-flow visible-absorption absolute rate technique. Rate data were also obtained at 295 K and atmospheric pressure using a relative rate method, employing FTIR spectroscopy for monitoring the loss of reactants. Agreement between the rate constants determined from the two experimental techniques was excellent. The rate coefficients obtained for the reaction of NO$_3$ radicals with both alcohols and ethers are significantly higher than would be expected on the basis of carbon-hydrogen bond strengths in these molecules. This effect could arise from stabilisation of the transition states for the reactions arising from an interaction between the nitrate radical and the oxygen atom of the hydroxy or ether groups.

Table 2: Rate constants for the reactions of NO_3 with VOCs at 298 K.

VOC	Method	k (cm^3 s^{-1})	Reference *
Alkanes			
methane	DSF-VA	$< 4 \times 10^{-19}$	Wayne (94)
ethane	DF-VA	8.3×10^{-19}	Wayne (94)
n-butane	DF-VA	4.5×10^{-17}	Wayne (94)
n-hexane	DF-VA	1.4×10^{-16}	Ljungström (91)
2-methyl propane	DF-VA	1.1×10^{-16}	Wayne (94)
Oxygenated Organics			
acetaldehyde	DF-VA	2.5×10^{-15}	Wayne (94)
acetone	DSF-VA	8.5×10^{-18}	Wayne (90)
methanol	DF-VA	9.0×10^{-16}	Ljungström (94)
ethanol	DF-VA	1.6×10^{-15}	Ljungström (94)
2-propanol	DF-VA	4.0×10^{-15}	Ljungström (94)
dimethyl ether	DF-VA, RR	2.6×10^{-16}	Ljungström (93)
diethyl ether	DF-VA, RR	2.8×10^{-15}	Ljungström (93)
methyl t-butyl ether	DF-VA, RR	6.4×10^{-16}	Ljungström (93)
di-n-propyl ether	DF-VA, RR	6.5×10^{-15}	Ljungström (93)
di-i-propyl ether	DF-VA, RR	5.5×10^{-15}	Ljungström (93)
t-amyl methyl ether	DF-VA, RR	2.2×10^{-15}	Ljungström (93)
ethyl t-butyl ether	DF-VA, RR	1.1×10^{-14}	Ljungström (93)
methyl formate	DF-VA	3.6×10^{-18}	Ljungström (92)
methyl acetate	DF-VA	7.0×10^{-18}	Ljungström (92)
methyl propionate	DF-VA	3.3×10^{-17}	Ljungström (92)
methyl butyrate	DF-VA	4.8×10^{-17}	Ljungström (92)
ethyl formate	DF-VA	1.7×10^{-17}	Ljungström (92)
ethyl acetate	DF-VA	1.3×10^{-17}	Ljungström (92)
ethyl propionate	DF-VA	3.3×10^{-17}	Ljungström (92)
propyl formate	DF-VA	5.4×10^{-17}	Ljungström (92)
propyl acetate	DF-VA	5.0×10^{-17}	Ljungström (92)
1,2-epoxyethane	DF-MS	3.9×10^{-16}	Schindler (91)
1,2-epoxybutane	DF-MS	1.5×10^{-15}	Schindler (91)
1,2-epoxyhexane	DF-MS	5.7×10^{-15}	Schindler (91)
3,4-epoxy-1-butene	DF-MS	7.3×10^{-15}	Schindler (91)
3,4-epoxy-3-methyl-1-butene	DF-MS	1.8×10^{-14}	Schindler (91)

Table 2: continued

VOC	Method	k (cm^3 s^{-1})	Reference *
Alkenes			
ethene	DSF-VA	1.7×10^{-16}	Wayne (94)
	EC-RR	9.3×10^{-17}	Becker (88)
1-butene	DF-VA	1.1×10^{-14}	Wayne (90)
	EC-RR	7.8×10^{-15}	Ljungström (90)
	EC-RR	1.3×10^{-14}	Becker (88)
cis-2-butene	DF-MS	3.6×10^{-13}	Schindler (92)
trans-2-butene	DF-MS	3.8×10^{-13}	Schindler (88)
2-methyl propene	EC-RR	3.2×10^{-13}	Becker (88)
	DF-MS	3.8×10^{-13}	Schindler (92)
2-methyl-2-butene	DF-MS	7.8×10^{-12}	Schindler (92)
2,3-dimethyl-2-butene	DF-MS	4.0×10^{-11}	Schindler (92)
	DF-MS	4.5×10^{-11}	Le Bras (89)
	PR-KS	4.5×10^{-11}	Nielsen (88)
1-pentene	DF-VA	1.8×10^{-14}	Wayne (92)
1-hexene	DF-VA	1.5×10^{-14}	Wayne (92)
cyclopentene	DF-VA	5.9×10^{-13}	Ljungström (92)
cyclohexene	DF-VA	6.3×10^{-13}	Ljungström (92)
1-methyl cyclohexene	DF-VA	1.5×10^{-11}	Ljungström (92)
1,3-butadiene	DF-MS	1.0×10^{-13}	Le Bras (89)
	EC-RR	1.9×10^{-13}	Becker (88)
2,3-dimethyl-1,3-butadiene	DF-MS	2.0×10^{-12}	Le Bras (89)
2-methyl-1,3-butadiene	DF-MS	7.8×10^{-13}	Le Bras (90/91),
(isoprene)			Schindler (90/91)
	EC-RR	6.0×10^{-13}	Becker (90)
α-pinene	EC-RR	6.5×10^{-12}	Becker (88)
β-pinene	EC-RR	2.8×10^{-12}	Becker (88)
Δ^3-carene	EC-RR	8.1×10^{-12}	Becker (88)
d-limonene	EC-RR	1.1×10^{-11}	Becker (88)
Aromatics			
toluene	RR	6.1×10^{-17}	Carlier (89)
o-xylene	RR	3.7×10^{-16}	Carlier (89)
m-xylene	RR	2.5×10^{-16}	Carlier (89)
p-xylene	RR	4.2×10^{-16}	Carlier (89)
1,2,3-trimethyl benzene	RR	2.5×10^{-15}	Carlier (89)
1,2,4-trimethyl benzene	RR	2.1×10^{-15}	Carlier (89)
1,3,5-trimethyl benzene	RR	1.4×10^{-15}	Carlier (89)

* These references are to EUROTRAC Annual Reports. The year in parentheses is the year of the report not the publishing year.
DSF-VA: discharge stopped flow - visible absorption
DF-VA: discharge flow - visible absorption, p < 5 Torr
DF-MS: discharge flow - mass spectrometry
EC-RR: environmental chamber - relative rate
RR: relative rate, atmospheric pressure of air
PR-KS: pulse radiolysis - kinetic spectroscopy

The rate data for reaction of NO_3 with aliphatic esters show that the presence of the ester group in an organic molecule has little influence on the reactivity compared to the parent alkane. The reactivity trends exhibited by the nitrate radical for reactions with alcohols, ethers and esters are similar to those shown for the analogous reactions of hydroxyl radicals. The major products identified from the NO_3 radical-initiated oxidation of alcohols, ethers and esters under atmospheric conditions were esters, carbonyls and alkyl nitrates. Similar products arise from the reactions of OH radicals with these molecules under atmospheric conditions.

The rate constants for reactions of NO_3 with oxygenated organic compounds are too low for these reactions to contribute to the overall atmospheric removal of these species, although they could provide a non-negligible source of HNO_3 under night-time conditions. Schindler has measured rate constants for reaction of NO_3 with saturated epoxides using a discharge-flow mass-spectrometry technique. The reactivity of the epoxides is around two orders of magnitude higher than that for the analogous alkanes, indicating the influence of ring-strain in these reactions.

Dimethyl sulfide is the most abundant of all the biogenic organo-sulfur compounds, and the available data show that the H atom abstraction reaction with NO_3 could contribute significantly to the conversion of NO_x to HNO_3 in polluted coastal regions (Hjorth, Le Bras/Poulet).

Reaction of NO_3 with alkenes

Rate constants for the gas-phase reactions of NO_3 with a variety of alkenes, dialkenes, cycloalkenes and terpenes have been measured using both relative (Becker, Ljungström) and absolute rate techniques (Nielsen, Le Bras/Poulet, Ljungström, Schindler, Wayne), (Table 2). The various data sets are, in general, in good agreement and help to identify patterns of reactivity that may have diagnostic and predictive value. Wayne has carried out work on correlations of rate constants with ionisation potentials for reactions of NO_3 with alkenes. The correlations provide accurate predictions of room temperature rate coefficients and Arrhenius parameters for the reaction of NO_3 with a wide range of alkenes, which may be used to estimate unknown rate constants or indicate experimental inconsistencies. The increase in rate constant with increasing degree of alkyl substitution at the double bond strongly suggests attack by the electrophilic NO_3 radical at the double bond site. Arrhenius parameters, obtained from rate coefficient measurements as a function of temperature, are consistent with this view. Rate constants for reaction of NO_3 with cyclic alkenes are slightly higher than those for structurally similar acyclic alkenes. This may reflect the loss of strain energy on formation of the radical adducts.

A number of product studies on the reactions of NO_3 with alkenes have been carried out using mainly FTIR spectroscopy (Becker, Hjorth, Le Bras/Poulet, Ljungström, Schindler). The product and intermediate distributions for all the alkenes investigated followed a similar pattern and can be represented in terms of the following simplified mechanism:

$$NO_3 + R^1CH = CHR^2 \quad \rightarrow \quad R^1CH(ONO_2)CHR^2$$
$$R^1CH(ONO_2)CHR^2 + O_2 \quad \rightarrow \quad R^1CH(ONO_2)CHR^2O_2$$
$$R^1CH(ONO_2)CHR^2O_2 + NO_2 \quad \leftrightarrow \quad R^1CH(ONO_2)CHR^2O_2NO_2$$
$$R^1CH(ONO_2)CHR^2O_2 + NO \quad \rightarrow \quad R^1CH(ONO_2)CHR^2O + NO_2$$
$$R^1CH(ONO_2)CHR^2O \quad \rightarrow \quad R^1CHO + R^2CHO + NO_2$$
$$R^1CH(ONO_2)CHR^2O + NO_2 \quad \rightarrow \quad R^1CH(ONO_2)CHR^2ONO_2$$
$$R^1CH(ONO_2)CHR^2O + O_2 \quad \rightarrow \quad R^1CH(ONO_2)COR^2 + HO_2$$

Under atmospheric conditions, the nitro-oxyalkyl peroxy radicals will probably form mainly the corresponding nitro-oxy alkoxy radicals. Thermal decomposition, yielding carbonyl compounds and NO_2, and reaction with O_2 giving carbonyl nitrates, appear to be the dominant reactions under most atmospheric conditions. The extent to which carbonyl nitrates can act as temporary reservoirs for NO_x will largely depend on their photolysis rates or reactions with OH radicals.

Product studies on the reaction of NO_3 with isoprene showed that the addition occurs mainly at the 1-position and that 3-methyl-4-nitroxy-2-butenal is the main reaction product (Hjorth). Results from an FTIR spectroscopy study on product formation in the reaction of NO_3 radicals with cyclic alkenes provided evidence that dicarbonyls are important products arising from ring opening processes (Ljungström).

Evidence for epoxide formation from NO_3 reactions with alkenes has been reported (Hjorth, Schindler). These results show that oxirane formation occurs via a radical adduct, in which rotation about the C-C bond can occur before elimination of NO_2. Reactions of NO_3 with isoprene and 2-butene at low pressure and at low O_2 concentrations gave mainly oxiranes, whereas in air at atmospheric pressure, oxirane yields were negligible. However, even at atmospheric pressures, the reaction of NO_3 with 2,3-dimethyl-2-butene gave an oxirane yield of around 20 %. Thus, it is apparent that, at least with some alkenes, oxirane formation may be important under tropospheric conditions.

Combination of the available rate constant data with estimates of the atmospheric concentrations of O_3, NO_3 and OH indicates that the reaction of NO_3 with substituted alkenes and terpenes may be the dominant tropospheric oxidation process for these species. Formation of carbonyl nitrates in the reactions could be important in the long-range transport of odd nitrogen.

Reactions of NO_3 with aromatic compounds

Carlier has determined rate constants for the reactions of NO_3 with a series of aromatic compounds using a relative rate method (Table 2). Agreement with literature values is reasonable. The gas-phase reactions of NO_3 radicals with benzene, toluene and xylene isomers have been studied in N_2O_5/NO_2/aromatic/air mixtures at room temperature and atmospheric pressure (Hjorth). Products were identified by *in situ* FTIR spectroscopy, and after sampling, on charcoal column or a XAD-2 trap by gas chromatography coupled to mass spectrometry.

Benzene was found to be virtually unreactive, with only trace yields of nitrobenzene detected. Reaction with toluene and xylenes gave benzaldehydes and benzyl nitrates as major products, which indicate that the reaction with NO_3 proceeds mainly by hydrogen abstraction from a methyl group. Minor yields of aromatic nitroderivatives were observed in the reactions suggesting that the nitrate radical may also react by addition to the aromatic ring system.

2.2.3 Reaction of NO₃ with peroxy radicals

Reactions of NO_3 with HO_2 and simple RO_2 radicals (CH_3O_2, $C_2H_5O_2$ and $CH_3C(O)O_2$) have been investigated within LACTOZ (Table 3), as possible chain propagation steps in the mechanism suggested for the night-time oxidation of VOCs (see above).

Table 3: Rate constants for the reactions of NO_3 with HO_2 and RO_2 radicals at 298 K.

RO₂	Method	k (cm³ s⁻¹)	Reference [a]
HO₂		$NO_3 + HO_2 \rightleftharpoons NO_2 + OH + O_2$, k_1	
		$NO_3 + HO_2 \rightleftharpoons HNO_2 + O_2$, k_2	
	DF-EPR	$3.6 \times 10^{-12}/k_1$	Le Bras (88)
		$0.9 \times 10^{-12}/k_2$	
	MP	$3.5 \times 10^{-12}/k_1 + k_2$	Wayne (88)
	FP-LA	$4.7 \times 10^{-12}/k_1 + k_2$	Zellner (88)
	DF-MS/RF	$2.5 \times 10^{-12}/k_1$	Schindler (91)
		$1.9 \times 10^{-12}/k_2$	
	EC-FTIR/TDL	$< 8.0 \times 10^{-13}/k_1$ [b]	Hjorth (92)
CH₃O₂		$NO_3 + RO_2 \rightleftharpoons NO_2 + RO + O_2$	
	MP	2.3×10^{-12}	Burrows,
			Moortgat,
			Le Bras (89)
	DF-LIF	1.3×10^{-12}	Le Bras (93)
	DF-LIF	1.0×10^{-12}	Wayne (93)
CD₃O₂	DF-MS	1.4×10^{-12}	Moortgat (93)
C₂H₅O₂	DF-LIF	2.5×10^{-12}	Wayne (94)
	DF-LIF	2.2×10^{-12}	Le Bras (94)
CH₃C(O)O₂	SFR-FTIR	$\sim 2 \times 10^{-11}$	Wayne (94)

[a] These references are to EUROTRAC Annual Reports. The year in parentheses is the year of the report not the publishing year.
[b] This value has not been confirmed so far.
DF-EPR: discharge flow - electron paramagnetic resonance
MP: modulated photolysis
FP-LA: flash photolysis - laser absorption
DF-MS/RF: discharge flow-mass spectrometry / resonance fluorescence
EC-FTIR/TDL: environmental chamber - Fourier transform IR spectroscopy/tuneable diode laser spectroscopy
DF-LIF: discharge flow-laser induced fluorescence
SFR- FTIR: slow flow reactor - Fourier transform IR spectroscopy

Several studies of the $NO_3 + HO_2$ reaction have been carried out (Le Bras/Poulet, Wayne and Cox, Zellner, Schindler, Hjorth) using four different techniques. The discharge flow-EPR method has been used to determine rate constant data at room temperature for the two possible reaction channels (Le Bras/Poulet):

$$NO_3 + HO_2 \quad \rightarrow \quad NO_2 + OH + O_2 \qquad\qquad\qquad (a)$$
$$HNO_3 + O_2 \qquad\qquad\qquad\qquad (b)$$

Good agreement with the value of the overall rate constant was obtained in the other studies (Wayne and Cox, Zellner, Schindler). The predominance of channel (a) was confirmed by Schindler and is in agreement with other reported work [4]. In contrast, channel (a) was found insignificant in an indirect study of the dark reaction of $HO_2NO_2/N_2O_5/^{13}C^{18}O$/air mixtures at room temperature and atmospheric pressure (Hjorth). However, this result has not been confirmed so far.

The $NO_3 + CH_3O_2$ reaction has been investigated at 298 K in discharge-flow reactors, with LIF analysis of CH_3O in the reaction system:

$$NO_3 + CH_3O_2 \quad \rightarrow \quad NO_2 + CH_3O + O_2 \qquad\qquad (a)$$
$$NO_3 + CH_3O \quad \rightarrow \quad NO_2 + CH_3O_2 \qquad\qquad\qquad (b)$$

The two studies performed (Le Bras/Poulet, Wayne) yielded k_a and k_b values which were in agreement. The rate constant obtained for the $NO_3 + CD_3O_2$ reaction in a discharge-flow mass-spectrometric study (Moortgat) agrees with these k_a values.

Similar discharge-flow LIF studies were recently performed for the $NO_3 + C_2H_5O_2$ reaction which was also found to convert RO_2 ($C_2H_5O_2$) into RO (C_2H_5O) (Wayne, Le Bras/Poulet). The rate constant values obtained in both studies were in agreement.

The rate constant of the reaction of NO_3 with another peroxy radical, the peroxyacetyl radical ($CH_3C(O)O_2$) was also measured in a slow discharge flow reactor at 300–423 K (Wayne), using thermal decomposition of PAN as a source of the organic radicals.

The results obtained for reactions of NO_3 with RO_2 (R = H, CH_3 and C_2H_5) show that rate constants at 298 K are comparable $(1 - 3.6 \times 10^{-12}$ cm^3molecule$^{-1}s^{-1}$ at 298 K), and they all proceed through RO_2/RO conversion. These reactions can consequently be rather fast and make important the proposed chain mechanism of VOC oxidation during nighttime in moderately polluted atmosphere. In addition, if the high rate constant found for the $NO_3 + CH_3C(O)O_2$ reaction was confirmed $(k \approx 2 \times 10^{-11}$ cm^3molecule$^{-1}s^{-1}$ at 298 K); this reaction could affect PAN concentrations [5].

2.2.4 Reactions of NO$_3$ with inorganic molecules

The equilibrium constant for NO$_2$ + NO$_3$ + M $\leftrightarrow$ N$_2$O$_5$ + M has been measured at room temperature by direct monitoring of the species by optical absorption in O$_3$/N$_2$O$_5$/NO$_2$/NO$_3$/air mixtures (Table 4, Hjorth). This is an important parameter in the calculation of the NO$_y$ atmospheric partitioning.

The rate constant for the thermal decomposition of NO$_3$ (NO$_3$ + M $\rightarrow$ NO + O$_2$ + M) has been derived from a computer simulation of the same experiments (Hjorth). The deduced lifetime of NO$_3$ would be of 10 min at 295 K and 1 atmosphere pressure.

Several other reactions of NO$_3$ with inorganic molecules have been investigated and found to be negligible or too slow to be of atmospheric significance. These include reactions of NO$_3$ with itself and ozone (Wayne) with HCl, HBr and HI (Le Bras/Poulet, Wayne), the reaction of NO$_3$ with NO$_2$ as a source of NO (Ljungström) and the reaction of electronically excited NO$_3$ with N$_2$ as a source of N$_2$O (Burrows, Zellner).

Table 4: Rate constants for the reactions of NO$_3$ radicals with inorganic species.

Reaction	k (cm^3 s^{-1})	T (K)	Method	Reference *
NO$_3$ + NO$_2$ + M $\leftrightarrow$ N$_2$O$_5$ + M (1)/(−1)	$K_e = k_1/ k_{-1} = (3.73\pm0.41) \times 10^{-11}$ (cm^3 molecule^{-1})	298	EC	Hjorth (90)
NO$_3$ + M $\rightarrow$ NO + O$_2$ + M	$(1.6\pm0.7) \times 10^{-3}$ (s^{-1}) (1013 mbar in air)	295	EC	Hjorth (90)
NO$_3$ + NO$_3$ $\rightarrow$ NO$_2$ + NO$_2$ + O$_2$	2.7×10^{-16}	298	LP-LA DSF-LA	Wayne (89)
NO$_3$ + NO$_2$ $\rightarrow$ NO + NO$_2$ + O$_2$	$(5.2\pm1.5) \times 10^{-14}$ exp $[(-1477\pm260)/T]$ 3.7×10^{-16}	296–332 298	EC-FTIR/TDL	Ljungström (91)
NO$_3$ + O$_3$ $\rightarrow$ NO$_2$ + 2 O$_2$	$\leq 1 \times 10^{-19}$	298	DSF-LA	Wayne (93)

* These references are to EUROTRAC Annual Reports. The year in parentheses is the year of the report not the publishing year.
EC: environmental chamber
LP-LA: laser photolysis-laser absorption
DSF-LA: discharge stopped flow-laser absorption
DF-EPR: discharge flow-electron paramagnetic resonance
EC-FTIR/TDL: environmental chamber - Fourier transform IR spectroscopy/tuneable diode laser spectroscopy

2.3 Reactions of O_3 with alkenes

2.3.1 Introduction

Ozone-alkene reactions are important sinks for both ozone and alkenes, and provide important sources of OH and other radicals, carbonyl compounds, hydrogen peroxide, organic hydroperoxides and acids in the troposphere. Atmospheric photo-oxidation chemistry cannot be adequately described without considering the ozonolysis of alkenes, in particular biogenic compounds, such as isoprene (see also Section 4.8 and Fig. 13) and terpenes. In the presence of NO_x, the ozonolysis of biogenic hydrocarbons eventually can lead to ozone generation rather than its loss. The Criegee reaction mechanism, presently accepted for the ozonolysis of alkenes is shown in Fig 4, the key steps being reactions (1a) and (1b) – the two decomposition channels of the primary ozonide, (2) – the decomposition of the excited biradical (several pathways possible), (3) – its collisional deactivation and (4) – reactions of the stabilised biradical:

$$O_3 + R^1R^2C=CR^3R^4 \;\rightarrow\; [\text{primary ozonide}] \;\rightarrow\; R^1R^2C=O + R^3R^4COO^* \qquad (1a)$$
$$\rightarrow R^3R^4C=O + R^1R^2COO^* \qquad (1b)$$
$$R^1R^2COO^* \qquad\qquad \rightarrow \text{decomposition products (CO, CO}_2\text{, radicals, } etc.) \qquad (2)$$
$$R^1R^2COO^* + M \qquad\qquad\qquad \rightarrow R^1R^2COO + M \qquad (3)$$
$$R^1R^2COO + (\text{aldehydes, H}_2O, SO_2, etc.) \qquad \rightarrow \text{products} \qquad (4)$$

The value of the rate constants for the reactions of O_3 with alkenes (k_1) at room temperature are fairly well documented. However, corrections for OH reactions have to be made, resulting in lower rate constants compared to the currently accepted values. The temperature dependence of the rate constants is known only for a limited number of alkenes and corrections for OH reactions are also required in this case.

The knowledge of the chemistry of the Criegee biradicals R^1R^2COO is very poor: the branching ratios k_{1a}/k_{1b}, the rate constants k_4, and the product distributions of reactions (2) and (4) are largely uncertain.

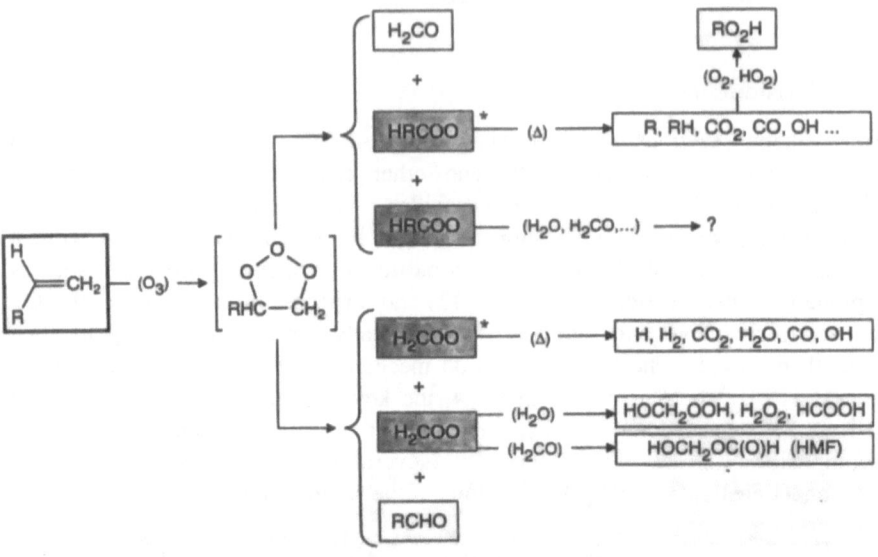

Fig. 4: Reaction scheme for the gas-phase ozonolysis of a simple alkene.

2.3.2 Results

Research within LACTOZ has addressed some of the open questions with respect to reaction rate constants, product formation and the general reaction mechanism. Overall reaction rate constants, k_1, have been measured as a function of temperature for a number of cyclic alkenes (Sidebottom), (Table 5a). The observed variation of Arrhenius parameters with the number of carbon atoms in the ring suggests that the ozonolysis rate is strongly influenced by the release of the ring strain, which is connected to the change of the C=C double bond in the cyclic alkene into the single bond in the intermediate trioxolane. In agreement with this view, k_1 is larger for cyclopentene and cycloheptene. Careful studies in remeasuring the rate constant for the fundamental reaction of O_3 with ethene have been carried out (Becker). The value obtained is about 30 % smaller than the previously recommended value [6], (Table 5b). In addition, rate constants were measured at room temperature for the O_3 reaction with unsaturated alcohols and 2-propenyl-oxirane, both being products of the oxidation of isoprene (Becker).

Table 5a: Rate constants for O_3 reactions at different temperatures and atmospheric pressure in air.

Compound	A (cm^3 s^{-1})	E_a/R (K)	k (cm^3 s^{-1})	Reference *
cis-2-butene	3.1×10^{-15}	939	1.3×10^{-16}	Sidebottom (93)
2-methyl-2-butene	5.2×10^{-15}	734	4.5×10^{-16}	Sidebottom (93)
cyclopentene	1.6×10^{-15}	348	4.9×10^{-16}	Sidebottom (93)
1-methyl-1-cyclopentene	-	-	6.7×10^{-16}	Sidebottom (93)
1-chloro-1-cyclopentene	4.4×10^{-16}	1008	1.5×10^{-17}	Sidebottom (93)
cyclohexene	2.6×10^{-15}	1062	8.5×10^{-17}	Sidebottom (93)
1-methyl-1-cyclohexene	5.3×10^{-15}	1039	1.7×10^{-16}	Sidebottom (93)
4-methyl-1-cyclohexene	2.2×10^{-15}	951	8.2×10^{-17}	Sidebottom (93)
1-nitro-1-cyclohexene	-	-	1.2×10^{-18}	Sidebottom (93)
cycloheptene	1.3×10^{-15}	493	2.4×10^{-16}	Sidebottom (93)
cis-cyclooctene	7.8×10^{-16}	217	3.7×10^{-16}	Sidebottom (93)
cis-cyclodecene	1.1×10^{-15}	1080	2.9×10^{-17}	Sidebottom (93)

* These references are to EUROTRAC Annual Reports. The year in parentheses is the year of the report not the publishing year.
k-values: at 298 K

Table 5b: Rate constants for O_3 reactions at 295 K and atmospheric pressure in air.

Compound	k (cm^3 s^{-1})	Reference [a]
2-methyl-3-butene-1,2-diol	4.8×10^{-18}	Becker (94)
3-methyl-3-butene-1,2-diol	6.3×10^{-17}	Becker (94)
2-propenyl-oxirane	3.3×10^{-18}	Becker (94)
ethene [b] (298 K)	1.2×10^{-18}	Becker (94)

[a] These references are to EUROTRAC Annual Reports. The year in parentheses is the year of the report not the publishing year.
[b] corrected for OH radical reactions.

Two groups (Becker, Sander) have presented some evidence for the need to complete the generally accepted Criegee mechanism (1)-(4) by additional reaction pathways:

-       The absence of methylacetate as a product in the ozonolysis of 2-methyl-propene is interpreted in terms of a "non-Criegee" mechanism (Sander);
-       The observed pressure independence of the product distribution in the ozone-ethene reaction between 100 and 1030 mbar indicates that the branching ratio k_3/k_4 may not be pressure dependent, in this pressure region. The pressure independence of the H_2 yield has been shown, even down to 1 mbar (Becker). These findings are not in agreement with reaction step (3) by which excited Criegee intermediates are stabilised.

The yield of OH radicals in the ozonolysis of a number of cycloalkenes has been determined by scavenging OH with cyclohexane (Sidebottom). The derived OH yield of about 40 % was found to be similar to that of cis-2-butene, providing support for the suggestion that OH yields from ozone-alkene reactions depend largely on the degree of alkyl groups substitution at the double bond site (Table 6a). From the number of alkene molecules by one O_3 molecule, the following OH yields were determined (Moortgat): ethene (5 %), cis-2-butene (40 %), trans-2-butene (60 %), isobutene (60 %) and isoprene (10–15 %).

Table 6a: OH radical yields of O_3 reactions at 298 K and atmospheric pressure in air.

Compound	OH yield (%)	Reference *
ethene	< 10 (a)	Sidebottom (94)
	< 5 (b)	
	5 (a)	Moortgat (94)
cis-2-butene	40 (a)	Moortgat (94)
trans-2-butene	60 (a)	Moortgat (94)
isobutene	60 (a)	Moortgat (94)
isoprene	24 (b)	Schindler (93)
	10–15 (a)	Moortgat (94)
cyclopentene	46 (a)	Sidebottom (94)
	62 (b)	
cyclohexene	39 (a)	Sidebottom (94)
	38 (b)	
cycloheptene	44 (a)	Sidebottom (94)
	46 (b)	
cis-cyclooctene	39 (a)	Sidebottom (94)
	35 (b)	
cis-cyclodecene	40 (a)	Sidebottom (94)
	35 (b)	

* These references are to EUROTRAC Annual Reports. The year in parentheses is the year of the report not the publishing year.
(a): based on reactions stoichiometry
(b): based on product analysis

Special emphasis has been laid on the formation of H_2O_2, hydroperoxides and hydroxy-hydroperoxides. Systematic studies of their yields for a number of alkenes have been made with and without the addition of water vapour (Becker, Gäb, Moortgat). In the case of ethene, the formation of 2-hydroxyethyl hydroperoxide clearly can be related to the reaction OH + ethene. It was also possible to compare the product yields of hydroperoxides from the alkene + O_3 reactions in the gas phase with that in the aqueous phase (Gäb), (Table 6b). The product distribution is different and, in general, the yields of α-hydroxy-hydroperoxides in the aqueous phase are 10 times higher than in the gas phase. In the presence of formaldehyde, the simplest Criegee intermediate, CH_2OO, forms hydroxylmethyl formate (Becker, Moortgat), which eventually decomposes into formic acid.

Table 6b: Hydroperoxide formation yields (%) of O_3 reactions with alkenes (mol % of reacted alkene).

Hydroperoxide	Ethene	Propene	Isoprene	α-Pinene	β-Pinene
H_2O_2	0.4 (a)	0.3 (a)	0.7 (a)	0.3 (a)	0.6 (a)
	- (b)	- (b)	< 0.1 (b)	< 0.1 (b)	< 0.1 (b)
CH_3O_2H	- (a)	4.1 (a)	5.0 (a)	0.8 (a)	- (a)
	- (b)	- (b)	- (b)	- (b)	- (b)
$HOCH_2O_2H$	0.2 (a)	0.2 (a)	0.5 (a)	0.1 (a)	0.3 (a)
	27.9 (b)	2.7 (b)	11.1 (b)	- (b)	0.2 (b)
$CH_3CH(OH)O_2H$	- (a)	- (a)	- (a)	- (a)	- (a)
	- (b)	11.1 (b)	- (b)	- (b)	-(b)
$HOCH_2CH_2CH_2O_2H$	< 0.1 (a)	- (a)	- (a)	- (a)	- (a)
	- (b)	- (b)	- (b)	- (b)	- (b)
total	0.6 (a)	4.6 (a)	6.2 (a)	1.2 (a)	0.9 (a)
	27.9 (b)	13.8 (b)	24.2 (b)	11.3 (b)	16.2 (b)

(a): no addition of H_2O, (b): aqueous-phase, "-": not detected
Gäb EUROTRAC Annual Report 1994.

Table 7: Product yields in the gas-phase ozonolysis of isoprene (units are mol-%: mole of product formed per mole of isoprene reacted).

Products		Becker (94) [a]		Moortgat (92)
	(900 mbar air)	(+ H_2O)[b]	(+ C_6H_{12})[c]	
methacrolein	24 ± 2	28 ± 2	28 ± 2	22 ± 3
methyl vinyl ketone	11 ± 1	15 ± 2	12 ± 2	11 ± 2
formaldehyde	53 ± 3	62 ± 5	62 ± 7	56 ± 6
remaining carbonyl compounds	47 ± 15			
CO	24 ± 2	25 ± 2	30 ± 4	21 ± 2
CO_2	18 ± 4	21 ± 9	~ 24	16 ± 3
formic acid	~ 3 [c]	1 – 25 [c]	1 – 25 [c]	3.5
ketene	~ 1	~ 1	~ 1	1.4
propene	2.3 ± 1.0	3 ± 1		< 9
H_2	2.0 ± 0.2	~2	~2	
epoxides	< 4	< 4		
CH_3OH	1.1 ± 0.3			
CH_3OOH	5		5	
$HOCH_2OOH$	0.5		0.6	
H_2O_2	~ 1		0.9	
OH	Schindler (93)[a]:	24 [d]	Moortgat (94)[a]:	10 – 15

[a] These references are to EUROTRAC Annual Reports. The year in parentheses is the year of the report not the publishing year.   [b] H_2O-vapour, 2–13 mbar;   [c] Addition of 400–1000 ppm C_6H_{12};
[d] Dependent on reaction time;   [e] Measured with CO as scavenger.

A considerable amount of work has been devoted to product studies of the O_3 reaction with isoprene (Becker, Moortgat), (Table 7). Ozonolysis of alkenes, in particular biogenic compounds, presents a significant source of OH radicals, H_2O_2 and a number of organic hydroperoxides. It is necessary to study the product yields also in the presence of NO_x to elucidate the potential for ozone formation. Complementary field studies on the contribution of OH production from ozone reactions and the formation of the different hydroperoxides are suggested. The gas-to-particle conversion during the ozonolysis of alkenes, especially biogenic compounds, should also be studied (blue haze problem).

2.4 Reactions of peroxy radicals

2.4.1 Introduction

Organic peroxy radicals, RO_2, are the first intermediate species formed during the oxidation of organic compounds in the atmosphere. In spite of their low reactivity, their reactions are important in determining the overall oxidation mechanism of organic compounds and the tropospheric ozone balance. They react principally with the nitrogen oxides NO and NO_2, and with other peroxy radicals, e.g. $R'O_2$ and HO_2. A great deal of work has been devoted in LACTOZ to studies of these reaction types, which has resulted in the collection of much new data and led to a much better understanding of the reactivity of peroxy radicals. A comprehensive review of RO_2 chemistry "Organic Peroxy Radicals: Kinetics, Spectroscopy and Tropospheric Chemistry", prepared by members of the LACTOZ group, has been published in Atmospheric Environment [1].

In polluted areas, reactions of RO_2 radicals with nitrogen oxides are of major importance. The reaction of RO_2 with NO results in the production of NO_2 and hence formation of ozone in the troposphere, and the reaction with NO_2 forms temporary reservoir peroxynitrate species. Much attention has been given in LACTOZ to the kinetics of these reactions, which has contributed to substantial improvement of the kinetic data base.

In remote atmospheres, where NO_x concentrations are very low, reactions of peroxy radicals with HO_2, or with other peroxy radicals, compete with the reaction with NO_x and result in the formation of various oxidised compounds such as aldehydes, alcohols, organic acids and hydroperoxides. Several studies performed as part of LACTOZ have provided important data for modelling these processes in the atmosphere.

Investigations performed within LACTOZ have mainly focused on the determination of structure-reactivity relationships for $RO_2 + NO_x$, $RO_2 + HO_2$, $RO_2 + RO_2$ and $RO_2 + R'O_2$ reactions. Reliable structure-reactivity relationships have been established for the RO_2 + NO reaction so that rate constants can be reliably estimated for any RO_2 reaction with NO. Significant progress has been made on the RO_2 + HO_2 reactions, and it is now clear that the rate constants increase markedly with the size of the peroxy radical but are practically insensitive to halogen substitution. The situation is not as clear for RO_2 + RO_2 self-reactions, despite the large number of results reported. An important result has emerged from the data, showing that the presence of various functional groups in the RO_2 radical gives rise to a significant increase in the rate constant of the self-reaction compared to the behaviour of similar unsubstituted alkylperoxy radicals. Thus, fairly good values of the rate constant can now be estimated according to the structure of R in a given peroxy radical. Cross-reactions of the type RO_2 + $R'O_2$ can play a significant role in the chemistry of hydrocarbon-rich atmospheres and new data have been obtained recently. They show, in particular, that reactions of the acetylperoxy radical are all probably quite fast and that these should be taken into account in tropospheric models and in mechanism reduction procedures.

In addition to kinetic and mechanistic studies, information on UV spectra has been obtained for all the peroxy radicals investigated.

Most of the work on peroxy radicals has been carried out by three of the LACTOZ groups (Cox/Hayman/Jenkin, Lesclaux and Moortgat).

2.4.2 RO_2 + HO_2 reactions

RO_2 + HO_2 reactions are generally fast and, owing to the fairly high concentration of HO_2 in the atmosphere, they play an important role under low concentrations of NO_x. Investigations of such reactions have represented a significant part of the LACTOZ project. The reactions that have been studied and the corresponding results are given in Table 8. Structure-reactivity relationships are fairly easy to establish for this particular class of reactions, as all rate constant values are within a factor of about three. There is no clear systematic effect of functional groups on the rate constants except for halogen substitution, which seems to lead to lower values for the rate constants. Apparently, the main factor influencing the rate constant is the size of the radical. The lowest rate constants are those of CH_3O_2 and of the corresponding halogen-substituted radicals, around 6×10^{-12} cm^3 molecule^{-1}s^{-1}. The rate constant increases rapidly with size to over 10^{-11} cm^3 molecule^{-1}s^{-1} for larger radicals and reaches values of 1.8×10^{-11} cm^3 molecule^{-1}s^{-1} for c-$C_5H_9O_2$ and c-$C_6H_{11}O_2$. The data suggest the following rate constants (units of cm^3 molecule^{-1} s^{-1}): 6×10^{-12} for CH_3O_2 and halogenated peroxy radicals, 1.0×10^{-11} for C_2—C_3 radicals and 1.5×10^{-11} for larger radicals (C_4 and over) and acylperoxy radicals. All reactions exhibit a strong negative temperature dependence. Exponential factors vary from 800K/T to 1600K/T, most of them being in the range 1000–1200K/T.

Table 8: Rate constants and branching ratios for the reactions: $RO_2 + HO_2$, $RO_2 + RO_2$ and $RO_2 + R'O_2$.

Rate Coefficient[a]	Branching Ratio[b]	Reference[c]
k (298 K) (cm^3 s^{-1})	k_a (298 K)/k(298 K)	

<table>
<tr><td colspan="3" align="center">RO₂ + HO₂ Reactions</td></tr>
</table>

$HO_2 + HO_2$		
$(2.4\pm0.2) \times 10^{-12}$ (1013 mbar O_2)		Cox (88)
$(2.8\pm0.3) \times 10^{-12}$ (1013 mbar air)		Cox (88)
$(3.8\pm0.4) \times 10^{-13}$ exp[(580±32)/T]		(1)
$(1.2\pm0.4) \times 10^{-33}$ [M] exp [(1150±97)/T]		
(248–573 K; 133–1013 mbar air)		
4.1×10^{-12} (1013 mbar air)		
$CH_3O_2 + HO_2$		
$(5.4\pm1.1) \times 10^{-12}$ (303 K; 13.3 mbar)		(2)
$(6.8\pm0.9) \times 10^{-12}$ (303 K; 1013 mbar)		(2)
$(4.8\pm0.2) \times 10^{-12}$ (933 mbar)		(3)
$(5.2\pm1.5) \times 10^{-12}$		Zellner (89)
$(4.4\pm0.7) \times 10^{-13}$ exp [(780±55)/T]		(1)
(248–573 K; 280–1013 mbar)		
6.0×10^{-12}		
$HOCH_2O_2 + HO_2$		
5.6×10^{-15} exp [(2300±1100)/T] (275–333 K)		Veyret (88)
1.3×10^{-11}		
$(1.2\pm0.3) \times 10^{-11}$		(4)
$C_2H_5O_2 + HO_2$		
$(1.6\pm0.2) \times 10^{-13}$ exp [(1200±40)/T		Lesclaux
1.1×10^{-11}		
$HOCH_2CH_2O_2 + HO_2$		
$(4.8\pm0.5) \times 10^{-12}$ (13.3 mbar)		Cox (89)
$(1.5\pm0.3) \times 10^{-11}$ (1013 mbar)		Hayman (94)

Table 8: continued

Rate Coefficient[a] k (298 K) (cm^3 s^{-1})	Branching Ratio[b] k_a (298 K)/k(298 K)	Reference[c]
CH$_3$C(O)O$_2$ + HO$_2$ $\underset{}{\overset{a}{\rightarrow}}$ CH$_3$C(O)O$_2$H + O$_2$		
$\underset{}{\overset{b}{\rightarrow}}$ CH$_3$C(O)OH + O$_3$		
$(4.3\pm1.2) \times 10^{-13}$ exp [(1040±100)/T]		(5)
(253–368 K; 800 mbar)		
$(1.3\pm0.3) \times 10^{-11}$		
1.0×10^{-12}	0.67 ± 0.07	
	0.73	Moortgat (91)
	$k_a = 1.15 \times 10^{-12}$ exp (550/T)	
	$k_b = 3.86 \times 10^{-16}$ exp(2640/T)	
	(233–333 K; 1013 mbar)	
CH$_3$C(O)CH$_2$O$_2$ + HO$_2$		
$(9.0\pm1.0) \times 10^{-12}$		Lesclaux (92)
CH$_2$=CHCH$_2$O$_2$ + HO$_2$		
$(5.6\pm0.4) \times 10^{-12}$ (390-430 K)		Lesclaux (94)
HOC(CH$_3$)$_2$CH$_2$O$_2$ + HO$_2$		
5.6×10^{-14} exp [1650/T] (303–398 K)		Lesclaux (94)
1.4×10^{-11}		
CH$_3$CH(OH)CH(O$_2$)CH$_3$ + HO$_2$		
$(1.5\pm0.4) \times 10^{-11}$		Hayman (94)
c–C$_5$H$_9$O$_2$ + HO$_2$		
2.1×10^{-13} exp[1323/T] (248–364 K, 270–1013 mbar)		
1.8×10^{-11}		
(CH$_3$)$_3$CCH$_2$O$_2$ + HO$_2$		
$(1.43\pm0.6) \times 10^{-13}$ exp [(1380±100)/T]		Lesclaux (92)
1.5×10^{-11}		
c-C$_6$H$_{11}$O$_2$ + HO$_2$		
2.6×10^{-13} exp[1245/T] (248–364 K, 270–1013 mbar)		
(CH$_3$)$_2$COHC(O$_2$)(CH$_3$)$_2$ + HO$_2$		
$\sim 2 \times 10^{-11}$		Hayman (94)

Table 8: continued

Rate Coefficient[a] k (298 K) (cm^3 s^{-1})	Branching Ratio[b] k_a (298 K)/k(298 K)	Reference[c]
C$_6$H$_5$CH$_2$O$_2$ + HO$_2$		
$(3.75\pm0.32) \times 10^{-13} \exp[(980\pm230)/T]$		Lesclaux (93)
$(1.0\pm0.2) \times 10^{-11}$		
CCl$_3$O$_2$ + HO$_2$		
$(4.81\pm0.46) \times 10^{-13} \exp[(706\pm31)/T]$		Lesclaux (94)
5.1×10^{-12}		
CHCl$_2$O$_2$ + HO$_2$		
$(5.6\pm1.2) \times 10^{-13} \exp[(700\pm64)/T]$		Lesclaux (94)
5.9×10^{-12}		
CH$_2$ClO$_2$ + HO$_2$		
$(3.31\pm0.61) \times 10^{-13} \exp[(830\pm60)/T]$		Lesclaux (93)
$(5.4\pm0.6) \times 10^{-12}$		
CH$_2$BrO$_2$ + HO$_2$		
$(6.7\pm3.8) \times 10^{-12}$		Lesclaux (94)

RO$_2$ + RO$_2$ Reactions

CH$_3$O$_2$ + CH$_3$O$_2$ $\overset{a}{\to}$ 2 CH$_3$O + O$_2$		
$\overset{b}{\to}$ **CH$_3$OH + HCHO + O$_2$**		
$(4.79\pm0.54) \times 10^{-13}$ [x]		Moortgat (88)
$(1.27\pm0.14) \times 10^{-13} \exp[(365\pm40)/T]$ (248–573 K)		Veyret (88)
4.4×10^{-13}	0.24	
$2.13\times10^{-13} \exp[(220\pm70)/T](268$–$353$ K; 1013 mbar) [x]		Cox (89)
$3.28\times10^{-13} \exp[(92\pm40)/T]$ (268–353 K;13.3 mbar) [x]		Cox (89)
4.5×10^{-13} (13.3 and 1013 mbar) [x]		
	$(y+\exp(1131\pm30/T)]/(19\pm5))^{-1}$	(6)
	(223–333 K)	
	0.30	
$(4.8\pm0.5) \times 10^{-13}$ (300 K; 320 mbar)		(7)
	0.25 ± 0.04 (1013 mbar)	Becker (90)

Table 8: continued

Rate Coefficient [a]	Branching Ratio[b]	Reference[c]
k (298 K) (cm^3 s^{-1})	k_a (298 K)/k(298 K)	
HOCH$_2$O$_2$ + HOCH$_2$O$_2$ $\overset{a}{\rightarrow}$ 2 HOCH$_2$O + O$_2$		
$\overset{b}{\rightarrow}$ CH$_2$(OH)$_2$ + HC(O)OH + O$_2$		
$(5.9\pm1.1) \times 10^{-12}$	0.89	Veyret (88)
C$_2$H$_5$O$_2$ + C$_2$H$_5$O$_2$		
	0.54±0.03	Becker (90)
2.49×10^{-13} exp[$-518/T$] + 9.39×10^{-16} exp(960/T)		
(220–330K, 1013 mbar)	0.65	Moortgat (91)
6.7×10^{-14}		
CH$_3$OCH$_2$O$_2$ + CH$_3$OCH$_2$O$_2$		
$(2.1\pm0.3) \times 10^{-12}$	0.67	Hayman
		(92/93)
HOCH$_2$CH$_2$O$_2$ + HOCH$_2$CH$_2$O$_2$		
$(2.3\pm0.3) \times 10^{-12}$ (1013 mbar/MMS) [x]	0.50	Hayman (93)
CH$_3$C(O)O$_2$ + CH$_3$C(O)O$_2$		
2.8×10^{-12} exp[(526±90)/T] (253–368 K; 800 mbar)		Veyret (88)
1.6×10^{-11}		
$(1.4\pm0.2) \times 10^{-11}$		Moortgat(94)
0.5±0.1 (1013 mbar)		
C$_2$H$_5$C(O)O$_2$ + C$_2$H$_5$C(O)O$_2$ $\overset{(O_2)}{\rightarrow}$ 2 C$_2$H$_5$O$_2$ +2 CO$_2$		
$(1.7\pm1) \times 10^{-11}$ (prelim.)		Veyret (90)
CH$_3$C(O)CH$_2$O$_2$ + CH$_3$C(O)CH$_2$O$_2$		
$(8.0\pm2.0) \times 10^{-12}$	0.75	Lesclaux (92)
CH$_2$ = CHCH$_2$O$_2$ + CH$_2$ = CHCH$_2$O$_2$		
$(6.8\pm1.3) \times 10^{-13}$	0.61	Hayman (92)
$(7.1\pm1.0) \times 10^{-14}$ exp[(700±100)/T]		Lesclaux (93)
$(7.3\pm1.5) \times 10^{-13}$		

Table 8: continued

Rate Coefficient[a] k (298 K) (cm^3 s^{-1})	Branching Ratio[b] k_a (298 K)/k(298 K)	Reference[c]
$(CH_3)_3CO_2 + (CH_3)_3CO_2 \rightarrow 2\ (CH_3)_3CO + O_2$		
1.0×10^{-11} exp [$-3890/T$]		Veyret (90)
2.1×10^{-17}		
$HOC(CH_3)_2CH_2O_2 + HOC(CH_3)_2CH_2O_2$		Lesclaux (94)
1.4×10^{-14} exp[$1740/T$] (303–398 K)	0.6 ± 0.2	
4.8×10^{-12}		
$CH_3CH(OH)CH(O_2)CH_3 + CH_3CH(OH)CH(O_2)CH_3$		
$(8.4\pm1.0) \times 10^{-13}$ [x]		Hayman (94)
$1\text{-}C_5H_{11}O_2 + 1\text{-}C_5H_{11}O_2$	> 0.16 (1013 mbar)	Becker (90)
$n\text{-}C_5H_{11}O_2 + n\text{-}C_5H_{11}O_2$	0.58 (overall)	Warneck (90)
$c\text{-}C_5H_9O_2 + c\text{-}C_5H_9O_2$		
2.9×10^{-13} exp[$-555/T$] (248–364 K, 1013 mbar)		Lightfoot (91)
4.5×10^{-14}		
$(CH_3)_3CCH_2O_2 + (CH_3)_3CCH_2O_2$		
3.02×10^{-19} $(T/298)^{9.46}$ exp [$4260/T$]		Veyret (90)
(248–373 K; 38–1013 mbar)		
$(1.04\pm0.09) \times 10^{-12}$	0.40 ± 0.01	
$c\text{-}C_6H_{11}O_2 + c\text{-}C_6H_{11}O_2$		
2.0×10^{-13} exp[$-487/T$] (248–364 K, 1013 mbar)		Lightfoot (91)
3.9×10^{-14}		
$(CH_3)_2C(OH)C(O_2)(CH_3)_2 + (CH_3)_2C(OH)C(O_2)(CH_3)_2$		
$(1.1\pm0.25) \times 10^{-14}$	1.0	Hayman (94)
$C_6H_5CH_2O_2 + C_6H_5CH_2O_2$		
$(2.75 \pm 0.15) \times 10^{-14}$ exp [$(1680\pm140)/T$]	0.4	Lesclaux (93)
$(7.7\pm1.5) \times 10^{-12}$	0.74	Warneck (93)

Table 8: continued

Rate Coefficient[a] k (298 K) (cm^3 s^{-1})	Branching Ratio[b] k_a (298 K)/k(298 K)	Reference[c]
CF$_3$O$_2$ + CF$_3$O$_2$		Wayne (94)
4.0×10^{-12}		
CCl$_3$O$_2$ + CCl$_3$O$_2$		
$(3.31\pm0.55) \times 10^{-13}$ exp[(745$\pm$58)/T]		Lesclaux (94)
4.0×10^{-12}	1.0	
CHCl$_2$O$_2$ + CHCl$_2$O$_2$		
$(2.65 \times 10^{-13}$ exp[(800/T)]		Lesclaux (94)
3.9×10^{-12}	> 0.95	
4.0×10^{-12}		Wayne (94)
CH$_2$ClO$_2$ + CH$_2$ClO$_2$		
$(1.95\pm0.27) \times 10^{-13}$ exp [(880$\pm$50)/T]	> 0.85	Lesclaux (93)
$(3.7\pm0.4) \times 10^{-12}$		
1.0×10^{-12}		Wayne (94)
CHBr$_2$O$_2$+CHBr$_2$O$_2$		
2.5×10^{-12}		Wayne (94)
CH$_2$BrO$_2$ + CH$_2$BrO$_2$		
$(1.05\pm0.4) \times 10^{-12}$	1.0	Lesclaux (94)
2.0×10^{-12}		Wayne (94)
CH$_3$CHBrCH(O$_2$)CH$_3$ + CH$_3$CHBrCH(O$_2$)CH$_3$		
7.1×10^{-15} exp[1360/T]		Lesclaux (94)
7.0×10^{-13}		
RO$_2$ + R'O$_2$ Reactions		
C$_2$H$_5$O$_2$ + CH$_3$O$_2$	0.5	Lesclaux (93)
$(2.0\pm1.0) \times 10^{-13}$		

Table 8: continued

Rate Coefficient [a] k (298 K) (cm^3 s^{-1})	Branching Ratio[b] k_a (298 K)/k(298 K)	Reference[c]
$CH_3C(O)O_2 + CH_3O_2 \xrightarrow{a} CH_3C(O)O + CH_3O + O_2$		
$\xrightarrow{b} CH_3C(O)OH + HCHO + O_2$		
9.6×10^{-12}	0.87	Moortgat (91,94)
	$k_a = 3.21 \times 10^{-11} \exp[-440/T]$	
	$k_b = 2.68 \times 10^{-16} \exp[2510/T]$	
	(233–333 K, 1013 mbar)	
7.6×10^{-12}		Lesclaux (94)
$C_3H_5O_2 + CH_3O_2$		
$(3.4\pm1.5) \times 10^{-13} \exp[(430\pm160)/T]$		Lesclaux (94)
1.5×10^{-12}		
$CH_3C(O)CH_2O_2 + CH_3O_2$		
$(3.8\pm0.4) \times 10^{-12}$	0.3	Lesclaux (92)
$CH_2=CHCH_2O_2 + CH_3O_2$		
$(2.0\pm1.0) \times 10^{-12}$	0.45	Lesclaux (93)
$C_5H_{11}O_2 + CH_3O_2$		
1.1×10^{-12}		Lesclaux (94)
$c\text{-}C_6H_{11}O_2 + CH_3O_2$		
$(3.0\pm1.5) \times 10^{-14}$	0.3	Lesclaux (93)
$C_6H_5CH_2O_2 + CH_3O_2$		
$< 10^{-12}$		Lesclaux (94)
$CCl_3O_2 + CH_3O_2$		
$(4.0\pm2.0) \times 10^{-12}$	0.65	Lesclaux (93)
$CH_2ClO_2 + CH_3O_2$		
$(2.5\pm1.2) \times 10^{-12}$	0.65	Lesclaux (93)
$CH_3C(O)O_2 + C_2H_5O_2$		
1.0×10^{-11}		Lesclaux (94)

Table 8: continued

ate Coefficient [a]	Branching Ratio[b]	Reference[c]
(298 K) (cm^3 s^{-1})	k_a (298 K)/k(298 K)	
$C_3H_5O_2 + C_2H_5O_2$		Lesclaux (94)
) × 10^{-12}		
$C_5H_{11}O_2 + C_2H_5O_2$		Lesclaux (94)
5 × 10^{-13}		
c-$C_6H_{11}O_2 + C_2H_5O_2$		Lesclaux (94)
< 10^{-14}		
$C_2H_5C(O)O_2 + C_2H_5O_2 \overset{a}{\to} C_2H_5O_2 + C_2H_5O + CO_2$		
$\overset{b}{\to} C_2H_5C(O)OH + CH_3CHO + O_2$		
±1.5) × 10^{-12} (prelim.)	0.25	Veyret (90)
	$k_a = (1\pm1) \times 10^{-12}$	
	$k_b = (3\pm0.5) \times 10^{-12}$	
$(CH_3)_2CHC(CH_3)_2O_2 + s$-$C_3H_7O_2$		
< 10^{-17}		Warneck (91)
$CH_3C(O)O_2 + c$-$C_6H_{11}O_2$		Lesclaux (94)
) × 10^{-11}		
$CH_3C(O)CH_2O_2 + CH_3C(O)O_2$		
0±1.0) × 10^{-12}	0.5	Lesclaux (92)
overall RO$_2$ reactions resulting from OH + isoprene	0.7–0.8	Becker (93)

[x] $k_{observed}$ with $k < k_{observed} < 2\,k$

[a] rate coefficients at 298 K and temperature dependence when available.

[b] ratio of rate of RO formation / total rate at 298 K; temperature dependence given when available.

[c] These references are to EUROTRAC Ann. Reps. The year in parentheses is the year of the report not the publishing year.

MMS molecular modulation spectroscopy

(1) P. D. Lightfoot, B. Veyret, R. Lesclaux; *J. Phys. Chem.* **94** (1990) 708.

(2) M. E. Jenkin, R. A. Cox, G. D. Hayman, L. J. Whyte; *J. Chem. Soc. Faraday Trans. 2*, **84** (1988) 913.

(3) G.K. Moortgat, R.A. Cox, G. Schuster, J.P. Burrows, G. S. Tyndall; *J. Chem. Soc. Faraday Trans. 2*, **85** (1989) 809.

(4) J.P. Burrows, G. Moortgat, G. Tyndall, R. Cox, M. Jenkin, G.D. Hayman, B. Veyret; *J. Phys. Chem.* **93** (1989) 2375.

(5) G. K. Moortgat, B. Veyret, R. Lesclaux; *Chem. Phys. Lett.* **160** (1989) 443.

(6) O. Horie, J. N. Crowley, G. K. Moortgat; *J. Phys. Chem.* **94** (1990) 8198.

(7) F. G. Simon, W. Schneider, G. K. Moortgat; *Int. J. Chem. Kinetics* **22** (1990) 791.

(8) I. Barnes, K. H. Becker, L. Ruppert; *Chem. Phys. Lett.* **203** (1993) 295.

It should be noted that the trends observed for the rate constants, particularly regarding the size of alkylperoxy radicals and halogen substitution, are generally opposite to those observed for the RO_2 + NO reactions. This is an important result, since the two types of reactions are in competition in the ozone formation processes. RO_2 + NO are propagating reactions, whereas RO_2 + HO_2 are terminating reactions, the products of the latter reactions being non-radical, generally hydroperoxides. An alternative reaction channel, producing a carbonyl compound and H_2O has been identified for the $HOCH_2O_2$, $CH_3OCH_2O_2$ and CH_2ClO_2 reactions, with a branching ratio of 40–50 %, but this reaction channel seems to be minor in the other reactions investigated. The formation of an organic acid along with ozone in reactions of acylperoxy radical reactions with HO_2 has been confirmed.

2.4.3 RO_2 + RO_2 reactions

Self-reactions of peroxy radicals may occur in the atmosphere under low NO_x concentrations, particularly since the reactions are fast, with rate constants which can reach values as high as 10^{-11} $cm^3 molecule^{-1}s^{-1}$. In addition, self-reactions must be well characterised for laboratory studies of other reactions of peroxy radicals, particularly those with HO_2 and $R'O_2$. The general mechanism for RO_2 self-reactions is the following:

$$RO_2 + RO_2 \quad \rightarrow \quad 2\,RO + O_2 \tag{1a}$$
$$\rightarrow \quad ROH + R_{-H}O + O_2 \tag{1b}$$

ROH is an alcohol, $R_{-H}O$ an aldehyde or a ketone and α the branching ratio of the non-terminating channel (1a): $\alpha = k_{1a}/k_1$. A possible third channel, forming the peroxide ROOR, has never been observed. The non-terminating channel generates an alkoxy radical which can further react with O_2 to yield a carbonyl compound and HO_2, decompose, or isomerise to generate a different peroxy radical. Rate constants for self-reactions have been determined by either ignoring subsequent reactions, thus giving an observed rate constant, k_{obs}, or modelling the complete reaction mechanism for a determination of the real rate constant. All rate constants measured during the LACTOZ project are collected in Table 8. The determination of the elementary rate constant requires knowledge of the branching ratio α, which has been determined for several reactions (Table 8). The values of α are fairly insensitive to the structure of alkylperoxy radicals and to the presence of functional groups (with the exception of halogen substitution) and generally fall in the range 0.4–0.7 at 298 K. In contrast, α seems to be close to 1 for all halogen substituted radicals, for all investigated temperatures.

Primary peroxy radicals (RCH_2O_2)

The self reactions of the simplest primary alkylperoxy radicals have rate constant values at 298 K of about 10^{-13} $cm^3 molecule^{-1}s^{-1}$. These rate constants for non–linear alkyl groups apparently increase with chain branching, e.g. 1.2×10^{-12} $cm^3 molecule^{-1}s^{-1}$ for neopentylperoxy. These values seem to be the lower and upper limits for primary alkylperoxy radicals and more data would be required to refine the structure-reactivity relationships for this class of radicals.

The most striking feature of the results is the dramatic increase of the rate constant following substitution of an H atom in the alkylperoxy radicals with halogen atoms, an OH group, a carbonyl group, unsaturated bond or aromatic ring. Apparently, this effect is independent of the position of the substituent (α or β) relative to the peroxy function. In the case of the methylperoxy radical, for example, the presence of a F atom, Cl atom(s), or an OH group, increases the rate constant by more than a factor of 10. The effect of bromine, in contrast, seems smaller. The enhancement is even larger with the presence of a carbonyl group (*e.g.* acetonylperoxy) or an aromatic ring (*e.g.* benzylperoxy). Rate constants for such primary peroxy radicals bearing a particular functional group generally fall in the range $2 - 8 \times 10^{-12}$ cm^3 molecule^{-1} s^{-1}.

The acetylperoxy radical, and probably other acylperoxy radicals RC(O)O$_2$, exhibit even higher rate constants: 1.7×10^{-11} cm^3 molecule^{-1} s^{-1}. Acylperoxy radicals are certainly the most reactive peroxy radicals, as emphasised below for other type of reactions.

A negative temperature dependence of the rate constant seems to be a general characteristic for self-reactions of primary alkylperoxy radicals, with most exponential factors being in the range 0–1000 K/T.

Secondary peroxy radicals (**RCHO$_2$R'**)

Only few data are available for self-reactions of secondary alkylperoxy radicals. One of the principal contributions of the project has been the study of the cyclo-alkylperoxy radicals c-C$_5$H$_9$O$_2$ and c-C$_6$H$_{11}$O$_2$ radicals. The corresponding rate constants are 40 times smaller than the only linear secondary peroxy radical studied so far, i-C$_3$H$_7$O$_2$, and exhibit a slight positive temperature dependence. More data would be necessary for a better description of such reactions but, due to their relatively low rate constants ($< 5 \times 10^{-14}$ cm^3 molecule^{-1} s^{-1}), their contribution to atmospheric chemistry is negligible.

It should be emphasised, however, that the presence of a functional group may result in an even more dramatic increase of the rate constant than in the case of primary radicals. This has been shown in the cases of radicals resulting from the addition of Br atoms or OH radicals to 2-butene (Moortgat, Hayman), the rate constant approaching 10^{-12} cm^3 molecule^{-1} s^{-1} for these particular cases. More data are needed to confirm whether this effect occurs for the presence of other functional groups.

Tertiary peroxy radicals (**RR'R''CO$_2$**)

No new data have been reported for tertiary alkylperoxy radicals. It has been confirmed that the rate constant of the t-butylperoxy radical self-reaction is very slow, with a fairly high activation energy. However, as above, a dramatic increase of the rate constant has been observed in the presence of an OH group at the β position with respect to the peroxy function (Hayman). However, the reaction is still slow (5.7×10^{-15} cm^3 molecule^{-1} s^{-1}) and has no relevance to atmospheric chemistry.

2.4.4   RO$_2$ + CH$_3$O$_2$ cross-reactions

CH$_3$O$_2$ is the most abundant alkylperoxy radical in hydrocarbon-rich atmospheres. It is of interest to assess the role of RO$_2$ + CH$_3$O$_2$ cross-reactions since a series of rate constants has been obtained for such reactions, either resulting from indirect determinations (secondary reactions of RO$_2$ + RO$_2$ reactions) or from direct measurements. Rate constants for this class of reactions are reported in Table 8.

It is observed that the rate constants are generally intermediate between the rate constants for self-reactions of RO$_2$ and CH$_3$O$_2$. A geometric mean such as

$$k(RO_2 + CH_3O_2) = 2 \sqrt{k(RO_2 + RO_2) \times k(CH_3O_2 + CH_3O_2)}$$

appears to provide a good estimate of the cross-reaction rate constant. However, it is observed that when the RO$_2$ self-reactions are fast, $k(RO_2 + CH_3O_2)$ is much closer to $k(RO_2 + RO_2)$ than to $k(CH_3O_2 + CH_3O_2)$. This is the case, for example, for acetyl, acetonyl and chloromethylperoxy radicals. If this behaviour is indeed general, a better approximation would be to set $k(RO_2 + CH_3O_2)$ equal to $k(RO_2 + RO_2)$. This would be incorrect for slow reactions, but as far as atmospheric chemistry is concerned, slow reactions are unimportant.

2.4.5   RO$_2$ + CH$_3$C(O)O$_2$ cross-reactions

A few cross-reactions of RO$_2$ radicals with the acetylperoxy radical have been investigated:

CH$_3$O$_2$, C$_2$H$_5$O$_2$ and c-C$_6$H$_{11}$O$_2$.

The reactions are very fast, with rate constants close to 10^{-11} cm^3 molecule^{-1} s^{-1}. This is even true for the secondary radical c-C$_6$H$_{11}$O$_2$, in spite of the much lower rate constant observed for the self-reaction (4×10^{-14} cm^{3}molecule^{-1}s^{-1}). This may be important since the acetylperoxy radical is also an abundant radical in the atmosphere and, if all cross-reactions of this type are as fast as those reported above, they should be taken into account in modelling the chemistry of remote atmospheres.

2.4.6   Reactions of RO$_2$ radicals with nitrogen oxides

As emphasised above, reactions of RO$_2$ radicals with NO are important propagating reactions, resulting in the formation of NO$_2$ which, on photolysis, yields O$_3$. The reactions of a series of peroxy radicals, derived from both hydrocarbons and halocarbons, have been investigated. The rate constants are given in Table 9 (see also Fig. 5a). Two significant trends can be observed. The rate constants increase markedly with halogen substitution (probably related to the electron accepting nature of these substituents), whereas they decrease with the size of the organic peroxy radical (probably due to the opposite effect). It is interesting to note that the very low rate constant measured for the reaction of the $(CH_3)_3CC(CH_3)_2CH_2O_2$ radical (1.8×10^{-12} cm^3 molecule^{-1} s^{-1}) is almost one order of magnitude smaller than the highest rate constants observed for halogenated peroxy radicals. As emphasised above, the trend observed with the size of radical is opposite to that observed for $RO_2 + HO_2$ reactions. Thus, the latter, which is a terminating reaction, is favoured for this particular class of radicals. Note that the acetylperoxy radical reaction is one of the fastest reactions.

Table 9: Rate constants for the reactions of RO_2 radicals with NO_x;
$RO_2 + NO$, k_1
$RO_2 + NO_2$, k_2

RO_2	Total Pressure (mbar)	Temperature (K)	k_2 (10^{-12} cm^3 s^{-1})	k_1 (10^{-12} cm^3 s^{-1})	k_2/k_1	Reference [a]
CH_3O_2	10.3/N_2	298		7.8 ± 1.7		Zellner (90)
		298		8.8 ± 1.4		Nielsen (93)
$C_2H_5O_2$	10.3/N_2	298	1.8 ± 0.1	7.2 ± 1.3		Zellner (89)
		298		8.5 ± 1.2		Nielsen (93)
$HOC_2H_4O_2$	1013/air	298		9 ± 4 [b]		Becker (91)
i-$C_3H_7O_2$	2.7/He	290		5.0 ± 1.2		Peeters (91)
t-$C_4H_9O_2$	2.7/He	290		4.0 ± 1.1		Peeters (91)
s-$C_4H_9O_2$	2.7/He	290		4.1 ± 1.0		Peeters (91)
$CH_3C(O)O_2$	1013/N_2	298	9.6 ± 0.9	21 ± 7		Becker (89)
						Veyret (89)
$(CH_3)_3CCH_2O_2$		298		4.7 ± 0.4		Nielsen (93)
$(CH_3)_3CC(CH_3)_2CH_2O_2$		298		1.8 ± 0.2		Nielsen (93)
$C_6H_5C(O)O_2$	1013/N_2	315			0.62	Becker (91)
CF_3CHFO_2	2.0/He	290		15.6 ± 4		Peeters (94)
$CF_3CH_2O_2$		290		12.3 ± 3.2		Peeters (93)
$CF_2ClCH_2O_2$		290		11.8 ± 3.0		Peeters (93)
$CF_3CCl_2O_2$		290		14.5 ± 3.9		Peeters (93)
CH_2FO_2		298		12.5 ± 1.3		Nielsen (93)
CH_2ClO_2		298		18.7 ± 2.0		Nielsen (93)
CH_2BrO_2		298		10.7 ± 1.1		Nielsen (93)
CHF_2O_2		298		12.6 ± 1.6		Nielsen (93)
$CFCl_2CH_2O_2$		298		12.8 ± 1.1		Nielsen (93)
$CF_2ClCH_2O_2$		298		11.8 ± 1.0		Nielsen (93)
		290		11.8 ± 3.0		Peeters (94)
$CH_3SCH_2O_2$	1000	298	(9.2 ± 0.9)			Nielsen (94)

[a] These references are to EUROTRAC Annual Reports. The year in parentheses is the year of the report not the publishing year.
[b] reaction channel $RO_2 + NO$ Æ $RO + NO_2$

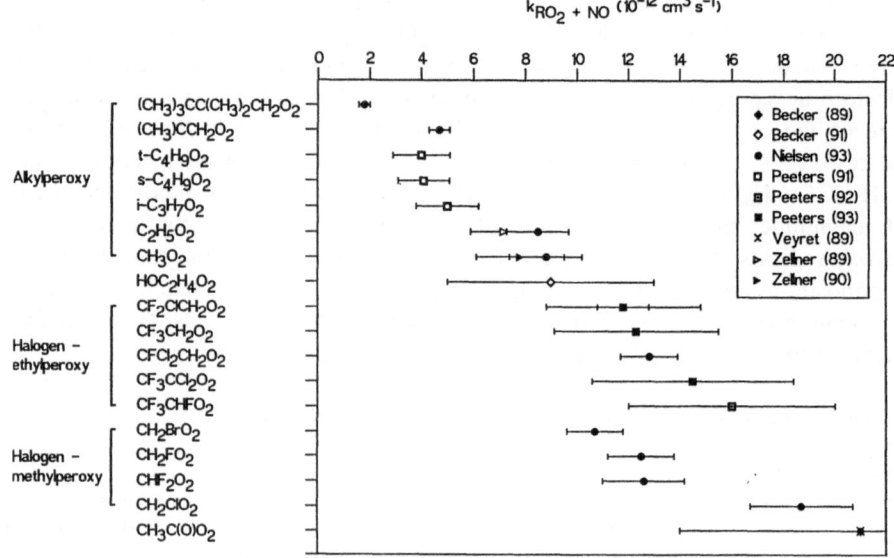

Fig. 5a: Rate constants for the reactions $RO_2 + NO$.

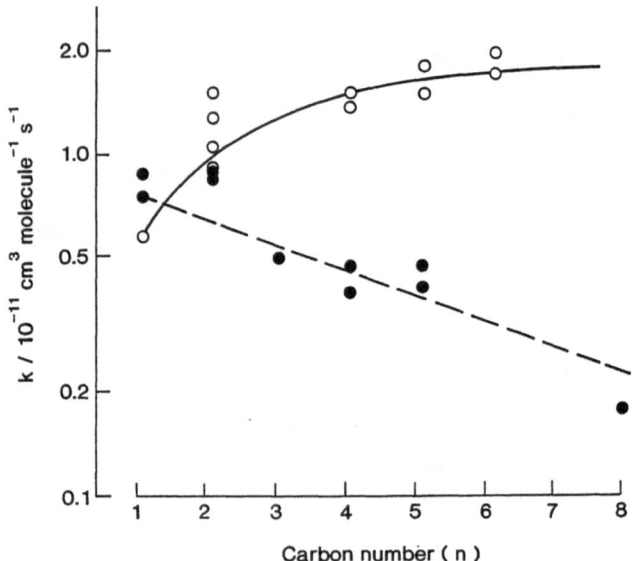

Fig. 5b: Room Temperature data for reactions of alkyl peroxy and β-hydroxy alkyl peroxy radicals with HO_2 (open circles) and NO (closed circles) as a function of carbon number.

Table 10: Branching ratios for nitrate formation at about 1000 mbar in air;
$RO_2 + NO \rightarrow RO + NO_2$, k_1 　　　　$RO_2 + NO \rightarrow RONO_2$, k_2

Precursor Radical R	Temperature (K)	$k_2/(k_2+k_1)$	Reference [a]
propane			
1-propyl	295	0.019	Becker (90)
2-propyl	295	0.049	Becker (90)
n-butane			
1 butyl	295	0.041	Becker (90)
isobutane			
2-methyl-1-propyl	295	0.075	Becker (90)
t-butyl	295	0.180	Becker (90)
isobutene			
2-methyl-1-hydroxy-2-propyl	300	0.40[b]	Peeters (89)
	295	0.057[c]	Becker (90)
2-butanone			
2-keto-3-butyl	293	0.0021	Becker (89)
	323	0.0025	Becker (89)
isopentane			
3-methyl-1-butyl	295	0.043	Becker (88)
2-methyl-1-butyl	295	0.039	Becker (88)
3-methyl-2-butyl	295	0.074	Becker (88)
2-methyl-2-butyl	295	0.044	Becker (88)
n-pentane			
1-pentyl	295	0.061	Becker (90)
n-hexane			
1-hexyl	295	0.121	Becker (90)
cyclohexane			
cyclohexyl	273	0.113	Becker (89)
	293	0.074	Becker (89)
	323	0.046	Becker (89)
	343	0.025	Becker (89)
n-heptane			
2-, 3- and 4-heptyl	295	0.265[d]	Kerr (88)
1-heptyl	295	0.195	Becker (90)
n-octane			
1-octyl	295	0.360	Becker (90)
2-, 3- and 4-octyl	295	0.370	Becker (90)
isoprene	295	0.10[d]	Becker (90)

[a] These references are to EUROTRAC Annual Reports. The year in parentheses is the year of the report not the publishing year.

[b] 2 Torr total pressure He, value seems to be too large.

[c] Represents the total nitrate yield for the OH radical initiated oxidation in the presence of NO_x.

[d] Mean branching ratio for 2-, 3- and 4-heptylperoxy radicals; temperature dependence obtained by $k_2/k_1 = 1.4 \times 10^{-3} \exp(1640/T)$.

It should be noted that in nearly all experiments carried out within LACTOZ, overall rate constants have been measured. However, the channel, $RO_2 + NO \rightarrow RONO_2$ relative to the $RO_2 + NO \rightarrow RO + NO_2$ channel, gains importance for larger RO_2 radicals. The branching ratios for nitrate formation have been shown to increase to more than 30 % (Table 10).

When the absolute reaction rate constants are not available, the ratio $k(RO_2 + NO_2)/k(RO_2 + NO)$ is an important parameter, giving the relative importance of the terminating and the non-terminating reaction channels. This ratio has been determined for a few typical radicals (Becker).

The reactions of RO_2 radicals with NO_2 are also very important. They are terminating reactions, in competition with the propagating $RO_2 + NO$ reactions (Fig. 5b), and they form peroxynitrates which can contribute to the transport of nitrogen oxides if they are sufficiently stable. The most important reaction influencing the formation of ozone is the reaction of the acetylperoxy radical with NO_2 forming PAN. The kinetics and thermodynamics of the forward and reverse reactions have been investigated in detail (Becker, Lesclaux), resulting in the currently recommended rate constant values. Examination of the absolute values of rate constants for reactions of RO_2 radicals with NO_2 at the high pressure limit, determined from LACTOZ investigations and also those reported in the literature, shows that they all appear to be very close to each other, *e.g.* $(7 \pm 2) \times 10^{-12}$ cm^3 molecule^{-1}s^{-1}. Thus, this value can be recommended for all $RO_2 + NO_2$ reactions. However, fall-off effects must be taken into account for the reactions of smaller radicals (Table 11).

Table 11: Kinetic parameters for the thermal decomposition of peroxynitrates, RO_2NO_2 (+M) → RO_2 + NO_2 (+M).

R	$k_{298K, 1\,bar\,N2}$ (s^{-1})	$E_{a, 1\,bar}$ (kJ/mol)	$A_{1\,bar}$ $(10^{16}\,s^{-1})$
C_6H_{11}	4.8	87.4±6.4 [a]	1 [b]
CH_3CH_2	3.7 [c]	86.2±4.0 [c, d]	0.47 [c]
$CH_3C(O)CH_2$	2.6 [c]	85.6±15 [c, d]	0.26 [c]
CF_2ClCH_2	2.3	90.9±5.9 [a]	2 [b]
$CFCl_2CH_2$	2.1	91.2±5.3 [a]	2 [b]
$C_6H_5CH_2$	1.8	89.8±7.2 [a]	1 [b]
CH_3	1.65	83.8±5.9	0.080
$HOCH_2$	1.0 [e]	85.3 [b]	0.09 [e]
CH_2I	0.8	91.8±6.8 [a]	1 [b]
CH_3OCH_2	0.45	93.5±5.4 [d]	1.11
CH_2Cl	0.26 [c]	91.7±2.9 [c, d]	0.31 [c]
$CH_3C(O)OCH(CH_3)$	0.23	94.9±7.3 [a]	1 [b]
$C_6H_5OCH_2$	0.22	95.0±6.8 [a]	1 [b]
CCl_3	0.19 [c]	96.8±1.4 [c, d]	1.8 [c]
H	0.083	89.2±2.2 [d]	0.036
CCl_2F	0.066 [c]	100.3±1.8 [c, d]	2.5 [c]
CF_3	0.043	97.7±1.0 [d]	0.57
$CClF_2$	0.040 [c]	98.7±1.8 [c, d]	0.8 [c]
$(CH_3)_2NC(O)$	0.00123	111.5±5.8 [d]	4.3
$CH_3OC(O)$	0.00084	107.0±4.5 [d]	0.48
$C_6H_5OC(O)$	0.00052	110.0±7.7 [a]	1 [b]
$CH_3C(O)$	0.00040 [c]	112.9±1.9 [c, d]	2.5 [c]
$C_2H_5C(O)$	0.00035	115.9±2.2 [d]	7.2
$C_6H_5C(O)$	0.00031	116.4±3.8 [d]	7.9
$ClC(O)$	0.00017 [c]	115.1±3.2 [c, d]	2.6 [c]
$CCl_3C(O)$	0.00012 [c]	120.6±5.2 [c, d]	16.4 [c]
$CFCl_2C(O)$	0.00012	118.0±5.2 [a]	6 [b]
$CF_2ClC(O)$	0.00010	118.6±5.2 [a]	6 [b]
$CF_3C(O)$	0.00008	119.1±5.0 [a]	6 [b]

[a] derived from k and estimated A, error reflects uncertainties of k and A;
[b] estimated by analogy from measured A or E_a of similar molecules
[c] p_{tot} = 800 mbar;
[d] 2σ error;
[e] p_{tot} = 600 mbar;
(Becker, EUROTRAC Annual Report 1994)

Appendix to Table 11: Predictions of rate constants k for the thermal decomposition of peroxynitrates.

1. Select d_x values for ^{13}C NMR shift of RX (X: OCH_3, Cl, F, H, OH, CH_3).
2. Enter the equation below with the proper numbers of a_x, b_x, c_x and d_x from the table.

$$k\,[s^{-1}] = A \cdot \exp\left(-\frac{E_a}{RT}\right) = 3 \times 10^{-3} \cdot \exp\left(\frac{a_x\delta_x + b_x}{R(c_x\delta_x + d_x)}\right) \cdot \exp\left(-\frac{a_x\delta_x + b_x}{RT}\right)$$

$$E_{a,1\,atm}\,[J/mol] = a_x\,\delta_x + b_x$$

X	OCH_3	Cl	F	H	OH	CH_3
a_x (Jxmol^{-1}·ppm^{-1})	307.5	226.5	358.0	153.4	277.1	170.9
b_x (Jxmol^{-1})	65 160	76 530	57 050	85 030	69 300	82 530
c_x (Kxppm^{-1})	0.7011	0.5437	0.8212	0.3458	0.6136	0.3927
d_x (K)	199.95	224.47	180.23	245.53	209.75	239.72

(Becker)

2.4.7 Thermal decomposition of peroxynitrates

Studies on the thermal stability of peroxynitrates have been extended (Becker). From the results of measurements of thermal decomposition rate constants for 28 peroxynitrates (Table 11), the following conclusions can be drawn:

- acyl peroxynitrates are thermally much more stable than any of the alkylperoxynitrates;
- the thermal stability of peroxynitrates is governed by the bond strength RO_2-NO_2 which is highly dependent on the electron density at the carbon atom next to the -O_2NO_2 group. Electron withdrawing substituents increase the thermal stability of peroxynitrates compared to substituents which are less electron withdrawing. For alkyl peroxynitrates inductive effects are the most important;
- the electronegativities of the atoms next to the C atom of the >CO_2NO_2 group will therefore determine the thermal stability of RO_2NO_2. For acyl peroxynitrates mesomeric effects are important and cannot be neglected.

Because of the importance of electron density, the activation energies for thermal decomposition of peroxynitrates are highly correlated with ^{13}C-NMR signal shifts of the corresponding R-X compounds (X= OCH_3, Cl, F, H, OH, CH_3). From these correlations an expression has been derived which can be used to estimate rate constants and activation energies for the thermal decomposition of peroxynitrates (see appendix to Table 11). Rate data for the decomposition of $CH_2IO_2NO_2$ indicate that these equations are not valid in cases where the substituents are heavy atoms. Using the equations, activation energies for the thermal decomposition of all measured peroxynitrates can be estimated within error limits to ± 6 kJ mol^{-1} (experimental error limits: 2–7 kJ mol^{-1}), with the exception of those peroxynitrates containing heavy atom substituents. All rate constants, which are calculated from the equation (see appendix to Table 11), agree with the measured values within a factor of 6, and 70 % of them agree with the measured values within a factor of 3. The experimental errors of the rate constants are usually between 10 and 50 %. Table 12 gives the relevant parameters of the equilibrium constant for $CH_3C(O)O_2 + NO_2 \Leftrightarrow CH_3C(O)O_2NO_2$. Knowledge concerning the thermal stability of peroxynitrates is now sufficiently advanced that the stability of peroxynitrates, for which no experimental data is available, can be estimated with reasonable confidence. Possible exceptions are peroxynitrates with a sulfur bonded to the peroxy group. Evidence has been found that these species might be stable enough, $e.g.$ $CH_3S(O_2)O_2NO_2$ (Hjorth) to play a role as NO_x carriers in the marine troposphere. Determination of the rate constants for photolysis and reaction with the OH radical of several representative peroxynitrates should be considered in order to compare the rates of these reactions with those for thermal decomposition of the more stable peroxynitrates.

Table 12:   Equilibrium constant* K_e $(cm^3$ $molecule^{-1})$;
$CH_3C(O)O_2 + NO_2 \leftrightarrow H_3C(O)O_2NO_2$.

$$K_e = 0.9 \times 10^{-28} \exp[(14.000\pm200)/T]$$

(200–300 K)

$$= (2.3\pm0.3) \times 10^{-8}$$

(298 K)

ΔS°_{298}	$= 174.0$	± 11 $J K^{-1} mol^{-1}$
ΔH°_{298}	$= -118.8$	± 4.0 kJ mol^{-1}
ΔH°_o	$= E_o = -115.1$	± 4.0 kJ mol^{-1}

* Equilibrium constants for other peroxynitrates can be derived from decomposition rate constants given in Table 11, taking into account that all measurements of k $(RO_2 + NO_2)$ give values very close to each other, k_∞ $(RO_2 + NO_2) \sim (7.0 \pm 3) \times 10^{-12}$ cm^3 s^{-1}.
Becker (89/90) and Veyret (89/90)

2.5 Reactions of oxy radicals

2.5.1 Introduction

Oxy radicals (RO) are important intermediates in all VOC oxidation chains. They are formed in the chain propagating channels of the reactions of RO_2 radicals with NO and in the self-reactions of RO_2. Three principal reaction pathways for RO radicals have been identified under atmospheric conditions:

- reaction with O_2, forming a carbonyl compound and HO_2;
- dissociation, involving C-C bond splitting and formation of a carbonyl compound and a new alkyl radical;
- isomerisation via an intramolecular H atom shift forming a hydroxyalkyl radical.

Table 13: Rate constants for decomposition / isomerisation of alkoxy radicals and branching ratios b at 298 K in air at atmospheric pressure, $\beta = (k_{dec} + k_{iso}) / (k_{dec} + k_{iso} + k_{O_2} [O_2])$.

RO	$k_{dec} + k_{iso}$ (s^{-1})	β	Reference [a]
CH3O	< 50	$< 4 \times 10^{-3}$	Zellner (90)
	dominant reaction with O_2		Lesclaux (93)
HOCH$_2$O	dominant reaction with O_2		
	and H atom elimination		Lesclaux (93)
CH$_2$ClO	dominant reaction with O_2		
	(small HCl elimination)		Lesclaux (93)
CHCl$_2$O	dominant decomposition		Lesclaux (93)
CCl$_3$O	dominant decomposition		Lesclaux (93)
propoxy [b]	$\leq 1 \times 10^3$	≤ 0.01	Zellner (93) [d]
i-propoxy	$\leq 3.0 \times 10^2$	≤ 0.01	Zellner (92) [d]
CH$_3$C(O)CH$_2$O	dominant decomposition		Lesclaux (93)
	dominant decomposition		Hayman (93)
CH$_3$CH(O$\cdot$)CH$_2$Cl	$\leq 2 \times 10^3$	≤ 0.05	Zellner (92) [d]
CH$_2$ = CHCH$_2$O	$< 10^3$	0	Hayman (91)
	dominant reaction with O_2		Lesclaux (93)
n-butoxy [d]	3.8×10^3	0.09	Zellner (92) [d]
i-butoxy	1.1×10^3	0.03	Zellner (92) [d]
HOC(CH$_3$)$_2$CH$_2$O	dominant decomposition		Lesclaux (93)
CH$_3$CH(O)OC$_2$H$_5$	$> 10^5$	> 0.75	Kerr (92)
n-pentoxy [a]	7.1×10^3	0.16	Zellner (92) [d]
(CH$_3$)$_3$CCH$_2$O	$> 3 \times 10^5$	~ 1 [c]	Lesclaux (90)
2-pentoxy	7.5×10^4	~ 0.9 [c]	Warneck (90)
3-pentoxy	5×10^3	~ 0.13 [c]	Warneck (90)
neopentoxy	competition between O_2		Lesclaux (93)
	reaction and decomposition		
c-C$_5$H$_9$O	dominant decomposition		Lesclaux (93)
n-hexoxy [d]	3.3×10^4	0.48	Zellner (92) [d]
2-hexoxy	99 % 1,5-isomerization		Kerr (93)
3-hexoxy	74 % 1,5-isomerization		Kerr (93)
	7 % reaction with O_2		
	19 % decomposition		
c-C$_6$H$_{11}$O	3×10^3	0.07	Zellner (93) [d]
	competition between O_2		
	reaction and decomposition		Lesclaux (93)
benzyloxy	dominant reaction with O_2		Lesclaux (93)
n-heptoxy [d]	4.4×10^4	0.53	Zellner (92) [d]
n-oxtoxy [d]	4.7×10^4	0.55	Zellner (92) [d]

[a] These references are to EUROTRAC Annual Reports. The year in parentheses is the year of the report not the publishing year.
[b] includes all oxy radicals originating from n-alkane; isomer unspecified.
[c] based on an assumed ratio coefficient for RO + O_2 of 8×10^{-15} cm^3 s^{-1}.
[d] based on measurements at 50 mbar of O_2, but extrapolated to 1 bar of air.

Table 14: Rate constants for the reactions of alkoxy radicals with O_2 at 298 K.

RO	k (cm^3 s^{-1})	Reference [a]
CH_3O	1.8×10^{-15}	Zellner (90)
C_2H_5O	1.1×10^{-14}	Zellner (89)
	8.8×10^{-15}	Zellner (92)
CH_2ClCH_2O	$(8\pm2) \times 10^{-15}$	Zellner (92)
propoxy [b]	' '	' '
i-propoxy	' '	' '
$CH_3CH(O^{\bullet})CH_2Cl$	' '	' '
n-butoxy [b]	' '	' '
i-butoxy	' '	' '
n-pentoxy [b]	' '	' '
n-hexoxy [b]	' '	' '
n-heptoxy [b]	' '	' '
n-octoxy [b]	' '	' '

[a] These references are to EUROTRAC Annual Reports. The year in parentheses is the year of the report not the publishing year.
[b] includes all oxy radicals originating from n-alkane; structure of isomer unspecified.

In the first of these processes, the carbonyl compound generated contains the same number of carbon atoms as the parent VOC. Formation of HO_2 radicals leads to subsequent NO to NO_2 conversion associated with HO_x radical recycling. The second process (dissociation) gives rise to breakdown of the carbon chain, and as a consequence the resulting carbonyl compound is of smaller size than the present VOC and reflects oxidative breakdown of the VOC. The third process (isomerisation) may lead to several NO to NO_2 conversion as the carbon atoms in the molecule are oxidised by internal H abstraction. The alkyl radicals generated in both dissociative and isomerisation reactions lead to formation of RO_2 radicals and hence enhanced NO to NO_2 conversion. Thus, the three alternative reaction pathways for oxy radicals have a large influence on the total net O_3 generation.

Despite the importance of RO radical reactions, study of their kinetic parameters is not very advanced. This is because, apart from the simplest RO's such as CH_3O and C_2H_5O, convenient production and detection techniques are not available for these species. As a consequence, rate data must be derived from relative product yields or by using indirect techniques. The results obtained within LACTOZ are summarised in Tables 13 and 14.

A large data base for the reactions of RO radicals has been provided by the work of Zellner. This group has used time resolved detection of both OH and NO_2 in the pulse-initiated oxidation of VOCs to determine branching ratios and absolute rate coefficients for the different reaction channels of RO radicals. However, in these studies the integrated oxidation chain for a particular VOC was investigated and hence the results for the RO reactions are weighted averages over a number of different RO radicals. Although detailed information is lost, this approach has the advantage that it can provide oxy radical lumped oxidation mechanisms for VOCs and provides information on total NO to NO_2 conversion. The major finding of this work is that the rate coefficients for reactions of different oxy radicals with O_2, except for CH_3O, are essentially independent of the structure of the radical ($k \approx 8 \times 10^{-15}$ cm^3molecule$^{-1}s^{-1}$). This implies that for the conditions of the lower troposphere the lifetime of RO is limited to 25 μs. Alternative reaction channels of RO such as decomposition and isomerisation are only important if their rate coefficients exceed $\approx 10^4$ s^{-1}. For n-alkane oxidation this is only the case for compounds $\geq C_5$ (Zellner).

Information on the reaction pathways for oxy radicals has also been derived from kinetic and mechanistic studies of RO_2 self-reactions. Lesclaux has shown that primary oxy radicals such as CH_3O, C_2H_5O, allyloxy and benzyloxy predominantly react with O_2 under tropospheric conditions. However, for cyclo-$C_6H_{11}O$ and neo-pentoxy radicals a competition between fragmentation and reaction with O_2 was observed. Decomposition of the oxy radicals cyclo-C_5H_9O, $CH_3C(O)CH_2O$ and $HOC(CH_3)_2CH_2O$ was found to be the major reaction channel. Information on the behaviour of the acetonoxy ($CH_3C(O)CH_2O$) and methoxy-methoxy (CH_3OCH_2O) radicals using RO_2 reaction kinetic studies has been provided (Jenkin). For CH_3OCH_2O an unusual reaction pathway, namely the elimination of an H atom, was identified in addition to the reaction with O_2.

Two groups (Kerr, Warneck) have provided reactivity data for oxy radicals using product distribution studies. From investigations on the oxidation of diethylether Kerr has derived relative reaction probabilities for the 1-ethoxyethoxy ($CH_3CH(O)OC_2H_5$) radical. It was found that under lower tropospheric conditions this radical mainly decomposes by CH_3 elimination to produce ethylformate as the dominant product (84 %). The atmospheric fate of 3-hexoxy and 2-hexoxy radicals, which are generated in the OH initiated oxidation of n-hexane, has also been investigated. It was shown that for both radicals that 1,5-H atom shift isomerisations are the dominant reaction pathways. In contrast, for the 2-hexoxy radical a significant fraction also reacts by decomposition and/or interaction with O_2. Surprisingly, these findings for the different reactivities of the hexoxy radicals are in very good agreement with suggested empirical predictions. Product distribution studies aimed at providing kinetic and mechanistic information on the reactions of oxy radicals have also been performed (Warneck). As a consequence of the large fractions of cyclic compounds observed in the oxidation of cyclo-hexane the cyclo-hexoxy radical must be stable with respect to C-C bond cleavage. Warneck also studied the behaviour of the different pentoxy radicals (1-, 2- and 3-pentoxy) which are formed in the OH initiated oxidation of n-pentane. For the 1- and 2-pentoxy radicals H-atom shift isomerisation reactions forming hydroxy-pentyl radicals are clearly dominant, whereas this is not the case for the 3-pentoxy radical.

Despite the collection of a reasonably large number of rate coefficients for reactions of RO radicals under atmospheric conditions within LACTOZ, the available data base is still too small to derive general conclusions and/or structure-reactivity relationships for these radicals. Moreover, since the branching ratios for the different reaction channels of RO radicals are expected to be strongly dependent on temperature, there is a remaining need for temperature dependent studies on the reactions of RO radicals.

2.6 Photochemical cross-sections and quantum yields

2.6.1 Introduction

Photochemical processes play a critical role in the chemistry of the atmosphere, since they control the daytime production of reactive free radicals, which initiate chemical transformations of many trace compounds. The photodissociation of atmospheric molecules occurs by absorption of solar ultraviolet (UV) and/or visible (VIS) radiation. The rate of photolysis is determined by the absorption cross-sections of the dissociated molecule and by the quantum yield for the various products channels at the absorbing wavelengths.

An important priority of the LACTOZ programme was placed on the high precision measurement of absorption cross-sections of molecules related to the production of tropospheric ozone. Emphasis was placed on NO_2, O_3 and on organic molecules formed in the photo-oxidation of VOCs which are responsible for generating odd hydrogen radicals (aldehydes and ketones) and the release of NO_x from reservoir molecules (PAN, organic nitrates). The species involved are generated in the following photolytic processes:

$$
\begin{array}{lcl}
O_3 + h\nu & \rightarrow & O_2 + O(^1D) \\
NO_2 + h\nu & \rightarrow & NO + O(^3P) \\
RCHO + 2O_2 + h\nu & \rightarrow & HO_2 + RO_2 + CO \\
RONO_2 + h\nu & \rightarrow & RO + NO_2
\end{array}
$$

For selected compounds, quantum yields were determined as a function of wavelength and pressure, resembling the conditions of the tropospheric environment.

2.6.2 Ozone (O_3)

The absorption cross-sections for O_3 have been measured with a resolution of 0.01 nm over the whole spectral range of the Hartley and Huggins bands (195–345 nm) at ambient temperature and at temperatures down to 218 K (Malicet). A negligible temperature effect (< 1 %) was observed at the maximum in the Hartley band near 250 nm. In contrast, a strong variation of the absorption cross-sections was observed in the Huggins band in the range 300 to 345 nm when the temperature was lowered from 295 to 218 K. This effect of temperature is shown in Fig. 6.

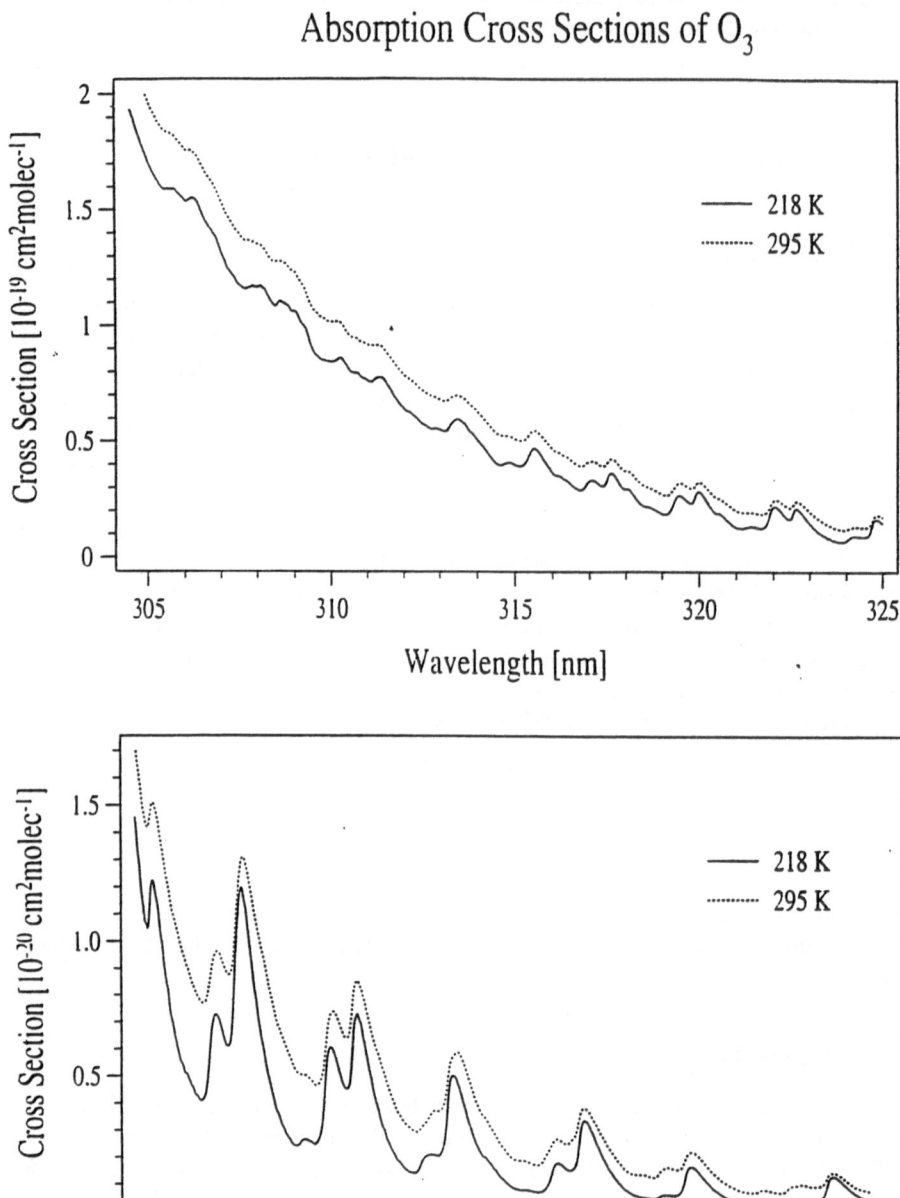

Fig. 6: O₃ Absorption cross-section between 305 and 345 nm at 295 and 218 K (Malicet).

2.6.3 Nitrogen dioxide (NO₂)

New high resolution (0.01 nm) measurements of the absorption cross-sections of NO_2 were performed in the 300–500 nm range at 293, 240 and 220 K using a coolable 5m long path absorption cell (Malicet). Experiments were carried out at very low NO_2 pressures in order to reduce absorption contributions of the dimer, N_2O_4. A definite temperature effect (up to 6 %) was observed in the structured region of the spectrum.

2.6.4 Formaldehyde (HCHO)

Absorption cross-sections of formaldehyde was obtained in the range 240–360 nm at 298 and 220 K by diode array spectroscopy with a spectral resolution of 0.022 nm (Moortgat). These values are consistently higher (up to 10 %) than those previously measured.

2.6.5 Carbonyl compounds

The absorption spectra and chemical degradation pathways of four selected carbonyl compounds, methylethylketone (MEK, $CH_3COC_2H_5$), methylvinylketone (MVK, $CH_3COCH=CH_2$), methacrolein (MACR, $CH_2=C(CH_3)CHO$) and methyl-glyoxal (MGLY, CH_3COCHO) have been investigated (Moortgat). MVK, MACR and MGLY are formed in the photo-oxidation of isoprene, initiated by the reaction of OH radicals and/or O_3, whereas MEK arises as a product in the photo-oxidation of butane. The measured spectra show that absorption of MEK extends to 330 nm, whereas the absorptions of MVK and MACR extend to 400 nm, and to 540 nm for MGLY (Fig. 7).

Quantum yields for decomposition were estimated from product analysis studies, following broad-band photolysis performed using sun lamps (290-360 nm). All these compounds showed increasing product yields at lower pressures in the total pressure range 50 to 760 Torr. The overall quantum yields of the primary processes are summarised in Table 15. In a separate study, the photolysis of MGLY was investigated and the CO quantum yield measured as a function of wavelength over the same pressure range (Moortgat). The quantum yields increase significantly from 0.01 at 410 nm to unity at 280 nm and displayed a strong pressure dependence, decreasing with increasing pressure.

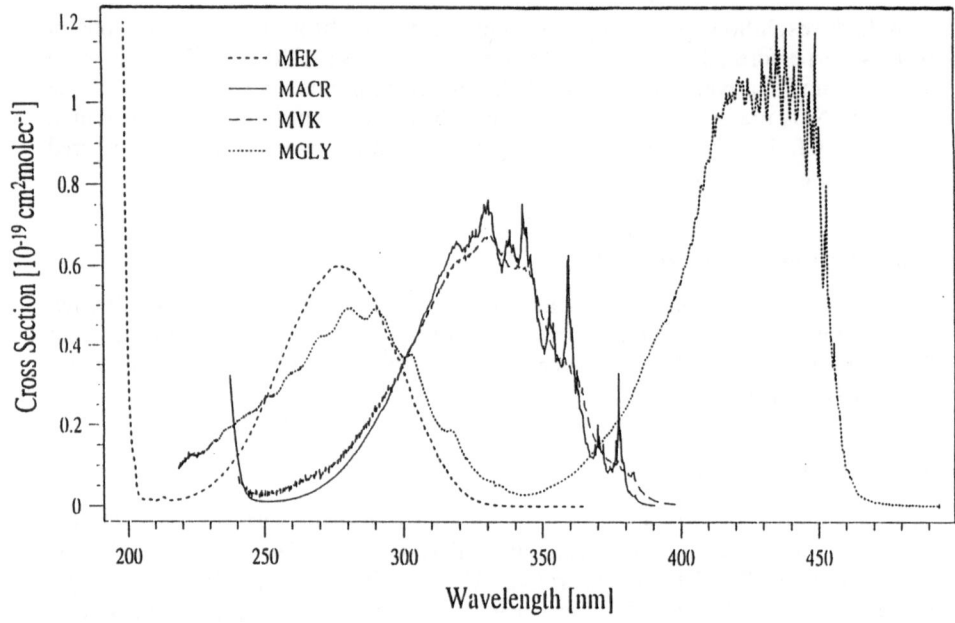

Fig. 7: UV visible absorption cross-sections of several carbonyl compounds. MEK = methylethylketone, MACR = methacrolein, MVK = methylvinylketone and MGLY = methylglyoxal (Moortgat).

Table 15: Calculated photolysis frequencies (J-values) for 30° and 70° solar zenith angle of organic nitrates and carbonyl compounds in the lower troposphere, using the given quantum yields ϕ.

Compound	ϕ	J value[a] (10^{-6}) s^{-1} 30°	70°
Nitrates (Becker)			
peroxyacetylnitrate	1	0.491	0.134
α-nitrooxy acetone	1	26.7	10.4
1-nitrooxy-2-butanone	1	15.1	5.42
3-nitrooxy-2-butanone	1	38.7	15.2
1,2-propandiol dinitrate	1	7.56	3.29
1,2-butandiol dinitrate	1	10.4	4.72
2,3-butandiol dinitrate	1	7.69	3.31
3,4-dinitrooxy-1-butene	1	4.52	1.81
1,4-dinitrooxy-1-butene	1	4.54	1.41
Carbonyl compounds (Moortgat)			
methylethylketone	0.34	3.19	0.932
methylvinylketone	0.05	20.7	13.8
methacrolein	0.03	12.6	8.37
methylglyoxal	b	39.6	29.3

[a] Calculated photolysis frequencies using radiation transfer model of MPI, Mainz.

[b] f = wavelength dependent quantum yield function (Moortgat).

The photolytic lifetimes for average tropospheric conditions were calculated on the basis of the cited quantum yields: 4 days for MEK, 14 h for MEK, 22 h for MACR and 35 min for MGLY. These results indicate that photolysis processes in the troposphere dominate (in the case of MGLY) or are in competition with removal reactions initiated by OH radicals.

2.6.6 Dicarbonyl compounds

Three unsaturated 1,4-dicarbonyl compounds, (butenedial, 4-oxo-2 pentenal and 3-hexene-2,5-dione), formed in the OH initiated oxidation of aromatic compounds, were studied in both UV and VIS regions (Becker). The results indicate that both reaction with OH radicals and photolysis are major atmospheric sinks for these species. In contrast, doubly unsaturated 1,6-dicarbonyl compounds (*e.g.* 2,4-hexadienedial) were found to photolyse extremely rapidly in the UV range at 254 nm, but negligibly in the visible range. These results may suggest that the latter compounds are not photolysed in the troposphere (Becker).

2.6.7 Nitrates and dinitrates

UV spectra of various keto-nitrates ($RCOCH(ONO_2)R$) and dinitrates ($RCH(ONO_2)CH(ONO_2)R$) (Table 15) were measured in the range 240–340 nm using diode array spectroscopy with spectral resolution of 0.7 nm (Becker). In the region of tropospheric interest (> 290 nm), the absorption cross-sections of keto-nitrates are approximately one order of magnitude higher than those of the dinitrates. The results indicate that, assuming a quantum yield of unity, photolysis of the saturated difunctional nitrates will generally be somewhat more important than loss via reaction with OH radicals in the troposphere. However, for un-saturated nitrates, loss via reaction with OH radicals will dominate over photolysis in the troposphere.

Photo-oxidation studies of the nitrates and dinitrates investigated revealed that NO_2 and PAN are important products (Becker). Thus, the photolysis of keto-nitrates and dinitrates will result in the release of NO_2 from these species. As a consequence, organic nitrates formed in the reactions of NO_3 with alkenes during night-time will act only as temporary reservoirs for NO_x.

2.6.8 Other photolytic studies

Nitrate radical (NO_3)

Photolysis of NO_3 at 662 nm in N_2 was performed in order to study the possible reaction of vibrationally excited $NO_3(^2A')$ with N_2 to yield N_2O and NO_2 (Burrows). An upper limit for the quantum yield (0.006) for N_2O formation was obtained. Zellner has observed a value of 0.014 in a recent flash photolysis study of this reaction.

Wavelength-dependent quantum yield studies of the photolysis of NO_3 were performed in the range 580–640 nm (Moortgat). The dominant photolysis channel was confirmed to be $NO_2 + O(^3P)$ at 298 K.

Peroxyacetylnitrate, PAN ($CH_3C(O)O_2NO_2$)

The UV absorption spectrum of PAN (Fig. 8) was determined in the range 235–330 nm using a diode array spectrometer (Becker). This study gave larger cross-sections for wavelengths > 300 nm than previously reported and gives photolysis rate coefficients at 298 K which are higher by about a factor 3 than currently recommended [7].

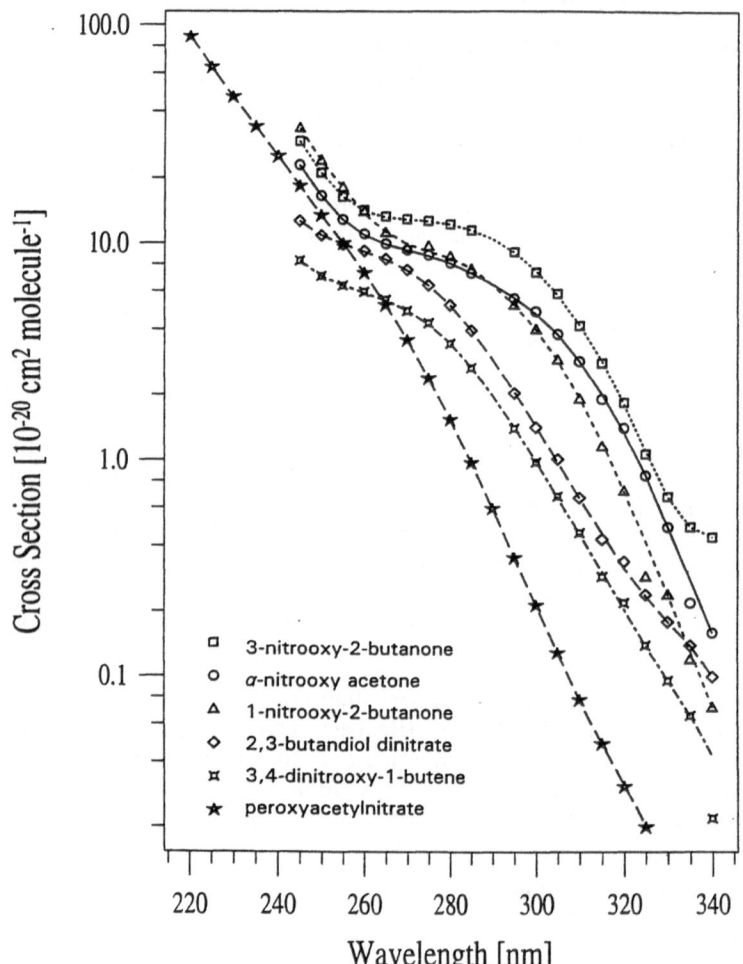

Fig. 8: UV absorption cross-sections of several organic nitrates and dinitrates (Becker).

Pyruvic Acid ($CH_3COCOOH$)

Photo-oxidation studies of pyruvic acid, a product found in the photo-oxidation of methylglyoxal, have been carried out (Warneck). Quantum yields of the primary reaction steps are given in Table 15.

Alkyl nitrites (**RONO**)

Excimer laser photolysis of ethyl nitrite was studied at 248 and 351 nm as a function of O_2 pressure (Zellner). In contrast to methyl nitrite, the photolysis produced not only C_2H_5O radicals but also methyl radicals with high yield. This study indicates that photolysis of higher alkyl nitrites is not an appropriate source for the laboratory generation of alkoxy radicals.

2.7 Oxidation of aromatic compounds

2.7.1 Introduction

Aromatic hydrocarbons (benzene, toluene, xylene isomers) are a major class of organic compounds associated with the urban environment and model calculations appear to indicate that they might contribute more than 30 % of the photo-oxidant formation in urban areas. However, the mechanisms currently used in models of tropospheric chemistry to describe the photo-oxidation of aromatic compounds are highly speculative. Studies within LACTOZ have involved both direct and smog chamber investigations and have provided data on many different aspects of the mechanisms for oxidation of aromatic compounds related to tropospheric ozone and oxidant formation. Most of this work was carried out by four of the LACTOZ groups (Becker, Devolder, Kerr, Zetzsch).

2.7.2 Results

The kinetics of the primary oxidation steps of the OH initiated oxidation of the alkylated benzenes, i) abstraction of an H atom from an alkyl side group, and ii) addition to the ring, as outlined in Fig. 9, are reasonably well established (Becker, Devolder, Kerr, Zetzsch). Branching ratios for abstraction/addition have been obtained for toluene and the xylene isomers, however, data for higher alkylated benzenes are poor. Reversible addition to the aromatic ring is the dominant pathway and accounts for approximately 90 % of the reaction.

Zetzsch has measured reaction rate constants for the reactions of the aromatic OH adduct with O_2, NO_2 and NO (Table 16) (units: $cm^3 molecule^{-1} s^{-1}$, at 298 K):

toluene-OH adduct

$k(O_2)\ = 5.5 \times 10^{-16}$
$k(NO_2) = 3.6 \times 10^{-11}$
$k(NO)\ < 3 \times 10^{-14}$

p-xylene-OH adduct

$k(O_2)\ = 8.0 \times 10^{-16}$
$k(NO_2) = 3.2 \times 10^{-11}$
$k(NO)\ < 10^{-14}$

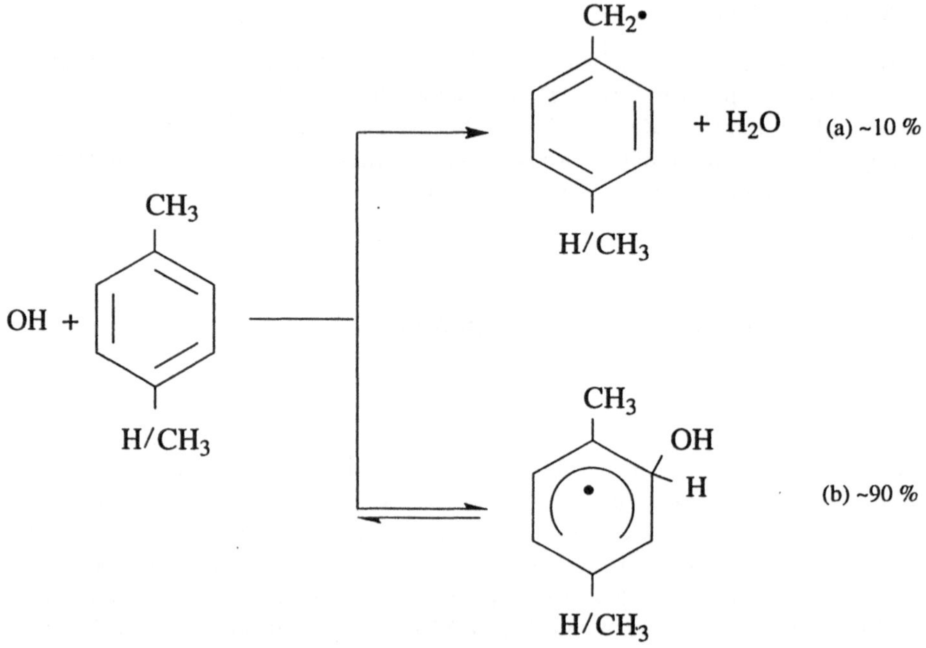

Fig. 9: Primary OH reaction with toluene and p-xylene.

These data show that under atmospheric conditions the adducts will react predominantly with O_2 to form hydroxycyclohexadienylperoxy radicals. The subsequent reactions of these peroxy radicals leading to ring-opening are still unclear. Several possible reaction pathways which result in formation of dicarbonyl and unsaturated dicarbonyl compounds are outlined in Fig. 10. Experiments on radical cycling (Zetzsch) indicate that the reaction of the aromatic-OH adducts with O_2 produce HO_2 radicals with high yield on a short timescale. The mechanism leading to this so-called "prompt HO_2" formation in the aromatic compound oxidation systems is presently unclear.

Careful attempts to detect hydroperoxides and peroxynitrates, expected products of the reactions of the hydroxycyclohexadienylperoxy radicals with HO_2 and NO_2, respectively, were unsuccessful. This has been taken, in conjunction with work on the effect of NO_x on the kinetics and products of the aromatic compound oxidation, as an indication that such peroxy radicals are very short-lived. It has been suggested, therefore, that the peroxy radicals are not able to oxidise NO as is the case for alkane and alkene atmospheric oxidation.

Table 16: Rate constants for the reactions of radicals derived from aromatic VOCs with O_2, NO and NO_2.

Radical	$k_{O_2} \times 10^{16}$ $(cm^3\ s^{-1})$	$k_{NO} \times 10^{14}$ $(cm^3\ s^{-1})$	$k_{NO_2} \times 10^{11}$ $(cm^3\ s^{-1})$	Reference *
benzene-OH				
(300 mbar N_2, 298 K)	2.1	< 1	2.5	Zetzsch (90)
(133 mbar Ar, 299 K)	1.6			Zetzsch (91)
(" , 314 K)	2.1			Zetzsch (91)
(" , 330 K)	3.0	< 3	2.5	Zetzsch (92)
(" , 354 K)	3.7			Zetzsch (91)
toluene-OH				
(1.3 mbar He, 353 K)			4	Devolder (89)
(130 mbar N_2, 298 K)	5	< 3	3.7	Zetzsch (90)
(133 mbar Ar, 330 K)	5.5	< 3	3.6	Zetzsch (92)
phenol-OH				
(130 mbar Ar, 298 K)	250	< 7	3.4	Zetzsch (90)
(130 mbar Ar, 330 K)	300	< 7	3.6	Zetzsch (92)
benzyl				
(1.3 mbar He, 297 K)	0.72×10^4	0.063	6.0	Devolder (91)
o-methylbenzyl				
(1.3 mbar He, 297 K)	1×10^4		5	Devolder (90)
"	0.93×10^4	0.105	5.7	Devolder (91)
m-methylbenzyl				
(1.3 mbar He, 297 K)	1.1×10^4		5	Devolder (90)
"	1.1×10^4	0.127	5.6	Devolder (91)
p-xylene-OH				
(130 mbar Ar, 330 K)	8.0	< 10	3.2	Zetzsch (92)
m-cresol-OH				
(130 mbar Ar, 330 K)	800	< 3	4.0	Zetzsch (92)
aniline-OH				
(130 mbar Ar, 330 K)	10	< 10	5.0	Zetzsch (92)
naphthalene-OH	1.1	-	-	Zetzsch (93)
(130 mbar Ar, 400 K)				

* These references are to EUROTRAC Annual Reports. The year in parentheses is the year of the report not the publishing year.

Yields have been determined for hydroperoxide formation from the alkyl side-groups ethylbenzene-2-hydroperoxide, 17 % C from OH + ethylbenzene; benzyl-hydroperoxide, 5 % C from OH + toluene (Bachmann). Low yields of methylhydroperoxide are reported for both systems, and ethylhydroperoxide was also observed in the reaction of OH with ethylbenzene.

Warneck has determined the product yields for the reaction OH + toluene under NO_x free conditions in air at atmospheric pressure and with very high partial pressures of toluene and H_2O_2, both on the order of 1000 ppm. Hydrogen peroxide was used as the photolytic source of OH radicals and the toluene concentration was kept high in order to minimise secondary OH reactions. High yields of cresols were observed which could be best explained by a mechanism involving H atom abstraction from the OH adduct by O_2. A similar study also gave high yields of cresols (Bachmann). Becker tried to confirm these findings under conditions more similar to those typical of the troposphere. It was also found that, at high aromatic compound concentrations and high levels of H_2O_2, cresol formation was important. Addition of NO_x to the system increased the cresol formation. The cresol yield was measured relative to that of benzaldehyde, which is a product of the abstraction channel. However, at low concentration cresol yields were minimal. From these results it has been concluded that under atmospheric conditions cresols in the case of toluene and alkylated phenols in the case of the xylene isomers will be only very minor products. This conclusion is further supported by the absence of quinone-type products in aromatic compound product studies; quinones were shown to be formed in high yields in the reactions of OH with phenolic-type compounds. These conclusions, however, do not necessarily include benzene which is more difficult to study and in which the oxidation mechanism may be different to that of the alkylated benzenes. Results from studies of Cl atom reactions with benzene indicate a 50 % yield for phenol formation from the reaction of O_2 with the benzene-OH adduct (Zellner).

Becker has performed detailed product studies on the OH reaction with toluene and p-xylene. At present, product formation can best be explained by the mechanism shown in Fig. 10 where muconaldehydes (hexa-2,4-diene-1,6-dial) are proposed as direct products of the reaction of the aromatic-OH adduct with O_2. Several of these muconaldehydes have been synthesised and their reactions with OH radicals investigated (Becker). These reactions result in the formation of unsaturated 1,4-dicarbonyl compounds, glyoxal, methylglyoxal and maleic anhydride, which have also been observed in the reaction of OH with toluene and p-xylene. It has been demonstrated that the unsaturated 1,4-dicarbonyl species react very rapidly with OH and photolyse, yielding products which possibly accelerate O_3 formation in smog chamber type studies. Many of the aromatic oxidation products especially the unsaturated 1,6-dicarbonyl species are known either to be toxic or are potentially toxic with both carcinogenic and mutagenic properties [8].

Becker has carried out smog-chamber type experiments on the photolysis of cyclohexane/NO_x mixtures in air which show that addition of toluene to the system produces O_3 much faster than the toluene-free system. However, the maximum value of O_3 reached in both cases remains approximately the same and NO_x is observed to be scavenged in the toluene system. At present the mechanism by which NO_x is scavenged in photoreactors is not clear. Formation of PAN and nitro aromatic compounds are observed, however, their yields are very low. The unidentified scavenging process could possibly be an artefact which results in a progressive slowing down of the O_3 formation rate in smog chamber experiments. The carbon balance found in aromatic compound product studies is presently only approximately 50 % C. Persistent product spectra, observed during the oxidation of toluene and p-xylene, can very probably be attributed to 3,3-dihydroxy hexane-2,4,5-trione in the case of p-xylene and in the case of toluene to a similar hydrated vicinal polyketone. Based on an estimated IR absorption cross-section for the carbonyl band of these compounds, it has been shown that these compounds could account for another 30 % C in the product studies of both aromatic compounds.

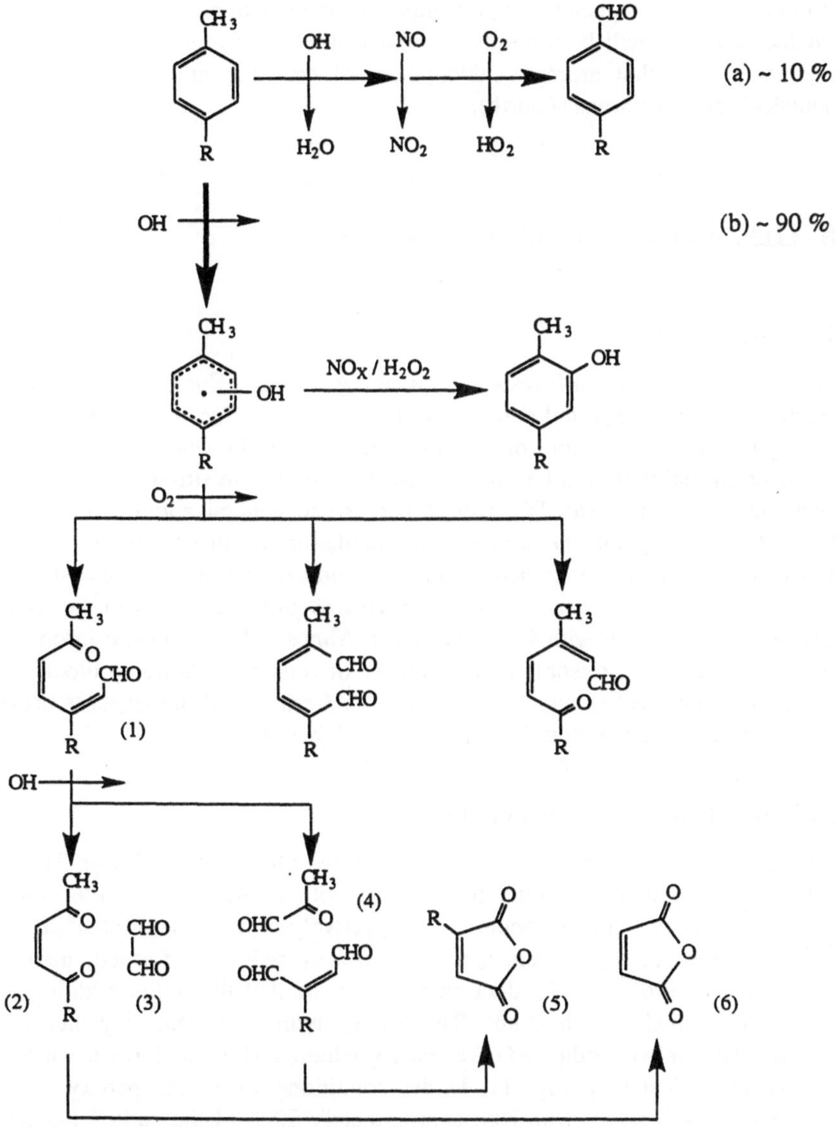

(a) Abstraction Channel   (b) OH-Addition-Channel
(1) 6-Oxo-2,4-heptadienal (R=H), 3-Methyl-6-oxo-2,4-heptadienal (R=CH$_3$)
(2) 4-Oxo-2-pentenal (R=H), 3-Hexene-2,5-dione (R=CH$_3$)
(3) Glyoxal   (4) Methylglyoxal
(5) Methyl maleic anhydride (R=CH$_3$), Maleic anhydride (R=H) = (6)

Fig. 10: Mechanism of atmospheric toluene (R=H) and p-xylene (R=CH$_3$) oxidation.

With the exception of phenolic-type compounds the reactions of NO_3 radicals with alkylated benzenes will be unimportant under atmospheric conditions. A product study has shown that aromatic aldehydes and aromatic nitrates are the main products of these reactions (Hjorth).

2.8 Oxidation of biogenic compounds

2.8.1 Introduction

Terpenes are emitted from forests and other vegetation in Europe, the importance of each individual compound depending on tree type and location. Isoprene is the main hydrocarbon emission of certain trees. Many biogenic VOCs show a significant diurnal variation because emission strengths are strongly dependent on temperature, humidity, *etc.* We collect here some general comments based on LACTOZ results about the atmospheric oxidation of these species. A more detailed discussion, accompanied by identification of the laboratories where the research was conducted, can be found in earlier sections dealing with the specific chemistry (*e.g.* reaction with OH, NO_3, *etc.*). Almost all mechanistic information that is available refers to isoprene. Isoprene is, of course, the building block of the terpenes, although its behaviour may well differ from that of the terpenes because of the location of the two double bonds and its higher volatility.

2.8.2 Oxidation mechanism of isoprene

Isoprene and the terpenes may be attacked by OH, NO_3 and O_3. The rate coefficients at room temperature for reaction with isoprene are of the order of 10^{-10}, 10^{-12} and 10^{-17} cm^3 molecule^{-1}s^{-1}, respectively. In the case of attack by OH, and in the presence of oxygen, a peroxy-hydroxy radical is formed; up to six isomers may be produced. Product studies indicate that the initial addition is to one of the terminal carbon atoms. The peroxy-hydroxy radical may then either react with NO or with other RO_2 radicals (including HO_2) to form a variety of products, as indicated in Fig. 11. Under conditions where the peroxy-hydroxy radical reacts exclusively with NO, approximately 50 % of the carbon balance is accounted for by three main products: methacrolein, methyl vinyl ketone and formaldehyde. Other carbonyl products and hydroxy-nitrates are thought to make up the carbon balance although there is presently no clear indication of the exact identity of these compounds or their yields.

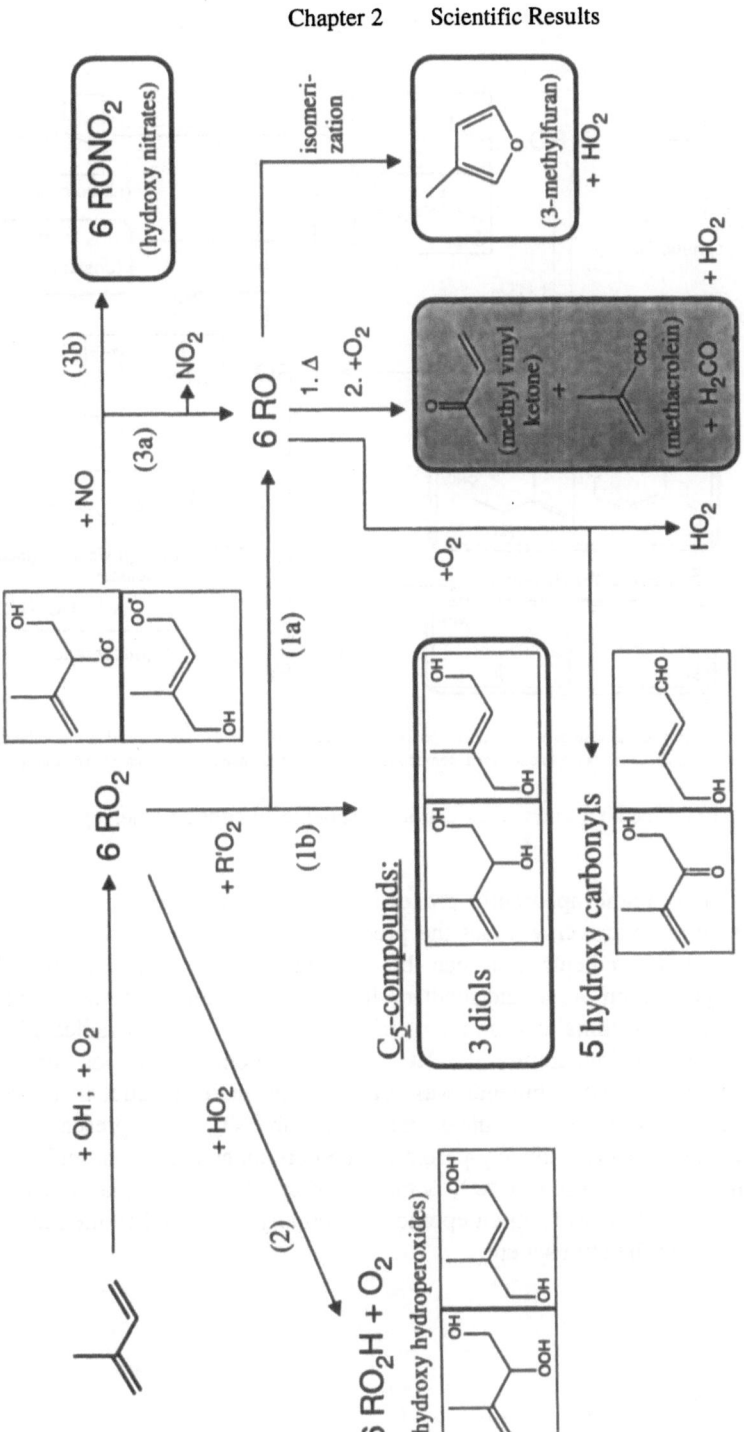

Fig. 11: Mechanism for the OH initiated oxidation of isoprene (simplified scheme).

Legend:

The numbers written in front of the products give the number of different compounds which can be expected (cis-/trans-isomerism in the case of 1,4-substituted compounds is neglected). The meaning of the different frames is:

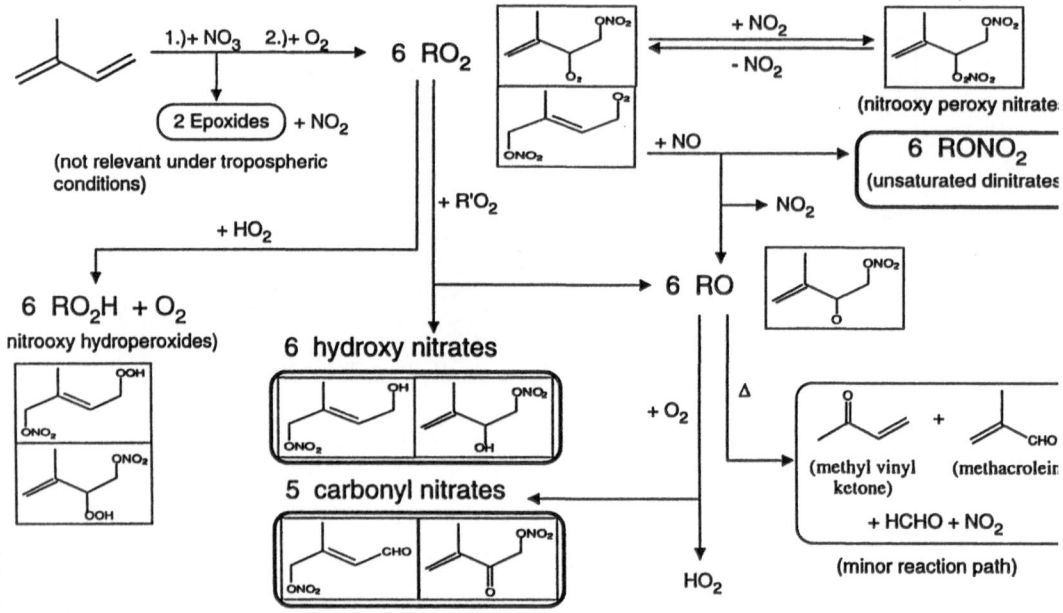

egend:
he numbers written in front of the products give the number of different compounds which can be expected (cis-/trans-isomerism in the case of 1,4
uted compounds is neglected). Dashed frames show example structures, compounds in solid frames have been identified in laboratory studies.

Fig. 12: Mechanism for the NO_3 initiated oxidation of isoprene (simplified scheme).

Attack of NO_3 on isoprene apparently proceeds in much the same manner, but there is considerable controversy about the precise reaction pathway because of the variety of peroxy radicals that can be formed. The products, such as 4-nitroxy-2-methyl-1-butan-3-one and methacrolein, are consistent with the initial addition of NO_3 to the terminal carbon atoms to form nitro-oxy-peroxy radicals in the presence of oxygen; apparently the NO_3 adds preferentially to position 1 (Fig. 12). 3-methyl-4-nitroxy-2-butenal was found as the main product in these experiments. The nitro-oxy-peroxy radicals can react with NO_2, in the presence of O_2, to yield thermally unstable nitroxy-peroxynitrate compounds. One particularly important feature of the addition of NO_3 is the extent to which the initial adduct, which might eliminate NO_2 to form an epoxide, is actually converted to the nitro-oxy-peroxy radicals in the atmosphere.

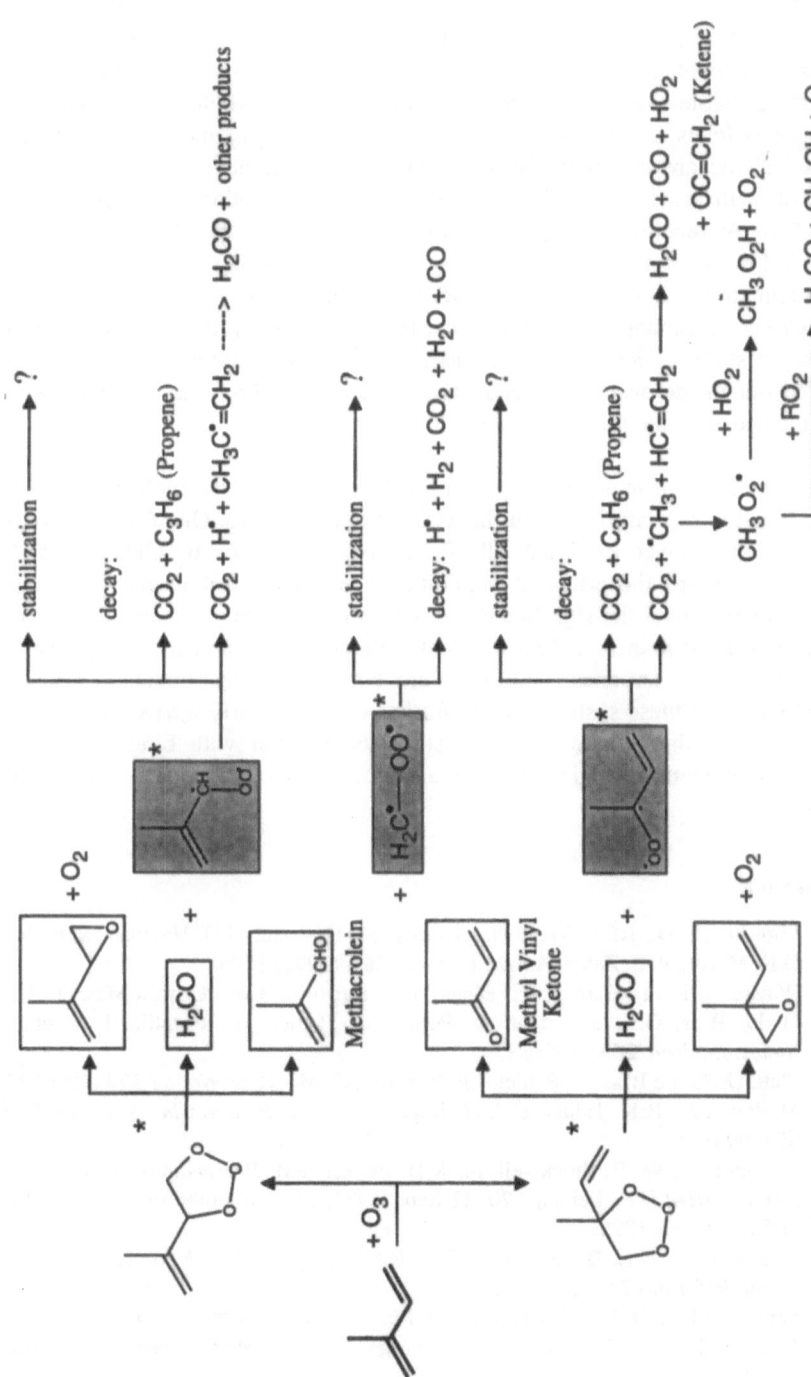

Fig. 13: Proposed mechanism for the gas-phase ozonolysis of isoprene.

Reaction of isoprene with ozone follows a rather different course. Ozone adds across one or other of the double bonds, and the ozonide may then fragment in several possible ways, to form methacrolein, methylvinylketone, epoxides and Criegee radicals as indicated in Fig. 13. Methacrolein, methylvinylketone and formaldehyde are the main carbonyl compounds that have been identified. The ratio of methacrolein to methylvinylketone yields is considerably higher than that found for the reaction of OH with isoprene. The fates of the Criegee radicals, and even if they are formed at all, are the subject of much disagreement, but the matter is of some importance, since these radicals might provide a source of OH at night. The Criegee mechanism cannot account for all the observed products found in the oxidation studies. Note that, necessarily, the laboratory studies on ozonolysis are carried out in the absence of NO_x, the presence of which may or may not alter relative and absolute product yields.

This summary will indicate that, even for the simplest of the molecules, isoprene, there remain uncertainties about the degradation pathways. One problem concerns poor carbon balance in almost all of the studies of attack by OH, NO_3 and O_3. Another concerns the effect of humidity on product distributions. Yet a further question hangs over the significance of the ozonolysis reactions as a source of OH radicals. Almost nothing is known about the mechanisms and specific pathways of reactions of the terpenes, and there are substantial experimental obstacles to investigation of these systems. Much further work is clearly warranted, in order to determine whether ozone is only lost in its reaction with biogenic VOCs or whether the reactions might constitute a source of atmospheric ozone when NO_x is present.

References

1.   Lightfoot, P.D., R.A. Cox, J.N. Crowley, M. Destriau, G.D. Hayman, M.E. Jenkin, G.K. Moortgat, F. Zabel; *Atmos. Environ.* **26A** (1992) 1805.
2.   Wayne, R.P., I. Barnes, P. Biggs, J.P. Burrows, C.E. Canosa-Mas, J. Hjorth, G. Le Bras, G.K. Moortgat, D. Perner, G. Poulet, G. Restelli, H. Sidebottom; *Atmos. Environ.* **25A** (1991) 1.
3.   Platt, U., G. Le Bras, G. Poulet, J.P. Burrows, G. Moortgat; *Nature* **172** (1990) 430.
4.   Mellouki, A., R.K. Talukdar, A.M. Bopegedera, C.J. Howard; *Int. J. Chem. Kinetics* **25** (1993) 25.
5.   Kirchner, F., W. R. Stockwell; in: K.H. Becker (ed), *Proceedings of the Workshop LACTOZ/HALIPP*, Leipzig, 20–22 Sept. 1994, EC Air Pollution Research Report N°54, Brussels 1995.
6.   Atkinson, R., D.L. Baulch, R.A. Cox, R.F. Hampson, J.A. Kerr, J. Troe; *J. Phys. Chem. Ref. Data* **21** (1992) 1125.
7.   Senum, G.I., Y.N. Lee, J.S. Gaffney, *J. Phys. Chem.* **88** (1984) 1269.
8.   Greenberg, A., C.W. Bock, P. George, J.P. Glusker; *Chem. Res. Toxicol.* **6** (1993) 701.

Chapter 3

Individual Reports from LACTOZ Contributors

3.1 Laboratory Studies of the Atmospheric Decay of Hydrocarbons in the NO$_x$ Free Atmosphere

K. Bächmann, B. Kusserow, J. Polzer, J. Hauptmann and M. Hartmann

Technische Hochschule Darmstadt, D-64287 Darmstadt, Petersenstr. 18, Germany

Summary

Analysis methods for the determination of hydroperoxides and hydroxyhydro-peroxides in reaction mixtures were developed and applied to photolysis mixtures of aliphatic and aromatic hydrocarbons, carbonyls, alcohols and halogenated hydrocarbons. HPLC with chemiluminescence, electrochemical and fluorescence detection was used for selective detection of hydroperoxides and hydroxohydro-peroxides and gas chromatography with mass spectrometric and flame ionisation detection for the analysis of all reaction products.

Laboratory experiments of the oxidation of hydrocarbons under NO$_x$ free conditions were made in 2 L and 10 L deactivated glass vessels by the photolysis of H$_2$O$_2$.

Hydroperoxides were found as reaction products of alkanes, alkylsubstituted aromatic hydrocarbons, ketones, chlorinated and iodated hydrocarbons. They were major products in the oxidation of alkanes.

Aims of the research

The atmospheric decay of hydrocarbons is studied intensively only for methane. The aim of our research was to develop analytical methods that could help to study the decay of hydrocarbons with longer chain. The hydroperoxides were of special interest.

Separation and detection methods for the specific and sensitive determination of hydroperoxides in laboratory experiments and in rain water must be developed. In addition, methods for the analysis of the whole product distribution were necessary. These methods should have been applied to the analysis of different reaction mixtures.

Principal findings during the project

During the project, we developed HPLC and GC methods for the determination of hydroperoxides.

HPLC

The separation of hydroperoxides with HPLC was developed with commercially available hydroperoxides and a post column derivatisation with luminol on an RP-2 phase.

Due to the lack of detection methods for organic hydroperoxides, we developed three different detection methods for HPLC analysis.

1. Post-column reaction with luminol
   The detection of hydroperoxides in this method is based on chemiluminescence. The system was optimised using t-buthylhydroperoxide, t-amylhydroperoxide and cumenehydro-peroxide. Cytochrome C was used as a catalyst, the eluent was 0.25 M borate buffer. Detection limits were 5 nmol of active oxygen, which was only sufficient for laboratory experiments.

2. Polarographic detection
   A polarographic detector was used with the method of Funk [1]. For electrochemical detection it is necessary to use an electrolyte as eluent. A new separation method was developed for this reason. The reduction of O_2, which could not be completely excluded, at the detector potential of -1400 mV led to distortion of the chromatogram. Because of this problem, the detection limit was 1 ppm. In the later research, this problem could be solved partially with a method which allowed the degassing of the sample. Detection limits were improved by a factor of 2.

3. Fluorometric detection
   In general, fluorometric detection achieves very good detection limits. Therefore we searched a method to derivatize hydroperoxides in a post column

reaction to fluorescent products. Kok *et. al.* [2, 3] used a flourometric system to measure H_2O_2 and the total hydroperoxide content. The reaction of hydroperoxides with p-hydroxyphenylacetic acid in the presence of a catalyst (peroxidase) leads to a fluorescent dimer of the acid.

Peroxidase is not useful for all hydroperoxides. Better results were achieved with microperoxidase. Detection limits of 0.2 µmol/L were achieved.

Solid-phase enrichment

A method for the enrichment of hydroperoxides on solid phases has been developed. The enrichment is performed on a small volume of RP-18 material, the solid phase is dried and the sample eluted with acetone. An enrichment factor of 50 can be reached.

Detection of 1-hydroxyhydroperoxides with chemiluminescence

The exposed method is based on the detection of hydrogen peroxide with the peroxyoxalate chemiluminescence method [4].

In the first step the 1-hydroxy hydroperoxides are separated by reversed-phase HPLC in acidic solution as described by Hellpointner [5]. After the separation, pH value as well as temperature are raised to enforce the decomposition of the labile 1-hydroxy hydroperoxides to hydrogen peroxide and the corresponding aldehyde. After mixing of eluent and reagent stream, the solution passes a flow cell placed in front of the window of a photomultiplier tube where the emitted light of the chemiluminescence reaction is registered. The main difficulties arise from high contents of hydrogen peroxide (*e.g.* rain samples).

Because of the high sensitivity of the chemiluminescence, limits of detection reached presently are 0.06 µmol/L (6 pmol absolute 0.4 ng HMHP).

In several rain samples 1-hydroxymethyl, 1-hydroxyethyl and 1-hydroxypropyl hydroperoxides could be determined.

Analysis of peroxy compounds in photolysis mixtures of hydrocarbons

A number of hydrocarbons were photolysed in a reaction vessel according to the system described by Prof. Warneck, MPI Mainz. Measurements were made with fluorometric and polarographic detection. The photolysis experiment were reproducible with a standard deviation of 3–6 %. Several hydroperoxides could be identified by "cross"-identifications by comparing the retention times of different photolysed hydrocarbons, aldehydes, ketones and alcohol's with the unknown peaks.

Determination of hydroperoxides by GC/MS and GC/FID

In our recent investigations, a low temperature GC/MS system (HP 5890/5970) with a vapour sample inlet system is applied to the determination of all oxidation products $> C_2$ including the alkylhydroperoxides. This system is described in the annual report 1993. We analysed photolysis mixtures of alkanes, aromatic hydrocarbons, alkenes, ketones and chlorinated hydrocarbons. Quantitations were made in the full scan mode and later with a flame ionisation detector. Carbon balances were made for alkanes and aromatics. They were 93–102 % (20 min photolysis) and 45–73 % (90 min photolysis), respectively.

Alkanes

We conducted photolysis experiments with methane, ethane, propane, n- and i-butane, n- and i-pentane and n-hexane. Hydroperoxides were the main products in all experiments. Other products were carbonyls, alcohols and organic acids. Peroxides were found only when hydrocarbons are present in high excess in the reaction mixture. In the photolysis of hydrocarbons > C4 no peroxides were found. We discuss the experiment with n-butane as an example.

Quantitations were made based on the "effective carbon number" concept of Sternberg et al. [6] with a FID. Response factors of the hydroperoxides were calculated as the corresponding alcohol's. Carbon balances are good, indicating that this increment method is suitable also for hydroperoxides.

The decrease in recovery with increasing reaction time is due to the forming of low boiling products, i.e. methanol, formaldehyde, CO and CO_2 which cannot be analysed with our analytical system.

The distribution of products of the attack on primary and secondary H-atoms is in good agreement with calculations of Atkinson [7].

Table 1: Reaction products in the photo-oxidation of butane. 54 ppm (= 24.07 µmol) n-butane 2.5 + 60 ppm (= 26.60 µmol) H_2O_2. We identified additionally: ethanol, propanal, methylhydroperoxide, formic acid and acetic acid.

	Reaction time 20 min	Reaction time 53 min
n-Butane	1.13 µmol[*]	2.57 µmol[*]
Butanal	0.03 µmol	0.05 µmol
Butanone	0.11 µmol	0.48 µmol
2-Butanol	0.03 µmol	0.09 µmol
Ethylhydroperoxide	–	0.07 µmol
1-Butanol	–	0.01 µmol
2-Hydroperoxy butane	0.89 µmol	1.67 µmol
1-Hydroperoxy butane	0.10 µmol	0.16 µmol
Reacted carbon (%)	102	98

[*] Loss due to oxidation.

Table 2: Percentage distribution of the attack on primary and secondary H of n-butane with OH radicals, experimental and calculated.

attack on	calculated	measured (20 min photolysis)
primary H	14 %	12 %
secondary H	86 %	88 %

Influence of the composition of the reaction mixture on the product distribution

At the beginning of our research we made experiments to determine the influence of the composition of the reaction mixture to the product distribution. First we kept the concentration of H_2O_2 constant and varied the concentration of n-butane at constant reaction times. As expected, the whole product concentration raised with increasing concentration of n-butane. The relative composition changes, the hydroperoxides are decreasing while the other oxidation products are increasing with increasing concentration of n-butane. Similar results were achieved when the concentration of n-butane is kept constant and the concentration of H_2O_2 is varied. With increasing the H_2O_2 / n-butane ratio the amount of hydroperoxides increases because the reaction of the hydroperoxy radical with ˙OOH predominates the self reaction.

We also changed the concentration of the reaction mixture without changing the H_2O_2 / n-butane ratio. No change in product distribution could be found.

Alkenes

No hydroperoxides could be found in photolysis experiments with 1-butene. Some substances in the experiments with 1-pentene and 1-hexene could not be identified. Main products were ketones.

Ketones

We photolysed 2 and 3-pentanone. We found ethylhydroperoxide in the experiment with 3-pentanone and methylhydroperoxide and propylhydroperoxide in the oxidation of 2-pentanone as product of the Norrish reaction. Other products were diketones and organic acids. Some main products at late retention time could not be identified. Tests with HPLC/fluorescence detection indicate that they are hydroperoxides.

Aromatic hydrocarbons

The decay of benzene, toluene, ethylbenzene and the xylenes was investigated. Hydro-peroxides were found in the experiments with alkyl substituted aromatics. Main products were organic acids. These experiments were described in the annual report 1993.

Chlorinated hydrocarbons

No hydroperoxide was found in the case of methylenchloride. Main product was phosgene. Chloracetylchloride was the main product in the photo-oxidation of 1,2-dichlorethane. 1,2-dichlorethylhydroperoxide could be identified with mass spectrometry and HPLC / fluorescence. Other products were chloracetaldehyde and chloracetic acid.

Achievements during the project

During the project we developed analytical methods for the selective and sensitive determination of hydroperoxides and hydroxyhydroperoxides. They can be used as analytical tools for laboratory experiments and partially for field measurements. A method for GC/MS and CG/FID allowed the analysis of all reaction products.

Acknowledgement

We are grateful to the BMBF for financial support.

References

1.   M.O. Funk, J. Baker, *J. Liq. Chromatog.* **8** (1985) 663–675.
2.   A.L. Lazrus, G.L. Kok, J.A. Lind, S.N. Gitlin, B.G. Heikes, R.E.Shetter, *Anal. Chem.* **58** (1986) 594–597.
3.   G.L. Kok, K. Thompson, A.L.Lazrus, *Anal. Chem.* **58** (1985) 1192–1194.
4.   M.M. Rauhut, L.J. Bollyky, B.G. Roberts, M. Loy, R.H. Whitman, A.V. Ianotta, A.M. Semsel, R.A. Clarke, *J. Amer. Chem. Soc.* **89** (1967) 6515.
5.   E. Hellpointer, S. Gäb, *Nature* **337** (1989) 661.
6.   J.C. Sternberg, W.S. Gallaway, D.T.L. Jones, in: N. Brenner, J.E. Callen, M.D. Weiss (eds), *Gas Chromatography*, Academic Press, New York 1962, pp. 231–276.
7.   R. Atkinson, *Int. J. Chem. Kinet.* **19** (1987) 799–825.

3.2   OH Initiated Oxidation of VOC under Variable NO$_x$ Conditions

K-H Becker, I. Barnes, A. Bierbach, K.J. Brockmann, F. Kirchner,
B. Klotz, H.G. Libuda, A. Mayer-Figge, S. Mönninghoff, L. Ruppert,
W. Thomas, E. Wiesen, K. Wirtz and F. Zabel

Physikalische Chemie / FB 9, Bergische Universität – Gesamthochschule Wuppertal,
D-42097 Wuppertal, Germany

Summary

During the reporting period the following investigations have been carried out:

- Determination of rate coefficients for the reactions of NO$_3$ radicals with unsaturated hydrocarbons
- Determination of alkene yields from the reaction of OH with alkanes
- Investigation of the thermal stability of peroxynitrates
- Determination of branching ratios for RO$_2$ + NO reactions: yields of alkyl nitrates
- Atmospheric chemistry of ketonitrates and dinitrates: OH reaction kinetics and UV absorption spectra
- Mechanistic study of the ozonlysis of ethene
- Investigation of the oxidation of isoprene, initiated by reaction with OH and with O$_3$
- Kinetic and mechanistic studies of the reaction of simple alkylated aromatics with OH
- Investigation of the reactions of unsaturated carbonyl compounds as products of the oxidation of aromatics under atmospheric conditions

Experimental

As described in previous LACTOZ reports the experiments were carried out in glass reaction chambers with volumes ranging from 37 to 1080 L at a total pressure of 1000 mbar synthetic air and at temperatures between 243 and 323 K. Most of the analyses of reactants and products for the different types of experiments were performed using long-path *in situ* Fourier transform IR spectroscopy (FTIR), however, other techniques such as gas chromatography, HPLC and TDL spectroscopy were also applied. UV absorption spectra were measured using a diode array spectrometer.

Principal results

Reaction of NO₃ radicals with unsaturated hydrocarbons

Rate coefficients have been determined for the reactions of NO_3 with a number of aliphatic mono- and dialkenes and monoterpenes with a relative kinetic method. The measured rate coefficients are listed in Chapter 2, Table 2. The products of the reactions were also investigated using *in situ* FTIR. The formation of thermally unstable nitro-oxy-peroxynitrate-type compounds was observed. The final products included aldehydes/ketones, nitro-oxy-aldehydes and -ketones and dinitrate compounds.

Alkene formation in the OH initiated oxidation of saturated hydrocarbons

The primary step in the OH initiated atmospheric oxidation of saturated hydrocarbons is abstraction of an H atom to form an alkyl radical and water. The predominant fate of the alkyl radical is addition of oxygen to form a peroxy radical. However, it is known for ethyl radicals that a pathway exists which leads to formation of ethene and HO_2 radicals. Two mechanisms have been discussed in the literature, a) a direct abstraction of an H atom by O_2 and b) a Lindemann mechanism involving an excited alkylperoxy radical:

$$CH_3CH_3 + OH \quad \rightarrow \quad CH_3CH_2 + H_2O$$
$$a) \quad CH_3CH_2 + O_2 \quad \rightarrow \quad CH_2{=}CH_2 + HO_2$$
$$b) \quad CH_3CH_2 + O_2 \quad <=> \quad CH_3CH_2O_2{*} \quad CH_3CH_2O_2{*} \quad \rightarrow \quad CH_2{=}CH_2 + HO_2$$

The formation yields for alkene production in the OH initiated oxidation of a number of saturated hydrocarbons (alkanes) have been determined. The 254 nm photolysis of H_2O_2 was employed as the OH source and GC-MS for detection of the products. The following alkanes were investigated: *n*-butane, *n*-pentane, 2,3-dimethylbutane and cyclopentane.

The alkene yields measured in the experiments are shown in Table 1. The yields have been corrected for loss due to reaction with OH radicals. For all the alkanes investigated formation of alkenes was observed, but the yields were always well below 1 %. The results show that the reaction of alkyl radicals with O_2 to form alkenes is not a significant process under normal tropospheric conditions.

Table 1: Formation yield for alkene production in the OH initiated oxidation of saturated hydrocarbons.

Saturated hydrocarbon	Alkene formation yield [%]
n-Butane	1-butene (0.18 ± 0.02)
	trans-butene (0.15 ± 0.03)
n-Pentane	1-pentene (0.014 ± 0.003)
	cis- and *trans*-2-pentene (0.053 ± 0.012)
2,3-Dimethylbutane	2,3-dimethyl-1-butene (0.62 ± 0.18)
	2,3-dimethyl-2-butene (0.16 ± 0.03)
Cyclopentane	cyclopentene (0.024 ± 0.003)

Thermal stability of peroxynitrates

Peroxynitrates are formed in the troposphere by the recombination of RO_2 radicals and NO_2. The relevant chemical reactions which determine the importance of peroxynitrates as temporary reservoirs of NO_x are the following:

$RO_2 + NO_2 (+ M)$	$\Longleftrightarrow$	$RO_2NO_2 (+ M)$	$(1,-1)$
$RO_2 + NO$	$\rightarrow$	$RO + NO_2$	$(2a)$
$RO_2 + NO (+ M)$	$\rightarrow$	$RONO_2 (+ M)$	$(2b)$
$RO_2NO_2 + h\nu$	$\rightarrow$	products	(3)

Within the framework of LACTOZ, rate coefficients k_{-1} were systematically measured as a function of temperature and total pressure for a variety of different groups R, see Chapter 2, Table 11. In addition, k_{2a} was estimated for $R = HOC_2H_4$, the ratios $k_1/(k_{2a}+k_{2b})$ were measured for several R, and UV absorption cross sections were determined for the most abundant peroxynitrate, PAN, see Chapter 2, Fig. 7. The following conclusions can be drawn from the results of this work:

- The thermal stability of peroxynitrates increases with the electron withdrawing effect of R thus implying, for example, that $CF_3O_2NO_2$ is considerably more stable than $CH_3O_2NO_2$.
- The strongest effect on the thermal stability is induced by a carbonyl group adjacent to the peroxy group, *e.g.* in $CH_3C(O)O_2NO_2$ (PAN). With respect to their importance in the lower troposphere, peroxynitrates can thus be divided into two groups: those with no adjacent carbonyl group (alkylperoxynitrates, substituted alkylperoxynitrates) with thermal lifetimes in the order of seconds at 298 K and 1 atm, and those containing a carbonyl group adjacent to the peroxy group (acetylperoxynitrate and substituted acetylperoxynitrates) with lifetimes in the order of hours at the same conditions. The effect of electronegative substituents or carbonyl groups distant from the peroxy group on the lifetime of a peroxynitrate is only minor — acetonyl peroxynitrate, for example, is very short-lived as compared to PAN.
- A linear relationship has been established between the Arrhenius parameters for the dissociation of a particular peroxynitrate RO_2NO_2 and the ^{13}C-NMR shifts in the corresponding compounds RX (X = H, Cl, CH$_3$). The physical background of this correlation is that both the ^{13}C-NMR shift of the C atom next to the peroxy group and the OO–N bond strength depend on the electron density at this C atom. Based on these correlations, thermal lifetimes of peroxynitrates of the type $C-O_2NO_2$ can be estimated from tabulated NMR data of the corresponding compounds RX with an uncertainty of about a factor of 5.
- Experimental results on the ratio $k_1/(k_{2a}+k_{2b})$ for acetylperoxy radicals are about 30 % lower than previously assumed leading to 30 % less PAN formation from acetylperoxy radicals.

- New measurements of the UV absorption spectrum of PAN resulted in larger cross sections for wavelengths ≥ 300 nm, leading to photolysis rate coefficients at 298 K which are higher by about a factor of 3 than a previous estimate [1].

Product yields of alkyl nitrates in the oxidation of hydrocarbons

Values of the branching ratios $k_{2b}/(k_{2a}+k_{2b})$ at 298 K have been determined for the OH initiated oxidation of a number of alkanes and alkenes. All of these branching ratios are listed in Chapter 2, Table 10, where the quoted errors are 2 σ and represent experimental precision only. Data for the branching ratios for the reactions of alkyl radicals with ≥ C$_4$ are known only for n-butane and n-heptane. From a knowledge of the branching ratios for the secondary alkylperoxy radicals and also of the ratio of primary to secondary alkylperoxy radicals, the branching ratios for primary alkylperoxy radicals formed in the reactions of OH with n-butane, n-pentane, n-hexane and n-heptane were also determined and are also listed in Chapter 2, Table 10. These results show clearly that the branching ratio increases with increasing chain length of the hydrocarbon.

Atmospheric chemistry of ketonitrates and dinitrates

Rate coefficients have been measured for the reaction of OH radicals with a number of organic ketonitrates (R–CO–CH(ONO$_2$)–R) and dinitrates (R–CH(ONO$_2$)–CH(ONO$_2$)–R) using a relative kinetic method. The compounds investigated included 1,2-propane-, 1,2-butane-, and 2,3-butanediol dinitrate, a-nitro-oxy acetone, 1-nitro-oxy-2-butanone, 3-nitro-oxy-2-butanone, 3,4-dinitro-oxy-1-butene and cis-1,4-dinitro-oxy-2-butene (LACTOZ 90). The rate coefficients are listed in Chapter 2, Table 1a. and this study represented the first determination of the OH rate coefficients for these compounds.

UV absorption spectra have been measured for the ketonitrates and dinitrates (LACTOZ 89). In the region of atmospheric interest (λ > 270 nm) the absorption cross sections of the carbonyl nitrates are approximately a factor of 10 higher than those of the dinitrates. From the measured cross sections photolysis frequencies have been calculated for the organic ketonitrates and dinitrates. Although the photolysis frequencies represent upper limits the results indicate that photolysis will generally be somewhat more important than loss via reaction with OH radicals for saturated difunctional nitrates. However, for unsaturated nitrates loss due to reaction with OH will dominate over photolysis as an atmospheric sink. The product studies show that photolysis of ketonitrates/dinitrates will result in the re-release of NO$_2$ and the formation of PAN-type compounds.

Ozonolysis of ethene

The currently accepted mechanism for the gas-phase reaction between ozone and ethene is represented by reactions (1)–(5) [2, 3]:

$O_3 + C_2H_4$	$\rightarrow$	$(O_3 \cdot C_2H_4 \rightarrow) CH_2O + CH_2O_2^*$	(1)
$CH_2O_2^* + M$	$\rightarrow$	$CH_2O_2 + M$	(2)
$CH_2O_2 + CH_2O$	$\rightarrow$	$HC(O)OCH_2OH$ (HMF)	(3)
$CH_2O_2^* \rightarrow HC(O)OH^*$	$\rightarrow$	$CO + H_2O$	(4a)
	$\rightarrow$	$CO_2 + H_2$	(4b)
	$\rightarrow$	$CO_2 + 2 H$	(4c)
$HC(O)OH^* + M$	$\rightarrow$	$HC(O)OH + M$	(5)

(* = highly excited, HMF = hydroxymethylformate)

The following results have been obtained within the framework of LACTOZ with respect to the above mechanism:

1) Usually, the loss of C_2H_4 and O_3 due to reaction (1) is enhanced by side reactions involving OH and HO_2 radicals. By using cyclohexane as an OH scavenger, $k_1 = (1.21 \pm 0.06) \times 10^{-18}$ cm^3 s^{-1} at 298 K was obtained from the first order decay of C_2H_4 in the presence of a large excess of O_3. This value is lower by ca. 30 % than the previously recommended value [3].

2) H_2 was measured for the first time in order to derive more reliable branching ratios for the reaction channels (4b) and (4c). With CO added as an OH scavenger, the H_2 yield relative to the total amount of C_2H_4 consumed increased from 12.1 % H_2 (no CO added) to 15.5 % H_2 ([CO] $\geq 5 \times 10^{16}$ molecules cm^{-3}).

3) Based on the experimental H_2 yields with and without CO added and the yields of CO and CO_2 in the absence of radical scavengers, the following product yields (corrected for side reactions and relative to the amount of C_2H_4 consumed in reaction (1)) were determined:
26.4 ± 0.5 % CO_2, 15.8 ± 0.3 % H_2, 10.5 ± 0.6 % H.

4) The pressure dependence of the yields of CO, CO_2, H_2, CH_2O, and $HC(O)OH$ was investigated in order to examine the pressure dependent steps (2) and (5). From the measured product yields, based on 64 experiments, the following conclusions are drawn:
- The H_2 yield (relative to the total amount of C_2H_4 consumed) is independent of total pressure between 1 and 1030 mbar.
- Within error limits, the yields of the products CO, CO_2, $HC(O)OH$, and CH_2O do not depend on total pressure between 100 and 1030 mbar.
- The results above both suggest that collisional deactivation of CH_2OO^* and $HC(O)OH^*$ (reactions (2) and (5)) are not effective. Ground state CH_2OO, if existing in this system, should be formed directly, implying a (at least) bimodal energy distribution of the Criegee radical formed in reaction (1).
- When total pressures are 10 mbar or lower, a change of the mechanism occurs, leading to significantly higher yields of CO and CO_2 and, simultaneously, lower yields of CH_2O.

Further mechanistic studies are under way to explain this result.

5) The products of the ozonolysis of ethene in the presence of added water vapour were analysed for the hydroperoxides H_2O_2, $HOCH_2OOH$ (hydroxymethyl hydroperoxide = HMHP) and HOC_2H_4OOH (2-hydroxyethyl hydroperoxide = 2-HEHP).

These experiments led to the following conclusions:

- H_2O_2, HMHP and 2-HEHP are formed in gas-phase reactions.
- The yields of H_2O_2 and HMHP increase with increasing partial pressure of H_2O, levelling off at high humidities. The yield of HMHP passes a distinct maximum of 4 % when the partial pressure of H_2O is increased.
- The major fraction of H_2O_2 is formed in a reaction other than
  $HO_2 + HO_2 \rightarrow H_2O_2 + O_2$.
- The formation of both H_2O_2 and HMHP is attributed to the reaction of CH_2OO with water vapour.
- For the reaction $CH_2OO + H_2O \rightarrow H_2O_2$ + products, a rate coefficient of $(2.4 \pm 1.4) \times 10^{-19}$ cm^3 s^{-1} was determined, relative to the rate coefficient of the reaction $CH_2OO + SO_2$ [2].
- 2-HEHP is formed via a reaction sequence initiated by the OH addition to ethene. The simplest explanation of the observed amounts of H_2O_2 and $HOCH_2OOH$ is the reaction of ground state Criegee radicals with water:
  $CH_2OO + H_2O \rightarrow ... \rightarrow H_2O_2 + HCHO; CH_2OO + H_2O \rightarrow ... \rightarrow HOCH_2OOH$.

Further experiments are necessary to establish the existence of ground state Criegee radicals and to elucidate their reactions.

Oxidation of isoprene

The products of the reaction with OH, especially in the absence of NO_x, and of the reaction with O_3 have been investigated. In addition rate coefficients for the reactions with OH and O_3 were measured for some recently identified products of the isoprene oxidation. The concentrations of the reactants were in the low ppm range (usually 1–10 ppm). OH radicals were generated either by continuous UV photolysis of H_2O_2 (254 nm) with or without NO present, or by photolysis of CH_3ONO.

These are the main findings of the investigations:

Isoprene + OH

In the presence of high NO_x concentrations (1–5 ppm), methacrolein (MAC, 19 ± 1 [%-molar yield]), methyl vinyl ketone (MVK, 29 ± 3) and formaldehyde (53 ± 5) were the main products in good agreement with the literature [4]. Formation of other carbonyl compounds and of nitrates could clearly be observed in the IR spectra of the products, but the yields were not quantified.

In the absence of NO_x, MAC (16 ± 2), MVK (15 ± 2) and H_2CO (33 ± 3) are also formed, but in significantly lower yields, as confirmed recently by Miyoshi *et al.* [5]. Remaining carbonyl compounds, which could not be identified, were evaluated from their IR absorption (18 ± 7) and in contrast to the NO_x experiments, strong absorptions of hydroxy- and/or hydroperoxy-groups were observed. Their yield was estimated to be 48 ± 20 % based on an average absorption coefficient. From this group of compounds two diols, 2- and 3-methyl-3-butene-1,2-diol, were unambiguously identified. Their yields are relatively low (5 and 2 %), but this first-time identification is a clear indication that the peroxy radicals formed in the OH initiated oxidation of isoprene follow the reaction mechanism described in the Chapter 2, Fig. 11.

From a comparison of the MAC and MVK yields in the presence to those in the absence of NO_x, an overall value has been estimated for the branching ratio of the $RO_2 + R'O_2$-reactions, occurring in the isoprene + OH reaction system:

$a = k_{1a} / k_1 = 0.7–0.8$

$RO_2 + R'O_2 \quad \rightarrow \quad RO + R'O + O_2$ (1a)

$\rightarrow \quad$ carbonyl comp. + alcohol + O_2 (1b)

Isoprene + O_3

Product yields were measured for isoprene / O_3 / air mixtures. In some experiments either water vapour (2–13 mbar) or cyclohexane (~ 500 ppm) was added to the mixture. While cyclohexane acts as a scavenger for OH radicals, water is expected to react with the proposed intermediate biradicals [6, 7]:

- As in the OH reaction, MAC, MVK and formaldehyde are the main products, but in substantially different yields. The ratio MAC / MVK ranges around 2.3 (OH: 0.6–1).
- Among the unidentified compounds there is a large fraction of carbonyl- and/or carboxyl compounds: ~ 50 %. As for the OH experiments this number was obtained by use of an average absorption coefficient.
- The yields of the three main products increase in the presence of water vapour, as well as in the presence of cyclohexane. The simultaneous presence of both scavengers leads to still higher yields for these carbonyl compounds. This trend, as well as the absolute yields, agrees well with literature data obtained under comparable conditions [8–10]. The explanation for this behaviour is not clear. One idea is that high concentrations of water vapour and cyclohexane suppress reactions of biradicals and OH radicals, respectively, with other products such as MAC, MVK and formaldehyde, so that higher yields are measured for these products. Also the IR spectrum of the non-identified products changes clearly with different concentrations of water vapour, but the nature of these products is still speculative. Nevertheless, it is evident that water vapour participates in the O_3 reaction.

- The formation of the low molecular weight compounds CO, CO_2, H_2, propene, ketene, methanol and methyl hydroperoxide can be explained by the decomposition of biradicals as shown in Chapter 2, Fig. 13. In view of the atmospheric relevance of the ozonolysis of alkenes, especially the influence of water vapour on the reaction mechanism and the product formation needs careful investigation. Under such conditions heterogeneous effects cannot be neglected and large-scale reaction chambers would be useful to account for such effects.

Kinetics of reactions of isoprene oxidation products with OH and O_3

Rate coefficients for the reactions of OH and O_3 with 2- and 3-methyl-3-butene-1,2-diol and 1,2-epoxy-3-methyl-3-butene, the latter having been reported very recently as product from isoprene + O_3 [11], were determined by use of a relative rate technique. In the O_3 experiments sufficient CO was added to scavenge > 90 % of OH radicals eventually being formed. For the O_3 reaction of the diols absolute measurements were carried out in addition by monitoring the decay of O_3 under pseudo-first order conditions. Absolute and relative rate data were in good agreement. The average values at 295 ± 2 K are:

k(2-methyl-3-butene-1,2-diol + O_3)   = $(4.8 \pm 0.6) \times 10^{-18}$ cm^3 s^{-1}
k(3-methyl-3-butene-1,2-diol + O_3)   = $(6.3 \pm 1.0) \times 10^{-17}$ cm^3 s^{-1}
k(1,2-epoxy-3-methyl-3-butene + O_3)   = $(3.3 \pm 0.3) \times 10^{-18}$ cm^3 s^{-1}

The rate coefficients for the OH reaction are listed in Chapter 2, Table 1a. All compounds investigated show high reactivity to both, OH and O_3, with the OH reaction dominating under "normal" tropospheric conditions. For OH, the experimental values are a factor of two higher than expected from the usual structure-reactivity estimation methods.

This high reactivity towards the important atmospheric photo-oxidants, probably valid also for similar isoprene oxidation products like unsaturated hydroxy carbonyls, hydroxy nitrates or hydroxy hydroperoxides, is an indication that these reactions need more consideration in laboratory experiments as well as photo-oxidation models.

Atmospheric chemistry of simple alkylated aromatics

Kinetics of the OH initiated oxidation of p-xylene and toluene

The rate coefficients for reaction of toluene and p-xylene with OH radicals have been determined and it was found that these rate coefficients are dependent on the O_2 concentration. Values of $(1.11 \pm 0.06) \times 10^{-11}$ and $(1.45 \pm 0.12) \times 10^{-11}$ cm^3 s^{-1} were obtained as rate coefficients for the reaction of p-xylene with OH radicals in 750 Torr N_2 and synthetic air, respectively. Intermediate values could not be determined, since, even in the presence of the smallest measurable partial pressure of O_2, the same rate coefficient as in synthetic air was obtained. The rate coefficients for the reaction of toluene with OH radicals rose from

$(4.74 \pm 0.11) \times 10^{-12}$ cm^3 s^{-1} in N$_2$ to a maximum value of $(6.61 \pm 0.17) \times 10^{-12}$ cm^3 s^{-1} in the presence of 12 Torr O$_2$. It has been observed that the measured rate coefficients also depend on the concentrations of the organic constituents of the reaction mixture. This observation has been confirmed by experiments using different starting concentrations of the aromatics and reference compounds. Under atmospheric conditions, reaction with O$_2$ will be the main fate of the initially formed OH-aromatic adducts.

Products of the OH initiated oxidation of toluene and p-xylene

Table 2 shows the products that have presently been identified and quantified for the p-xylene and toluene OH initiated oxidation systems. In the case of p-xylene 51.8 % of reacted carbon has been detected using FTIR and 51.7 % using the GC-MS technique. In the case of toluene 40.3 % of the reacted carbon has been detected by FTIR spectroscopy. Table 3 shows products which have been identified, but not yet quantified.

Table 2: Quantified products from the reaction of p-xylene and toluene with OH radicals using FTIR and GC-MS. Yields are given in percent of reacted carbon detected.

Technique	FTIR		GC-MS
Product	toluene[1]	p-xylene[1]	p-xylene[2]
CO		8.0 ± 1.4	
CO$_2$		4.0 ± 0.4	
Benzaldehyde	7.1 ± 1.8		
p-Tolualdehyde		6.4 ± 1.5	11.2 ± 4
2,5-Dimethylphenol		(8.0 by GC-FID)	5.8 ± 2
cis-3-Hexene-2,5-dione		8.3 ± 1.9	14.4 ± 4
Maleic anhydride	4.2 ± 1.1		
Methylglyoxal	4.4 ± 1.1	2.2 ± 0.5	
Acetic acid	5.2 ± 1.3	3.8 ± 0.3	20.0 ± 3
Glyoxal	3.7 ± 0.9	4.0 ± 1.7	
Formic acid	12.9 ± 3.2	5.1 ± 1.0	
Formaldehyde	2.3 ± 0.6	1.2 ± 0.2	
Ketene	0.5 ± 0.1	0.3 ± 0.1	
Total amount	40.3	51.8	51.7

[1] Identified by FTIR in a 1080 L reaction chamber using 1–2 ppm as initial concentration of the aromatic hydrocarbon.
[2] Identified by GC-MS in a 36.5 L chamber using 50–200 ppm as initial concentration of the aromatic hydrocarbon.

Because of a large discrepancy in the yields of phenol-type compounds in the aromatic oxidation system systems, a detailed study has been made of the processes leading to cresol formation in the oxidation of toluene. The studies were performed using different initial concentrations of toluene, NO$_2$ and H$_2$O$_2$ and it has been observed that there is an increase of the o-cresol yield with increasing concentrations of NO$_x$ and H$_2$O$_2$. At low starting concentrations of toluene, NO$_x$ and H$_2$O$_2$, the cresol formation was found to be low and will not be significant under tropospheric conditions.

Table 3: Products qualitatively identified in the reaction of toluene and p-xylene with OH radicals.

Toluene	p-Xylene
Acetylene[1]	
Methanol[1]	Methanol[1]
Methylhydroperoxide[1]	Methylhydroperoxide[1]
4-Oxo-2-pentenal[1]	4-Oxo-2-pentenal[1]
Benzylalcohol[2]	4-Methylbenzylalcohol[4]
Phenol[2]	2-Hydroxy-4-methylbenzaldehyde[4]
Cresol-isomers[2]	4-Methylbenzoic acid[4]
Benzoic acid[2]	Methylmaleic anhydride[1]
evidence of formation	
Methylmaleic anhydride[3]	
Methylhexadiendial isomers[3]	
Hydroxymethylhexadiendial isomers[3]	
Butenedial[5]	2-Methylbutenedial[5]

[1] Identified by FTIR in a 1080 L reaction chamber
[2] Identified by GC-MS after solvent trapping from the 1080 L chamber
[3] Identified by DF-MS in a flow reactor
[4] Identified by GC-MS from a stirred reactor
[5] Investigations have shown that butenedial (2-methylbutenedial) is the precursor for maleic anhydride (methylmaleic anhydride) formation

Atmospheric chemistry of unsaturated carbonyl compounds

The largest ring-fragmentation products identified in the OH initiated oxidation of aromatics so far are unsaturated 1,4-dicarbonyls (see Tables 2 and 3), and there is indirect evidence that the primary ring-opening products of aromatics are doubly unsaturated 1,6-dicarbonyls like 2,4-hexadienedial (CHO–CH=CH–CH=CH–CHO), also known as muconaldehydes, and its methylated derivates. Therefore, the atmospheric chemistry of such unsaturated carbonyl compounds has been investigated. The kinetic data determined for these compounds are listed in Chapter 2, Table 1a.

Mono unsaturated 1,4-dicarbonyls

Unsaturated 1,4-dicarbonyls (butenedial, 4-oxo-2-pentenal, 3-hexene-2,5-dione) were found to undergo an oxidative cyclisation process under VIS (λ_{max} = 360 nm) and UV photolysis (λ_{max} = 254 nm), leading to the formation of maleic anhydride. Additionally, butenedial and 4-oxo-2-pentenal were found to form 3H-furan-2-one and 5-methyl-3H-furan-2-one (a-angelicalactone), respectively. The major atmospheric sinks of butenedial, 4-oxo-2-pentenal, 3-hexene-2,5-dione and the furanones will be reaction with OH radicals and photolysis. OH product studies revealed the additional formation of smaller oxidation products like dicarbonyls and carbonyls. The potentially fast photolysis of these compounds is of great significance, it might represent an important radical source in the atmospheric oxidation of aromatics.

Doubly unsaturated 1,6-dicarbonyls

The kinetic data listed in Chapter 2, Table 1a show a very fast reaction of these compounds with OH radicals, while the NO_3 reactions were found to be slow. Furthermore, an extremely rapid UV photolysis was determined; irradiation with VIS lamps did not result in photolysis. The major atmospheric sinks of 2,4-hexadienedials will therefore be the reaction with OH radicals. The fast OH reactions and rapid UV photolysis can explain why it has not yet been possible to unambiguously identify 2,4-hexadienedials in aromatic oxidation systems, especially where the UV photolysis of H_2O_2 is used as the OH radical source.

The primary photo-oxidation products of doubly unsaturated 1,6-dicarbonyls are smaller carbonyls like butenedial, (methyl)glyoxal and their follow-up products. These compounds have been identified in the atmospheric oxidation of aromatics.

Kinetics and mechanism of the OH initiated oxidation of acetylformoin

Acetylformoin (3-hydroxy-hexane-2,4,5-trione: CH_3–CO–$CH(OH)$–CO–CO–CH_3) is a suspected major product of the OH initiated oxidation of 3-hexene-2,5-dione, itself a product of the OH initiated oxidation of p-xylene. This compound has been synthesised in its isomeric enediol form 3-hexene-3,4-diol-2,5-dione. The rate coefficients for the reaction with OH radicals was found to be $(2.7 \pm 0.7) \times 10^{-10}$ cm^3 s^{-1}. The residual FTIR spectrum of its OH initiated photo-oxidation was found to be identical to that obtained from the OH initiated oxidation of 3-hexene-2,5-dione, the major product has been tentatively identified as hexane-3,3-diol-2,4,5-trione (CH_3–CO–$C(OH)_2$–CO–CO–CH_3). These results suggest that acetylformoin readily enolises and then rapidly reacts with OH radicals to give products of the type observed for the OH initiated oxidation of 3-hexene-3,4-diol-2,5-dione.

Conclusions

Significant progress has been made in product studies of the reaction of NO_3 radicals with alkenes, however, further studies, in particular, with biogenic hydrocarbons are necessary. It was shown conclusively that the formation yields of alkenes from the reaction of OH radicals with alkanes is insignificant under atmospheric conditions. The investigations on the thermal stability of peroxy-nitrates are so advanced that it is now possible to estimate atmospheric lifetimes for nearly all types of peroxynitrate. Possible exceptions, that need further studies, are sulfur and iodine containing compounds. The mechanistic studies on the ozonolysis of alkenes have turned out to be more complicated than originally anticipated. Further investigations on different product channels and the mechanisms involving, $e.g.$ Criegee intermediates, are necessary. The investigations of the reaction pathways of simple alkylated aromatics have led to a new view of the ring-opening mechanism and point to the need for more detailed investigations on unsaturated carbonyl compounds which appear to be major secondary products. Also of major importance for further work is the identification of the fate of NO_x during the aromatic oxidation. Much more work is required to obtain a reliable and representative mechanism that can be used for modelling.

Acknowledgements

Financial support by the State NRW, BMBF, IFP and the EC is gratefully acknowledged.

References ´

1.   G.I. Senum, Y.-N. Lee, J.S. Gaffney, *J. Phys. Chem.* **88** (1984) 1269.
2.   R. Atkinson, A.C. Lloyd, *J. Phys. Chem. Ref. Data* **13** (1984) 315.
3.   Atkinson R., D.L. Baulch, R.A. Cox, R.F. Hampson., J.A. Kerr, J. Troe, *J. Phys. Chem. Ref. Data* **21** (1992) 1125.
4.   E.C. Tuazon, R. Atkinson, *Int. J. Chem. Kinet.* **22** (1990) 1221.
5.   A. Miyoshi, S. Hatakeyama, N. Washida, *J. Geophys. Res.* **99** (1994) 18779.
6.   K.H. Becker, K.J. Brockmann, J. Bechara, *Nature* **346** (1990) 256.
7.   K.H. Becker, K.J. Brockmann, J. Bechara, *Atmos. Environ.* **27A** (1993) 57.
8.   S.E. Paulson, R.C. Flagan, J.H. Seinfeld, *Int. J .Chem. Kinet.* **24** (1992) 103.
9.   G.K. Moortgat, J. Crowley, F. Helleis, O. Horie, W. Raber, C. Zahn, in: *EUROTRAC Annual Report 1992*, part 8 , EUROTRAC ISS, Garmisch-Partenkirchen 1993, pp. 149 – 160.
10. S.M. Aschmann, R. Atkinson, *Environ. Sci. Technol.* **28** (1994) 1539.
11. R. Atkinson, J. Arey, S.M. Aschmann, E.C. Tuazon, *Res. Chem. Intermed.* **20** (1994) 385.

3.3   Laboratory and Field Measurement Studies of the Tropospheric Chemistry of Nitrate and Peroxy Radicals

J. P. Burrows[1], T. Behmann[1], J. N. Crowley[2], D. Maric[1],
A. Lastätter-Weißenmayer[1], G. K. Moortgat[2], D. Perner[2] and
M. Weißenmayer[1]

[1]Institut für Umweltphysik (IUP), Universität Bremen, Postfach 330440, D-28334 Bremen, Germany
[2]Max Planck Institut für Chemie, D-55020 Mainz, Germany

Summary

Both laboratory and field measurement studies of RO_2 and NO_3 radicals were part of this LACTOZ study. The discovery of a rapid reaction between NO_3 and RO_2 and its involvement in a potentially important night-time oxidation mechanism for VOC are discussed. Investigation of the photo-excitation of mixtures of NO_2, NO_3 and O_3 around 662 nm showed that reaction of $NO_3(A^2E')$ with N_2 is not likely to be a source of N_2O and provided an upper limit for rate coefficient for the reaction of NO_3 with O_3. N_2O is produced by 185 nm photolysis of NO, NO_2 and even air, as well as heterogeneously from NO_2, NO_3 and N_2O_5 mixtures. Finally the development of a tropospheric ambient RO_2 detector, the analysis of RO_2 measurements and their impact on our understanding of the chemistry of the daytime and night-time planetary boundary layer are discussed.

Aims of the scientific research

The nitrate radical (NO_3) is formed in the atmosphere primarily by the reaction of nitrogen dioxide (NO_2) with ozone (O_3). At the outset of this project the potential importance of the role of NO_3 as an oxidant in the troposphere had just been recognised. In order to assess the latter accurate physico-chemical models, describing the behaviour of NO_3 in the troposphere, are needed. These require a detailed understanding of the elementary photochemical or chemical reactions and physical processes such as deposition or transport, which determine the tropospheric lifetime of NO_3.

HO_2 and organic peroxy radicals (RO_2) play a variety of roles and functions in the tropospheric chemistry:

i)   the reactions of HO_2 and RO_2 lead to the formation peroxides;

ii)   during the day the reaction of HO_2/RO_2 with NO producing NO_2 plays a key role in the photochemical production of O_3;

iii) HO_2 and possibly RO_2 in air masses with low NO_x participate in a daytime catalytic cycle destroying O_3.

As NO_3 is photolysed rapidly during the day, its maximum concentration occurs at night. The reaction of NO_3 with volatile organic compounds (VOC) is therefore anticipated to represent a night-time source of RO_2 radicals.

The studies undertaken within this project aimed to combine laboratory studies of the tropospheric reactions of NO_3 and RO_2 with relevant field measurement studies. The objective of the latter was the investigation of the validity of the current understanding of the tropospheric role of NO_3 and RO_2. The primary target species for the field studies was therefore RO_2.

Principal scientific findings

Laboratory studies of NO_3 and RO_2 reactions

Previous studies from this laboratory and elsewhere (notably by Atkinson and co-workers) had shown that NO_3 reacted rapidly with a variety of VOC (e.g. DMS or isoprene or terpenes) and inorganic radicals such as NO_2 and Cl. Similarly it was already known that the equilibrium between NO_2, NO_3 and N_2O_5 played an important role in determining the atmospheric fate of NO_3.

Reaction of NO_3 with RO_2

The reaction between NO_3 and CH_3O_2 was studied in the modulated photolysis at 254 nm of HNO_3, CH_4 and O_2 mixtures. The reaction of CH_3O_2 with NO_3 was shown to be rapid and its rate coefficient at 298 K estimated for the first time: the value for $k(CH_3O_2 + NO_3)$ reported was $(2.3 \pm 0.7) \times 10^{-12}$ cm^3 molecule^{-1}s^{-1} [1]. The recognition of a source of NO_3 from the liquid decomposition of HNO_3 led to a re-analysis of this system with the rate coefficient $k(CH_3O_2 + NO_3)$ being now estimated to be approximately 1.0×10^{-12} cm^3 molecule^{-1} s^{-1} [2].

A study of the reaction of $C_2H_5O_2$ with NO_3 was undertaken by photolysing mixtures of HNO_3, C_2H_6 and O_2. The reaction appeared to be fast but an unequivocal value for the rate coefficient was not established. Finally an upper limit for the reaction of NO_3 with O_3 was established with $k(NO_3+O_3)$ 2×10^{-19} cm^3 molecule^{-1}s^{-1} [3].

Further studies of the reaction between NO_3 and CH_3O_2 using a discharge flow coupled with mass spectrometry and studies and the photo-oxidants of HCHO, CH_3CHO and CO_2 are reported by Moortgat elsewhere in this report.

Liquid-phase decomposition of HNO_3

The liquid-phase decomposition of HNO_3 was investigated by observing the oxides of nitrogen produced when a flow of N_2 or O_2 was passed through freshly prepared HNO_3 [2]. In addition to HNO_3 both NO_2 and NO_3 were identified directly in the gas phase. These results indicate that concentrated liquid HNO_3

disproportionates to form N_2O_5 and H_2O. There may also be other channels which directly form NO_2.

As tropospheric aerosol typically has a substantial H_2O content, this reaction was not considered to be of significance for tropospheric chemistry. However stratospheric aerosol is dry and consequently this reaction may occur. Further studies are required to confirm this conclusion.

Atmospheric significance of NO_3

Two studies were undertaken concerning the atmospheric role of NO_3. The first was a theoretical study which focused on the NO_3 initiated oxidation of DMS and the consequences of a chain propagated by the reactions of NO_3 with RO_2 and HO_2 [4]. The second took place as part of a review of the physics, chemistry and atmospheric significance of NO_3 [5].

The DMS study indicated that a potentially significant source of night-time OH results from the reaction of NO_3 with DMS. The concept of a night-time chain initiated by NO_3 reactions with VOC was further developed in Wayne et al. [5].

Photochemical and chemical sources of N_2O

N_2O is one of the most important tropospheric "greenhouse" gases, and its transport into the stratosphere is considered to be the dominant source of stratospheric oxides of nitrogen. The tropospheric abundance of N_2O has been shown to be increasing by approximately 0.2 % per year in recent times. The anaerobic denitrification/reduction of NO_3^- and the nitrification NH_4^+ in soils have been postulated as important microbial sources of N_2O.

The interest in this investigation surrounded the quantification of the potential atmospheric importance of a photochemical source of N_2O involving the photo-excitation of NO_3 into its A^2E' state by (0,0) band absorption followed by its reaction with N_2 [6]. Mixtures of NO_2 and O_3 in air were photolysed by a 662 nm light source, comprising a Tungsten lamp and interference filter. N_2O was observed by means of a sampling gas chromatograph equipped with an ECD detector.

Overall an upper limit for the quantum yield for N_2O production, FN_2O, of 0.6 % was established [7]. However control experiments showed that N_2O was also produced in the dark experiments and in the effective absence of N_2. The F_{N_2O} value is at least a factor of 2 lower than that obtained by Zellner et al. [8].

In spite of the efforts to deactivate the surface of the relatively large reaction vessel, the observed N_2O production was best explained by a heterogeneous reaction involving NO_2, NO_3 or N_2O_5.

Although the NO_3 photoexcitation source was considered unlikely to be a significant atmospheric source of N_2O, the potential importance of a chemical heterogeneous source of N_2O, discovered in this study, at the earth's surface as heterogeneous source of N_2O, discovered in this study, at the earth's surface as compared to the biological source needs to be quantified. The chemical conversion of NO_2, NO_3 or N_2O_5 to N_2O has been up to the present ignored as a significant source of N_2O.

In a further study the following sources of N_2O were investigated

i)   the association reaction of $O(^1D)$ with N_2;
ii)   the photolysis of NO in N_2 at 185 nm;
iii)   the photolysis of NO_2 in N_2 at 185 nm;
iv)   the photolysis of synthetic air at 185 nm.

The rate coefficient for the reaction of $O(^1D)$ with N_2, $k(O^1D + N_2+M)$, at 298 K was found to be $(8.8 \pm 3.3) \times 10^{-37}$ cm^6 $molecule^{-2}$ s^{-1} in reasonable agreement with previous studies. This confirms that this reaction is not a significant atmospheric source of N_2O.

The photolysis at 184 nm of both mixtures of NO and NO_2 in N_2 resulted in significant formation of N_2O. Surprisingly photolysis of air also generated N_2O. This could only be explained by the involvement of excited states of O_2. Any N_2O formation mechanism require a long lived state of NO. The latter is most likely the $NO(^4P)$ state [7] and this mechanism generates a new stratospheric radical.

Peroxy radical measurements

As part of this study, two potential RO_2 measurement techniques have been investigated: the chemical amplification and tuneable diode laser absorption spectroscopy. The first approach has proven to be successful for the measurement of ambient mixing ratios of the sum of all peroxy radicals which react with NO to form NO_2.

The second approach utilises the low-lying electronic transitions and vibrational overtones in the 1 to 2 μm spectral region. This method has the spectroscopic advantage of high specificity. As a first step in the development of this technique a prototype spectrometer was developed which could operate in this region [9, 10]. As a second step the first observation of a high resolution spectrum of HO_2 around 1.5 μm was obtained [11]. Further work is currently underway to investigate the use of an external resonator laser system to enable highly sensitive intracavity measurements of HO_2 to be made.

Development of the chemical amplifier detector

The idea of using chemical amplification for the measurement of RO_2 was first proposed by Stedman. It utilises the chain reaction of HO_2 with NO followed by OH with CO. This initial concept was found to be difficult to calibrate ([12] and references therein). A calibrated chemical amplifier system was developed in this laboratory for the first time [13]. This system detects RO_x (*i.e.* the sum of RO_2, HO_2, RO and OH) which participate effectively in the NO/CO chain reaction. As the concentrations of RO and OH are much smaller than HO_2 and RO_2 in the troposphere, the detector effectively measures the sum of RO_2 and HO_2. This detector requires a detailed understanding of the chemical mechanism which generate NO_2 from RO_2 or HO_2. Prior to the advent of LACTOZ, the theoretical study reported in Hastie *et al.* [13] would not have been feasible.

RO_x measurement campaigns 1991–1994

In 1991 RO_x measurements in continental air were made in Mainz and in Israel. In autumn and winter the instrument was used to study RO_x in the remote northern and southern Atlantic oceans, aboard the German research vessel Polar Stern from 50° N to 40° S. In 1993 the instrument took part in the first FIELDVOC campaign in Brittany in May and June and participated in the OCTA Project measurements at Izana Tenerife in July and August. In 1994 RO_x measurements were undertaken as part of the second FIELDVOC campaign in Portugal and as part of the PRICE Project at the Schauinsland TOR station in southern Germany. The details of the measurements made, the results and their interpretation are reported elsewhere [14–18].

Ambient RO_x measurements – daytime

The boundary layer measurements of RO_x at land sites have resulted in the following conclusions about daytime RO_2 and tropospheric chemistry:

a) in the absence of NO_x and VOC, RO_x values are quite low with maximum mixing ratios at midday ≤ 30 ppt;
b) in the presence of NO_x and VOC, high values of RO_x are obtained: the highest recorded RO_x mixing ratio being » 150 ppt;
c) the rate of production of O_3 in polluted air appears to be reasonably well determined by $Sk_{RO_2+NO}[NO][RO_2]$, implying that the rate determining step for tropospheric O_3 production is the reaction of RO_2 with NO forming NO_2;
d) the $[NO]/[NO_2]$ appears to be explained by the reactions of NO with O_3, RO_2 with NO and the photolysis of NO_2 at desert sites (Israel) and in the free troposphere (Izana);
e) the $[NO]/[NO_2]$ at the coastal site Brittany can only be explained by the presence of an additional NO oxidising agent.

The last observation has resulted in the hypothesis that either halogen oxides (*e.g.* BrO) or Sulfur radicals may be active in tropospheric chemistry at least in coastal environments ([19] and references therein).

Analysis of the RO_x measurements from the remote parts of the North and South Atlantic can be summarised as follows:

a) RO_x is low typically being < 10 ppt at midday;
b) an unidentified source of RO_x is required in the North Atlantic between approximately 10 and 35° N;

d) a small additional loss process for RO_x is required in both hemispheres;
e) for remote maritime air the observed loss of O_3 is similar to that predicted by the reaction of RO_x with O_3.

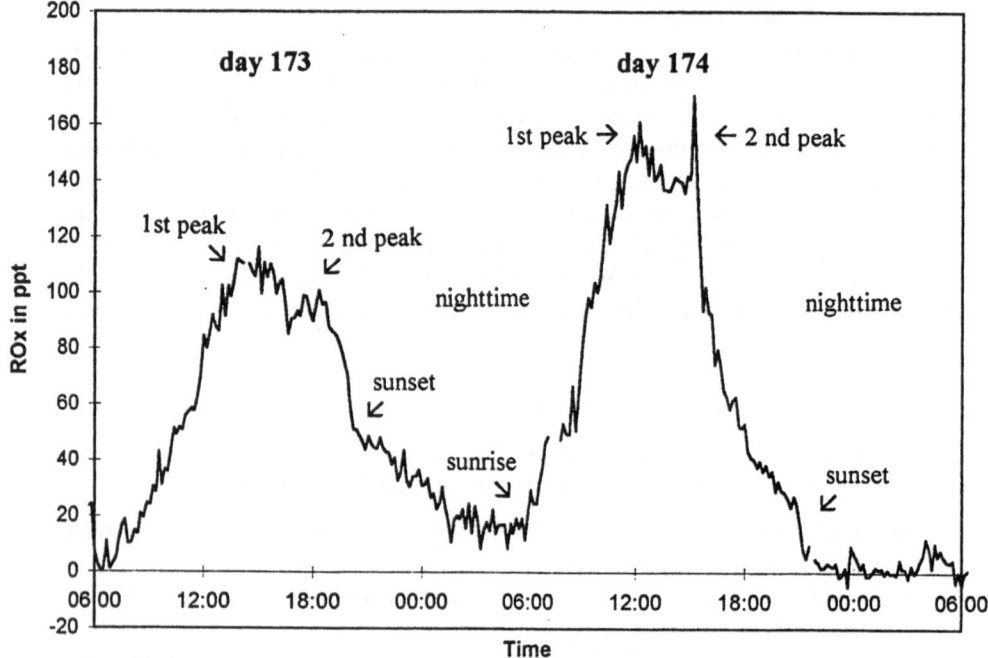

Fig. 1: RO_2 mixing ratio measured during the FIELDVOC Portugal Campaign 1994.

The observation b) is best explained by the existence of a biogenic maritime source of VOC in the upwelling region off the coast of Africa, or by an as yet unidentified anthropogenic pollution source. The observation d) may imply that loss of RO_x occurs on aerosol or via a homogeneous reaction with an as yet unidentified radical.

Ambient RO_x measurements – night-time

The night-time measurements of RO_x have shown the following:

a) in remote maritime air the mixing ratio of RO_x is less than 2 ppt;
b) in coastal environments RO_x significant levels *i.e.* above 2 ppt have been observed *e.g.* Brittany;
c) in forests unambiguous RO_x signals of up to 30 ppt have been observed.

An example of two days' data from the FIELDVOC campaign in Portugal in 1994 is shown in Fig. 1. Large night-time RO_2 signals are clearly visible on day 173. In addition a double maxima is observed in the daytime RO_2 on both days. The appearance of RO_x is linked to the presence of VOC, O_3 and NO_x. Qualitatively this is in agreement with the hypotheses of Platt *et al.* [4] and Wayne *et al.* [5].

Assessment

The original objective of the EUROTRAC project LACTOZ was to provide kinetic and mechanistic data to enable a quantitative description of the tropospheric ozone budget in numerical models. Following the review by the EUROTRAC-SSC in 1991 the priorities set for the second phase of LACTOZ were understanding

a)   Ozone production in the free troposphere;
b)   Complex VOC and NO_x chemistry in the boundary layer.

In addition it was pointed out at this time that *the inclusion of "box modelling" and specially designed "field measurements" in order to test or evaluate chemical mechanisms, in particular with measurements of photochemically generated free radicals such as OH and HO_2* was necessary.

The discovery of the rapid reactions of organic peroxy radicals with NO_3 and the recognition of their importance in tropospheric chemistry has been a significant achievement within this project. Similarly the development of a method for the detection of RO_x has provided a large number of measurements of RO_x to be made in the planetary boundary layer and a few to be made in the free troposphere (Izana, Tenerife). This has allowed box models of HO_2 and RO_2 chemistry to be developed and the inadequacies of our current understanding to be pointed out. In conclusion the aims and objectives of LACTOZ have been well served by the achievements of this project.

Acknowledgements

The work described in this study was funded in part by the University of Bremen, the Max-Planck-Institut für Chemie Mainz and the Max-Planck-Gesellschaft zur Förderung der Wissenschaft and by the following EU research contracts STI-056-J-C, ST2J-0193, EV4V-0093-C and EV5-CT91-0040 and by the German Ministry of Research and Technology under Project No. 325-4007-07 INT 037. We acknowledge the active co-operation of G. LeBras, G. Poulet and U. Platt in this project.

References

1. J.N. Crowley, J.P. Burrows, G.K. Moortgat, G. Poulet, G. LeBras; Room temperature rate coefficient for the reaction between CH_3O_2 and NO_3, *Int. J. Chem. Kin.* **22** (1990) 673–681.

2. J.N. Crowley, J.P. Burrows, G.K. Moortgat, G. Poulet, G. LeBras; Optical detection of NO_3 and NO_2 in pure HNO_3 vapour, the liquid phase decomposition of HNO_3, *Int. J. Chem. Kin.* **25** (1993) 795–803.

3. D. Maric, J. P. Burrows, G. K. Moortgat; A study of the formation of N_2O in the reaction of $NO_3(A^2E')$ with N_2, *J. Atmos. Chem.* **15** (1992) 157–169.

4. U. Platt, G. LeBras, G. Poulet, J. P. Burrows, G. K. Moortgat; Peroxy radicals from night-time reaction of NO_3 with organic compounds, *Nature* **348** (1990) 147–149.

5. R.P. Wayne, I. Barnes, P. Biggs, J. P. Burrows, C. E. Canosa–Mas, J. Hjorth, G. LeBras, G. K. Moortgat, D. Perner, G. Poulet, G. Restelli, H. Sidebottom; The nitrate radical: physics, chemistry and the atmosphere, *Atmos. Environ.* **25**A (1991) 1–206.

6. R. Zellner A. Hoffmann; Kinetics of the Reactions of NO_3 and $NO_3(A^2E')$" *2nd Franco-German Workshop on the study of the chemical reactions with tropospheric interest.*

7. D. Maric, J.P. Burrows; Formation of N_2O in the photolysis/photoexcitation of NO, NO_2 and air, *J. Photochem. Photobiol. A: Chem.* **66** (1992) 291–312.

8. R. Zellner, D. Hartmann, I. Rosner; N_2O formation in the reactive collisional quenching of $NO_3{}^*$ and $NO_2{}^*$ by N_2, *Ber. Bunsenges. Phys. Chem.* **96** (1992) 385–390.

9. T.J. Johnson, F.G. Wienhold, J.P. Burrows, G. W. Harris; High frequency modulation of diode lasers for the detection of atmospheric trace gases, in: *Proc. of the MIP 90, Microwaves and Optronics Conference on High Frequency Technology*, Network GmbH, Hagenburg 1990, pp. 211–216, (ISBN 3-924651-24-8).

10. T.J. Johnson, F.G. Wienhold, J.P. Burrows, G.W. Harris; FM spectroscopy at 1.3 mm using InGaAsP lasers: a prototype field instrument for atmospheric chemistry, *Appl. Opt.* **30** (1991) 407–413.

11. T.J. Johnson, F.G. Wienhold, J.P. Burrows, G.W. Harris, H. Burkhard; Measurements of line strengths in the HO_2 υ1 overtone band at 1.5 mm using an InGaAs laser, *J. Phys. Chem.* **55** (1991) 6499–6502.

12. D.H. Stedman, J.G. Walega, C.A. Cantrell, J.P. Burrows, G.S. Tyndall, in: W. Jaeschke (ed), *NATO ASI Series in Chemistry of Multiphase Atmospheric Systems*, Springer-Verlag, Berlin 1986, G6, pp. 352–366.

13. D.R. Hastie, M. Weißenmayer, J.P. Burrows, G.W. Harris; Calibrated Chemical Amplifier for Atmospheric ROx measurements, *Anal. Chem.* **63** (1991) 2048–2057.

14. T. Behmann, M. Weißenmayer, J.P. Burrows; Peroxy Radicals in night-time oxidation chemistry, in: G. Angeletti, G. Restelli (eds), *Proc. Sixth European Symp. on Physico-Chemical Behaviour of Atmospheric Pollutants*, ECSC-EC-EAEC, Brussels and Luxembourg 1994, pp. 259–265, (ISBN 92-826-7922-5).

15. M. Weißenmayer, J.P. Burrows, R. Gall, D. Hastie, A. Lastätter-Weißenmayer, M. Luria, M. Peleg, D. Perner, P. Russell; Direct insight into atmospheric ozone formation by peroxy radical observation, in: P.M. Borrell, P. Borrell, T. Cvitaš, W. Seiler (eds), *Proc. EUROTRAC Symp. '92*, SPB Academic Publ., The Hague 1993, p. 186.

16. M. Weißenmayer, J.P. Burrows, M. Schupp; Peroxy radical measurements in the boundary layer above the Atlantic Ocean, in: G. Angeletti, G. Restelli (eds), *Proc. Sixth European Symposium on Physico-Chemical Behaviour of Atmospheric Pollutants*, ECSC-EC-EAEC Brussels and Luxembourg 1994, pp. 575–582, (ISBN 92-826-7922-5).

17. M. Weißenmayer J.P. Burrows, R. Gall, D. Hastie, A. Lastätter-Weißenmayer, M. Luria, M. Peleg, P. Russel, D. Perner; Observation of peroxy radicals during ozone formation in polluted tropospheric air, *Nature* (1995) in press.

18. M. Weißenmayer, J.P. Burrows; Peroxy radical measurements in the clean maritime boundary layer, to be published.

19. T. Behmann, M. Weißenmayer, J.P. Burrows; Measurements of peroxy radicals in a coastal environment, *Atmos. Environ.* (1995) in press.

3.4 Tropospheric Degradation of Aromatics: Laboratory Kinetic Studies of some First Steps of the OH Initiated Oxidation

Pacal Devolder, Jean-Pierre Sawerysyn, Christa Fittschein, Abdel Goumri, Lahcen Elmaimouni, Bertrand Bigan and Christine Bourbon

Laboratoire de Cinétique et Chimie de la Combustion, URA CNRS 876, Université des Sciences et Technologies de Lille, 59655 Villeneuve d'ascq Cedex, France

Summary

We have performed laboratory measurements of rate constants of a few first steps of the tropospheric oxidation mechanism of some monocyclic aromatic hydrocarbons; the following reactions have been studied: OH radical with benzene or toluene (+M); benzyl radicals (from OH abstraction pathway) with O_2, NO and NO_2; on the other band, the branching ratios (abstraction/addition + abstraction) have been measured by a direct spectroscopic technique. All these measurements have been achieved thanks to the Discharge Flow (DF) technique with detection of OH by Resonance Fluorescence (RF) and other radicals by Laser Induced Fluorescence (LIF).

Aims of the research

Although the specific and significant impact of aromatics upon tropospheric chemistry in urban areas is well documented [1], the exact mechanism of their tropospheric oxidation remained largely uncertain at the beginning of LACTOZ. Taking toluene as an example, the first reaction steps are as represented on the scheme of Fig. 9 [2, 3].

The abstraction path, with a yield of roughly 10 % [3], leads to the intermediate benzaldehyde which is known to exhibit subsequent fast reactions [4]. The major addition path ($\approx$ 90 %), results in the formation of a radical adduct (of cyclo-hexadienyl type) which is suspected to lead ultimately to a variety of ring-containing or ring-cleavage products [1]. Whether or not there is an intermediate step of rearomatization (formation of o-cresol) is still a matter of controversy, Although very recent smog chamber experiments point to a negligible contribution of this pathway in real atmospheric conditions.

Our research aimed at the following objectives:

(i)   Measurements of the rate constants with O_2, NO and NO_2 of a few benzyl-type radicals derived from toluene or xylenes. No reaction rate with NO_2 was reported so far, whereas the data for O_2 and NO were very limited.

(ii) Measurements of the branching ratios between abstraction and addition.

These branching ratios were only inferred from end product analysis [3] and thus could be questioned since the carbon mass balance was usually very poor during the oxidation of aromatics (no more than 60 % of the initial reactant recovered in the collected end products). We have measured the branching ratios by a direct spectroscopic method based on a monitoring of the benzyl radical relative concentrations by LIF.

(iii) Since the low pressure discharge flow technique has been used throughout our investigation, it has been necessary to measure the reaction rate constants (OH + benzene (+ M)) and (OH + toluene (+ M)) in our specific conditions in a preliminary compulsory step. These experiments have provided the preliminary data to perform subsequent experiments.

Principal results

Rate constants for the intermolecular reactions (OH + toluene + He) *and* (OH + benzene + He); *rate constant of the adducts with* NO_2

These reactions have been investigated by the discharge-flow technique with detection of OH by resonance fluorescence. OH radicals were generated by the classical transfer reaction: (H + NO_2 → OH + NO). The experimental parameters were as follows: pressure of helium: 0.5/9.5 Torr, T = 298 or 353 K.

Both reactions are in the fall-off range in our experimental conditions; our global results (over the pressure range 0.5/10 Torr) can be fitted to derive the Troe parameters k_0 and k_∞ according to the classical Troe formula [5].

The corresponding results are presented in Table 1.

Table 1: Fall-off Troe parameters for OH + benzene (+ He) and OH + toluene (+ He) reactions.

		k_0/cm^6 molec^{-2} s^{-1}	k_∞/cm^3 molec^{-1} s^{-1}
benzene	(297 ± 3)K	(1.7 ± 0.5)10^{-29}	(10 ± 2)10^{-13}
	353 K	(1 ± 0.2)10^{-29}	(10 ± 1)10^{-13}
toluene	(295 ± 3)K	(4 ± 0.5)10^{-28}	(6.0 ± 0.7)10^{-12}
	353 K	(2 ± 0.5)10^{-28}	(4 ± 0.7)10^{-12}

The main reaction product (*i.e.* path b) with both benzene and toluene is an adduct (a cyclohexadienyl-type radical) which is known to exhibit back decomposition towards OH [6]; since these adduct radicals react very quickly with NO_2 [7], the OH decay kinetics proved to be very sensitive to NO_2 concentration in our experimental conditions, in agreement with numerical simulations. By taking into account the adduct unimolecular back decomposition rates of Zetzsch and co-workers [6], we have been able to estimate a few reaction rates of these adducts with NO_2 and O_2:

benzene (353K, 1 Torr He):
$$k(\text{adduct} + NO_2) = (4 \pm 2)\ 10^{-11}\ \text{cm}^3\ \text{s}^{-1}$$
toluene (353K, 1 Torr He):
$$k(\text{adduct} + NO_2) = 4 \times 10^{-11}\ \text{cm}^3\ \text{s}^{-1}$$
toluene (353K, 1 Torr He):
$$k(\text{adduct} + O_2) < 5 \times 10^{-14}\ \text{cm}^3\ \text{s}^{-1}$$

Rate constants for O_2, NO and NO_2 with a few benzyl-type radicals

The benzyl radical is the reaction product of the abstraction path (path a) of the reaction of toluene with OH. Similarly, methylbenzyl radicals are formed during the OH abstraction reactions with xylenes and trimethylbenzenes. We have measured the reaction rate constants with O_2, NO and NO_2 of a series of benzyl-type radicals, including a few fluorosubstituted ones in purpose of comparison.

The experimental technique was the fast discharge flow at low pressure (a few Torr) with detection of radicals by LIF.

The benzyl radicals were generated by chlorine atom attack on the methyl substituent since there is apparently no chlorine addition on the ring [8, 9]; chlorine atoms were generated in the upstream part of the flow tube thanks to the following reaction of transfer: $F + Cl_2 \rightarrow FCl + Cl$. The LIF signals were obtained by exciting in the visible band of benzyl radicals ($l \approx 450/465$ nm) and detected at maximum fluorescence intensity, around (480 ± 10) nm [10].

The experimental results are gathered in Tables 2 and 3.

Table 2: Rate constants (in 10^{-12} cm^3 s^{-1}) for benzyl-type radicals measured in this work.

Radical	O_2	NO	NO_2
benzyl	0.72	6.3	60
p-fluorobenzyl	0.78	8.9	50
m-fluorobenzyl	0.51	7.45	41
o-methylbenzyl	0.91	10.5	57
m-methylbenzyl	1.11	13	56

Table 3: Comparison between the room temperature values of the rate constant for (benzyl + O_2) by various techniques.

$k\ /\ 10^{-12}$ cm^3 s^{-1}	technique	pressure / Torr	reference
0.99 ± 0.07	FP/UV Abs	N_2, 160	[11]
1.12 ± 0.2	LP/LIF	N_2, 3/15	[12]
0.74 ± 0.2	DF/MS	He, 0.75/3	[13]
1.2	FP/UV Abs	N_2, 760	[14]
1.5 ± 0.2	PR/UV Abs	Ar, 1013	[9]
0.97 ± 0.1	LP/MS	He, 4	[15]
0.72 ± 0.14	DF-LIF	He, 1	this work

Our measurements are the first ones performed with the technique of discharge flow/LIF; also, concerning the reaction rates with NO_2, there are no other data with which our results can be compared. For the reaction rates with O_2 and NO, our results are systematically below the results of other techniques (obtained at higher pressures) Although the uncertainty ranges usually overlap (Table 3); these minor discrepancies could be assigned (at least for the reaction rate with O_2) to a fall-off behaviour [14].

Taking into account the experimental uncertainties and the small variations between the benzyl radicals, we propose the following generic values for the rate constants of a benzyl-type radical (in cm^3 s^{-1}): O_2: 0.9×10^{-12}; NO: 8×10^{-12}; NO_2: 5.8×10^{-11}.

Branching ratio measurements (R = abstraction/abstraction + addition) by a direct spectroscopic method

The branching ratios have been measured with essentially the same set-up of discharge flow/LIF used for kinetic measurements. A double fluorescence cell allows simultaneous monitoring of the benzyl radical (by LIF) and OH (by resonance fluorescence).

The benzyl radicals are formed by attack of the suitable precursor $R-CH_3$ in large excess by one of the three reactants: F, Cl or OH at the same concentration; with each reactant, the corresponding signal of the benzyl radical fluorescence: S^F, S^{Cl}, S^{OH} is probed by LIF:

$$
\begin{array}{lllll}
F + R-CH_3 & \rightarrow & R-CH_2 & \Rightarrow & S^F & (1) \\
Cl + R-CH_3 & \rightarrow & R-CH_2 & \Rightarrow & S^{Cl} & (2) \\
OH + R-CH_3 & \rightarrow & R-CH_2 & \Rightarrow & S^{OH} & (3)
\end{array}
$$

Stoichiometric conversions between the three reactants are achieved thanks to the following rapid transfer reactions:

$$F + H_2O \text{ (excess)} \rightarrow OH + HF; F + Cl_2 \text{ (excess)} \Rightarrow Cl + Fcl$$

Since the branching ratio for reaction 2 is one at room temperature [8, 9, 14], the branching ratio for OH abstraction (*i.e.* in reaction 3) is simply $R = S^{OH}/S^{Cl}$ (this is our so called direct method). Another indirect method can also been used if the branching ratio for reaction 1 is known: $R'' = S^F/S^{Cl}$; this happened for toluene and *p*-xylene [16]. The results of both methods used in the present work are gathered in Table 4, together with available literature data from end product analysis.

The good agreement between our values (from a spectroscopic method monitoring the first abstraction step) and the values derived from smog chamber data (from intermediate or final products) confirms that the "missing" final products responsible for the imperfect mass balance are probably associated with the dominant addition path.

Table 4: Room temperature branching ratios (abstraction/total).

	this work (direct technique)		product analysis	ref
	first method	second method		
toluene	0.06 ± 0.02	0.07 ± 0.03	0.12	[3]
			0.073	[17]
			0.065	[18]
o-xylene	0.11 ± 0.03		0.11	[3]
			0.059	[19]
m-xylene	0.07 ± 0.01		0.04	[19]
			0.04	[3]
p-xylene	0.08 ± 0.01	0.07 ± 0.02	0.08	[3]
			0.078	[19]
			0.08	[20]

Assessment of the achievement

Our work during the project is part of a concerted effort of various European groups to unravel the mechanism of tropospheric oxidation of aromatics. A lot of kinetic, mechanistic and analytic data have been produced, especially concerning (i) the first primary steps of the mechanism (abstraction and addition) (ii) the nature and the yield of intermediate or final oxidation products. Also, it appeared that "standard" reaction schemes for oxidation (*i.e.* like those for alkanes) are not applicable to aromatics; this specificity of the latter prevented us from presenting a clear picture of the oxidation mechanism and requires new investigations.

Acknowledgements

We thank the French programs "grands cycles biogeochimiques" (PACB), "environment" and the organisations CNRS, "Ministère de l'Environnement" for their support. We thank the CEC for a grant (STEP).

References

1.   R. Atkinson, S.M. Aschmann, *Int. J. Chem. Kinet.* **26** (1994) 929.
2.   J.A. Leone, J.H. Seinfeld, *Int. J. Chem. Kinet.* **16** (1984) 159.
3.   R. Atkinson, *J. Phys. Chem. Ref. Data, Monograph* No. 1 (1991) 1, and references therein.
4.   F. Kirchner, F. Zabel, K.H. Becker, *Chem. Phys. Lett.* **191** (1992) 169.
5.   R. Atkinson *et al.*, *J. Phys. Chem. Ref. Data* **21** (1992) 1125.
6.   R. Knispel, R. Koch, M. Siese, C. Zetzsch, *Ber. Bunsenges Phys. Chem.* **94** (1990) 1375.
7.   R. Zellner, B. Fritz, M. Preidel, *Chem. Phys. Lett.* **121** (1985) 412.
8.   T.J. Wallington, *Photochem. Photobiol.* A**45** (1988) 167.
9.   F. Markert, P. Pagsberg, *Chem. Phys. Lett* **209** (1993) 445.
10.   T.R. Choularton, B.A. Thrush, *Chem. Phys. Lett.* **125** (1986) 547.
11.   T. Ebata, K. Obi, I. Tanaka, *Chem. Phys. Lett.* **77** (1981) 480.

12. H.H. Nelson, J.R. McDonald, *J. Phys. Chem.* **86** (1982) 1242.
13. K. Hoyermann, J. Seeba, in: *Proc. 25th Int. Symp. on Combustion,* 1994.
14. F.F. Fenter, B. Noziere, F. Caralp, R. Lesclaux, *Int. J. Chem. Kinet.* **26** (1993) 171.
15. K. Bayes, private communication.
16. J. Ebrecht, W. Hack, H. G. Wagner, *Ber. Bunsenges, Phys. Chem.* **93** (1989) 619.
17. R. Atkinson, S.M. Aschmann, J. Arey, W.P.L. Carter, *Int. J. Chem. Kinet.* **21** (1989) 801.
18. E. Evmorfopoulos, S. Glavas, in: R.A. Cox (ed), *Air Pollution Research Report* **42**, CEC, E. Gouyot SA, Brussels 1992, pp. 55–60.
19. R. Atkinson, S. Aschman, J. Arey, *Int. J. Chem. Kinet.* **23** (1991) 77.
20. K.H. Becker, I. Barnes, A. Bierbach, E. Wiesen, *Joint CEC-EUROTRAC Workshop,* Leuven 1992.

3.5   Laboratory Studies of the Formation of Hydroperoxides in Ozonolysis of Anthropogenic and Biogenic Alkenes

Silke Wolff[1], Walter V. Turner[1], Siegmar Gäb[1], Simone Mönninghoff[2], Lars Ruppert[2] and Klaus Brockmann[2]

[1]Analytische Chemie, Bergische Universität-Gesamthochschule Wuppertal, 42119 Wuppertal, Germany
[2]Physikalische Chemie, Bergische Universität-Gesamthochschule Wuppertal, 42119 Wuppertal, Germany

Summary

An HPLC fluorescence method for quantitative and qualitative hydroperoxide analysis was improved, so as to determine linear hydroperoxides and hydroxy hydroperoxides up to a chainlength of seven carbon atoms.

The hydroperoxide products of ozonolysis of a number of alkenes were determined under three sets of ozonolysis conditions: 1. concentrations of 4 ppm in dry synthetic air, 2. concentrations of 140 ppm in synthetic air at 100 % relative humidity and 3. concentrations of 100 ppm in water.

The main products of ozonolysis in dry air were alkyl hydroperoxides, while in aqueous solution 1-hydroxyalkyl hydroperoxides were the only products observed. In humid air at the higher concentrations, both types of products were present, and similar results were obtained when ozone was bubbled through aqueous solutions of the alkenes.

A mechanism consistent with these results is postulated.

Aims of the research

Atmospheric ozonolysis is believed to be a major pathway for the degradation of the vast quantities of biogenic and anthropogenic alkenes emitted annually. Analysis of air and precipitation has indicated that H_2O_2, alkyl hydroperoxides and 1-hydroxyalkyl hydroperoxides (1-HAHPs) are present [1]. Since both ozonolysis and other oxidation processes can lead to hydroperoxides, it is of interest to know to what extent ozonolysis is involved in the formation of those hydroperoxides actually found in the atmosphere.

In continuation of a long-standing study, we set out under LACTOZ to determine the products of ozonolysis of 1-pentene, 1-hexene and 1-heptene at gas-phase concentrations of 140 ppm at 100 % relative humidity. It was necessary to develop our HPLC system further in order to do this. We later extended the study to the ozonolysis of lower alkenes and of some terpenes at ca. 4 ppm in dry air.

Ozonolysis in water was then examined in search of a method for distinguishing ozonolysis in atmospheric droplets from true reaction in the gas phase.

Principal scientific findings

The analytical system

The first step within this project was to improve the HPLC-fluorescence method [1] in order to determine not only hydrophilic but also lipophilic hydroperoxides of up to seven carbon atoms. Among the aspects investigated were column length, mixed solvent and gradient elution and the use of microperoxidase instead of horseradish peroxidase in the detection system [2, 3]. The use of microperoxidase let us add secondary, tertiary and other branched hydroperoxides to the list of peroxide species we can determine. For full analysis a number of hydroperoxides were synthesised as standards for retention time and quantitation: methyl hydroperoxide (MHP), ethyl hydroperoxide (EHP), 1- and 2-propyl hydroperoxides (PHPs), 1-butyl hydro-peroxide (1-BHP), hydroxymethyl hydroperoxide (HMHP), 1-hydroxyethyl hydroperoxide (1-HEHP), 2-hydroxyethyl hydroperoxide (2-HEHP), 1-hydroxypropyl hydroperoxide (1-HPHP), 1-hydroxybutyl hydroperoxide (1-HBHP), 1- hydroxypentyl hydroperoxide (1-HPentHP), 1-hydroxyhexyl hydroperoxide (1-HHHP), bis(hydroperoxy) methane (MBHP) and bis(hydroxy-methyl) peroxide (BHMP). We decided in the end not to use gradient elution, which brought baseline problems, but rather to use side-by-side HPLC systems: one with acidified water as eluent for the more hydrophilic hydroperoxides (H_2O_2 and the lower alkyl hydroperoxides) and a second, with a mixture of water and acetonitrile as eluent, for more lipophilic compounds. Neutralisation of the sample before chromatography allowed us to distinguish between alkyl and 1-hydroxyalkyl hydroperoxides (1-HAHPs), since it converted the latter to H_2O_2 and the corresponding aldehydes. The detection limit of the enhanced system is about 0.05 mmol/L for both linear and branched primary and about 1 mmol/L for secondary hydroperoxides.

The role of the ozonolysis conditions

Dry gas-phase ozonolysis was carried out on ethene, propene, isoprene, and a- and b-pinene [4] at concentrations of 4 ppb in a 1000 L reactor [5]. A certain volume of the resulting gas was drawn through acidified water, which was then examined for its hydroperoxide content. The results are shown in Table 1. Small amounts of H_2O_2 and HMHP were found in all the dry gas-phase experiments; it is still not certain whether these are artefacts, though it is reasonable, in a reaction that appears to have a high radical component, that some H_2O_2 be formed via HO_2 radicals. The only significant hydroperoxide detected was methyl hydroperoxide from those olefins with a methyl group on the double bond. A trace of 2-HEHP, readily distinguished from MHP by its retention time, must be attributed to reaction of OH radicals with ethene [6].

Table 1: Hydroperoxide distribution in dry gas-phase ozonolysis as % of alkene; a dash means "not detected".

	Ethene	Propene	Isoprene	α-Pinene	β-Pinene
H_2O_2	0.4	0.3	0.7	0.3	0.6
HMHP	0.2	0.2	0.5	0.1	0.3
MHP	–	4.1	5.0	0.8	–
1-HEHP	–	–	–	–	–
2-HEHP	< 0.1	–	–	–	–
other HAHPs	–	–	–	–	–
total	0.6	4.6	6.2	1.2	0.9

The experiments at 100 % relative humidity were carried out on 1-pentene, 1-hexene and 1-heptene at concentrations of 140 ppm in a 20 L borosilicate glass flask [7]. After an hour's reaction time the gaseous contents of the flask were drawn through acidified water; the walls of the flask were then washed with the same water. Analysis of the resulting solution for hydroperoxides gave the results in Table 2. Significant yields of H_2O_2, alkyl hydroperoxides and 1-HAHPs were obtained.

Table 2: Hydroperoxide distribution in humid gas-phase ozonolysis as % of alkene.

	1-Pentene	1-Hexene	1-Heptene
H_2O_2	2.8	2.4	2.5
HMHP	2.5	2.2	2.4
BHMP	0.7	1.2	1.3
MBHP	1.0	0.7	0.9
MHP	0.6	0.4	0.4
EHP	2.7	0.4	0.1
1-PHP	5.0	0.5	0.1
1-BHP	–	1.4	0.1
1-HPentHP	–	0.1	–
1-PentHP	–	–	0.7
1-HHHP	–	–	0.2
unknown	0.5	0.2	0.6
total	15.8	9.5	9.5

Ozonolysis in water was carried out by mixing a solution of the alkene in water (water/acetonitrile in the case of isoprene and the pinenes) with an aqueous solution of ozone The hydroperoxide yields are in Table 3. The only significant products were 1-HAHPs [4].

When the ozonolysis in water was carried out by bubbling ozone through an aqueous solution of the alkene, rather than by mixing prepared solutions of the two, the product distribution (not tabulated) resembled the "mixed-reaction" distribution seen in the humid gas-phase reactions.

Table 3: Hydroperoxide distribution in aqueous-phase ozonolysis as % of alkene.

	Ethene	Propene	Isoprene	α-Pinene	β-Pinene
H_2O_2	–	–	< 0.1	< 0.1	< 0.1
HMHP	27.9	2.7	11.1	–	0.2
MHP	–	–	–	–	–
1-HEHP	–	11.1	–	–	–
2-HEHP	–	–	–	–	–
other HAHPs	–	–	13.1	11.3	16.0
total	27.9	13.8	24.2	11.3	16.2

The products are consistent with the idea, as is some of the literature [8], that the early stages of the reactions under all the conditions examined here proceed by the Criegee ozonolysis mechanism (Fig. 1). Thus a primary ozonide decomposes rather unselectively to two carbonyl compounds (aldehydes or ketones) and two thermally excited carbonyl oxides (2 and 3), whose fate depends on the reaction medium. In aqueous solution the carbonyl oxides are largely relaxed by collision, whereafter they undergo high yield bimolecular reactions with water to 1-HAHPs [4]. In the gas phase, by contradistinction, the excited carbonyl oxides fragment thermally to a large extent before they can undergo bimolecular reaction [9] (Fig. 2). The major peroxide products come from the reactions of the fragment radicals with oxygen, the largest alkyl hydroperoxide corresponding in each case to the alkyl chain of the original alkene. The alkyl radical apparently first forms a peroxy radical, which extracts hydrogen from HO_2; either direct formation of smaller alkyl fragments from the carbonyl oxide or further fragmentation of the initial alkyl radical yields other alkyl hydroperoxides all the way down to MHP. The experiments described here were not designed to tell us anything about the possible intermediacy of complexes in the cleavage of primary ozonides in the gas phase, such as have been suggested by some others [10].

The products of ozonolysis in humid air indicate that the carbonyl oxide can react with water in the gas phase, when enough water is present. The course of this reaction is as yet unknown. The 1-HAHPs observed may result from the fraction of carbonyl oxide that is not relaxed by collision. Alternatively, they may come from bimolecular reaction of the excited state with water; in either case, the products observed or some intermediate complexes with water then undergo relaxation.

It seems as if part of the reaction takes place in the gas phase when ozone is bubbled through aqueous solutions of alkene. The alkenes examined here have a rather low Henry's Law constant, so that it should not be surprising if a considerable fraction enters the gas phase, where it reacts with ozone and whence polar products like hydroperoxides are absorbed into the water.

In general, the yields of hydroperoxides decrease as the starting alkene becomes larger. This may simply reflect the greater number of reaction pathways open to larger molecules: our analysis detects only hydroperoxides, and in some of the experiments only primary linear alkyl hydroperoxides.

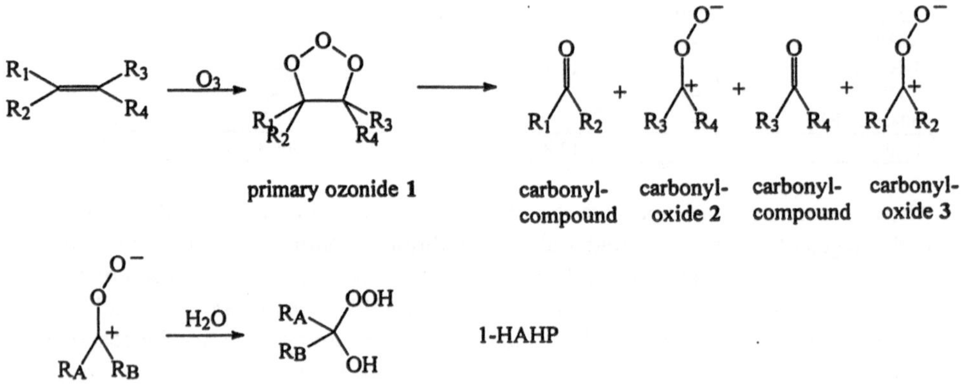

Fig. 1: Postulated mechanism for aqueous-phase ozonolysis.

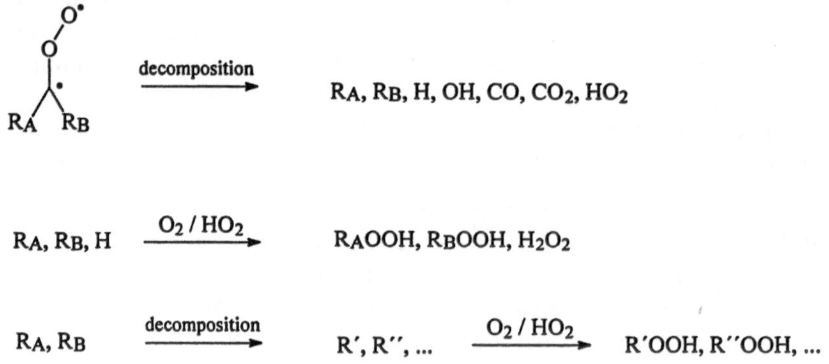

Fig. 2: Postulated mechanism of dry gas-phase ozonolysis.

The role of the alkene in ozonolysis

Because ethene is symmetrical, a single carbonyl oxide, CH_2OO, is formed, along with formaldehyde. Addition of water to the carbonyl oxide gives HMHP, which was observed in all cases where water was present. The trace of HMHP observed from ozonolysis in dry air may, as noted above, be an artefact, but fragmentation of CH_2OO has been postulated by others to give H radicals; these would be expected to give HO_2, then H_2O_2.

Propene gives two carbonyl oxides, CH_2OO and CH_3CHOO, and accordingly the reaction in water gives HMHP and 1-HEHP; as observed by others, the carbonyl oxide is preferentially formed on that end of the olefin bond with the methyl substituent [9]. Fragmentation of CH_3CHOO to methyl radicals in the dry gas phase leads to MHP.

The ozonolysis of isoprene is complicated by the presence of two unlike double bonds. The 11.1 % yield of HMHP in water shows that a significant fraction of the primary ozonide cleaves to CH_2OO. No other hydroperoxide could be detected under the conditions used to analyse the products of this reaction, but we have indirect evidence that other HAHPs are indeed present in the sample. Thus neutralisation of the sample brought about a great increase in the yield of H_2O_2, which would arise from HAHPs too large to be eluted from the HPLC column by water. Normally, hydroxy C-4 hydroperoxides would be eluted by water, so we prefer not to speculate as yet about the nature of the "hidden" product, which we hope to detect by using the H_2O/methanol or H_2O/acetonitrile eluents. Ozonolysis of isoprene in the dry gas phase gives MHP in accordance with our scheme.

Either direction of cleavage of the primary ozonide from α-pinene would produce a carbonyl oxide with ten carbon atoms. Fragmentation of this intermediate might well lead to a number of hydroperoxides we could detect, but only MHP was observed from gas-phase ozonolysis. As in the case of isoprene, neutralisation of the sample from aqueous ozonolysis before analysis indicates that significant amounts of HAHPs are formed in water. These are still under investigation.

β-Pinene might in principle give CH_2OO, but the absence of HMHP from the products in water shows that the cleavage is very selective to the substituted carbonyl oxide. The nature of the larger HAHPs revealed by neutralisation is still unknown.

In all the cases here studied under both sets of limiting conditions, the yields on ozonolysis in water were much higher than those in dry air; this doubtless reflects the high stabilisation of the excited intermediates in the condensed phase.

The product in highest yield in the ozonolysis of 1-pentene in humid air (where the concentration is also considerably higher) is 1-PHP;. MHP and EHP are also seen, the yields decreasing with the chain length. The HMHP probably results from the reaction of water with CH_2OO, though some of it may come from reaction of formaldehyde with H_2O_2, while the BMHP observed must arise from addition of

H_2O_2 to CH_2OO. It is odd that no 1-HBHP was observed, since the corresponding 1-HAHPs had been observed in the ozonolysis of all the lower linear alkenes and were seen in the case of 1-hexene and 1-heptene as well. The ozonolysis of 1-hexene and 1-heptene also gave linear alkyl hydroperoxides. The reason the yields of these decrease with the chain length is uncertain, the energetics of the fragmentation of the carbonyl oxides not having been fully examined.

Assessment

Ozonolysis in the gas phase leads to peroxide products and is especially interesting in that regard, because some of these peroxides can be formed even in the presence of NO_x. The studies described here provide information about the yields and identity of hydroperoxide products in gas-phase ozonolysis under two sets of conditions highly relevant to the troposphere. These products must be part of any mechanism that purports to explain atmospheric ozonolysis. Since peroxides are a source of radicals and a reservoir of oxidising power, their concentrations must be considered in quantitative descriptions of the tropospheric ozone budget.

Acknowledgements

Part of this work was carried out at the Institute of Ecological Chemistry, GSF-Gesellschaft für Gesundheits- und Umweltforschung, Freising/Attaching, Germany.

References

1.   E. Hellpointner, S. Gäb, *Nature* **337** (1989) 631–634.
2.   R.H.-H. Kurth, *Doctoral Thesis*, Technical University, Munich 1992.
3.   R. Rabong, *Master's Thesis*, Ludwig Maximilian University of Munich 1991.
4.   S. Wolff, *Master's Thesis*, Bergische Universität, Wuppertal 1994.
5.   K.J. Brockmann, *Doctoral Thesis*, Bergische Universität Wuppertal 1992.
6.   S. Mönninghoff, *Master's Thesis*, Bergische Universität, Wuppertal 1995.
7.   B. Thome, *Master's Thesis*, Technical University, Munich 1992.
8.   S.M. Aschmann, R. Atkinson, *Envir. Sci. Technol.* **28** (1994) 1539–1542.
9.   K.J. Brockmann, S. Gäb, W.V. Turner, S. Wolff, S. Mönninghoff, submitted for publication.
10. K.H. Becker, K.J. Brockmann, J. Bechara, *Nature* **346** (1990) 256–258; D. Cremer, E. Kraka, M.L. McKee, T.P. Radhakrishnan, *Chem. Phys. Lett.* **187** (1991) 491–493.

3.6 Gas-Phase Reactions of Interest in Night-time Tropospheric Chemistry

J. Hjorth, N.R. Jensen, H. Skov, F. Capellani and G. Restelli

The European Commission, Environment Institute, J.R.C., I - 21020 Ispra (VA), Italy

Summary

The nitrate radical, NO_3, is formed in the atmosphere by the reaction between NO_2 and O_3. During the day NO_3 is rapidly photolysed, but at night it builds up to several ppt. In recent years, the chemistry of NO_3 in air has been studied extensively and found to play an important role in the troposphere [1]. This contribution to LACTOZ has been focused on certain aspects of NO_3 chemistry: the identification of products and the understanding of mechanisms in the NO_3 initiated degradation of some volatile organic compounds (VOC) and the role of NO_3 in reactions leading to interconversion of NO_y species.

The organic compounds investigated were alkenes (propene, isobutene; *trans*- and *cis*-2-butene; 2-methyl-2-butene; 2,3-dimethyl-2-butene; isoprene [3–6], chlorinated butenes (2-chloro-1-butene; 3-chloro-1-butene; 1-chloro-2-butene; 2-chloro-2-butene) [7] and aromatics {benzene; *o*-xylene; *m*-xylene; *p*-xylene 4-fluorotoluene} [8].The following main products were found for the reaction between NO_3 and alkenes in air in the presence of NO_2: unsubstituted carbonyl compounds, nitroxycarbonyl compounds, nitroxy alcohols and dinitrates. For the reaction between NO_3 and isoprene it was found that the main pathway, for the initial step, is an addition reaction to the C_1-carbon atom. From the study of the reactions between NO_3 and chlorinated butenes the following conclusions were made: When a chlorine atom is attached to a double-bonded carbon, the channel leading to formation of carbonyl nitrates becomes less important while the one leading to non-nitrate carbonylic compounds becomes substantially more important. The reactions of NO_3 with aromatic hydrocarbons formed aromatic aldehydes and aromatic nitrates as the main products.

The equilibrium constant for the reaction $N_2O_5 + M \rightleftharpoons NO_3 + NO_2 + M$ was determined as 2.8×10^{10} molecule cm^{-3} at 298 K [9] while the rate of the reaction $NO_3 + NO_2 \rightarrow NO + NO_2 + O_2$ was determined as $(5.1 \pm 1.8) \times 10^{-16}$ cm^3 molecule^{-1} s^{-1} at 296 K [10]. Results of model simulation of experimental data indicated that the rate constant of the reaction $HO_2 + NO_3 \rightarrow OH + NO_2 + O_2$ should be below 1×10^{-12} cm^3 molecule^{-1} s^{-1} [11].

Aims of the scientific research

The aims of this investigation was to provide mechanistic, product and kinetic data for NO_3 reactions of atmospheric importance and to contribute to the background knowledge of the NO_3 reaction mechanisms. These kind of investigations are needed to be able to perform a quantitative description of the tropospheric ozone budget in numerical models.

Principal scientific findings during the project

Experimental setup

The product studies were performed by reacting the organic compounds with NO_3 radicals in an $N_2O_5/NO_2/NO_3$/organic/air mixture at atmospheric pressure and at 295 ± 2 K.

The experimental setup applied is shown in Fig. 1. The experiments were performed in a 480 L Teflon-coated evacuable reaction chamber as shown by Fig. 1. The chamber is equipped with multiple reflection mirror systems for on-line Fourier Transform Infrared Spectroscopy (FTIR) and on-line Tunable Diode Laser Spectroscopy (TDL) as well as UV/VIS spectroscopy. In addition, samples can be taken from the chamber and analysed by GC or by GC-MS.

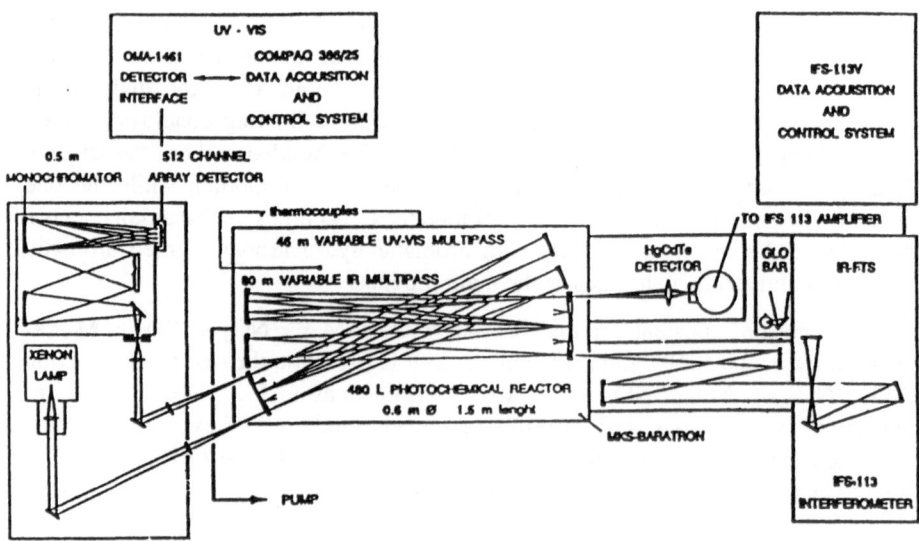

Fig. 1: Schematic picture of the experimental setup.

VOC + NO₃

In the investigation of products and intermediates from the reactions between NO_3 and the alkenes (propene; isobutene; *trans*- and *cis* butene; 2-methyl-2-butene; 2,3-dimethyl-2-butene and isoprene) it was found that all of them followed a similar pattern. The build-up of organic nitrate bands (845, 1280 and 1667 cm^{-1}) and peroxynitrate bands (790, 1300, 1725 cm^{-1}) were observed in the IR-spectra as showed in Fig. 2. The subsequent decay of the peroxynitrate bands was accompanied by a build-up of spectral features attributed to stable products (see Fig. 2). These stable products were identified as aldehydes, alcohol nitrates, carbonyl nitrates and dinitrates [3, 4].

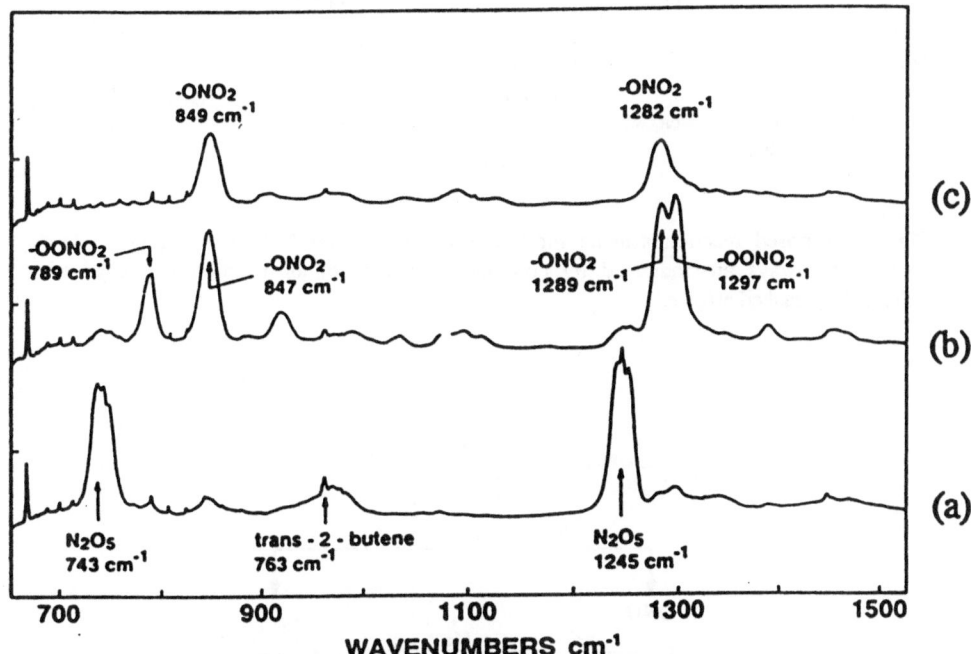

Fig. 2: Typical IR spectra recorded during the reaction of trans-2-butene with NO_3, showing the build-up and decay of the nitroxy-nitroperoxy intermediate and the formation of stable organic nitrates. (a) Initial spectrum (N_2O_5 is used as source of NO_3 radicals). (b) Spectrum recorded after 15 seconds reaction time. (c) Spectrum recorded after 1080 seconds reaction time.

Based on the end products identified and the intermediates observed in the IR spectra a reaction mechanism was proposed for these reactions [3, 4]. The reaction between NO_3 and isoprene was found to proceed mainly through a reaction pathway where NO_3 is added to the C_1-carbon in an initial step (see Fig. 3).

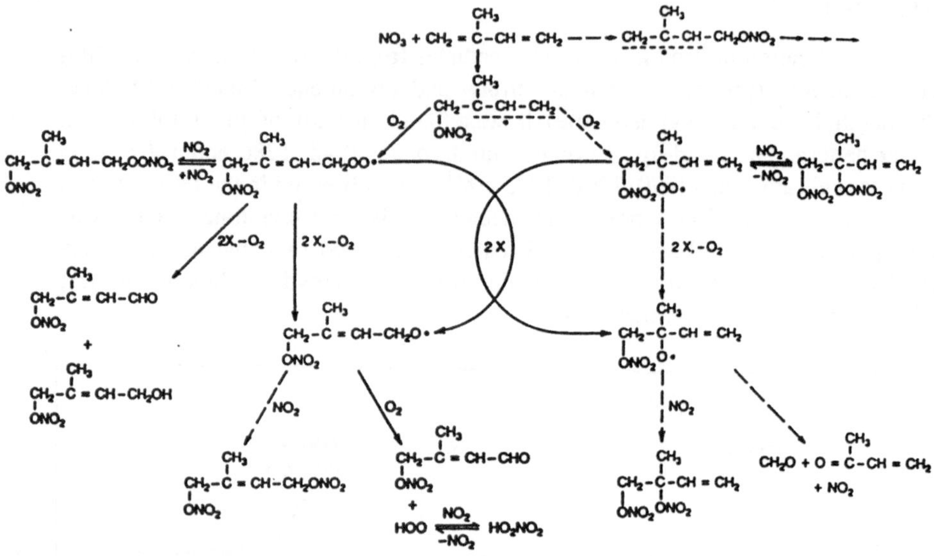

Fig. 3: Proposed reaction scheme for the reaction between NO_3 and isoprene. Main reaction pathways are indicated by continuous arrows; minor reaction pathways are indicated by dashed arrows.

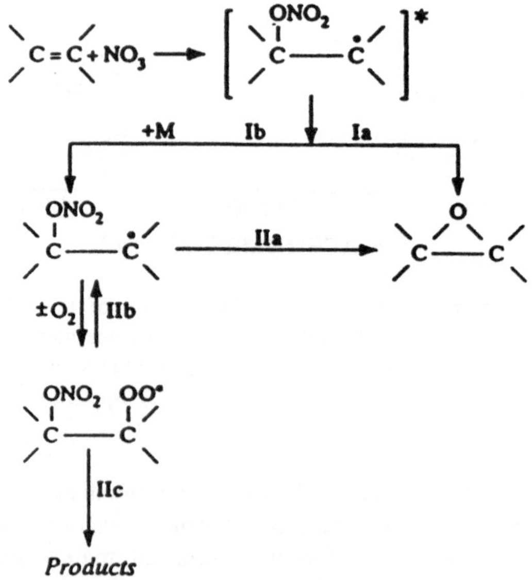

Fig. 4: Proposed scheme for the initial steps in the reaction between NO_3 and 2,3-dimethyl-2-butene in the presence of NO_2 and O_2.

In an investigation of the oxirane formation from the reaction between NO_3 and alkenes [5, 6] it was found that at low pressures and at low O_2 concentrations the oxirane formation pathway was the dominating pathway, but at atmospheric pressure (about 760 Torr) in air it was only a minor channel. The observations could be explained by the mechanism outlined in Fig. 4.

The reactions between the chlorinated butenes and the nitrate radical can be divided into two cases [7]:

Case 1: The chlorine substitution is next to the double bond:
For the two investigated compounds with the chlorine atom substituted next to the double bond, the intensity of the nitroperoxy bands passed through a maximum while the nitroxy bands continued to increase during the course of the experiments. A carbonyl band at approximately $1750\ cm^{-1}$ was seen at the same time. This indicates that either a carbonyl nitrate compound or a nitrate and a carbonyl compound were formed through decomposition of the nitroperoxy nitrate intermediate. Small amounts of acetaldehyde and chloroacetaldehyde were found among the products formed from the reaction with 1-chloro-2-butene. In the reaction with 3-chloro-1-butene, significant amounts of formaldehyde were formed.

Case 2: The chlorine substitution is a the double bond:
In the experiments with 2-chloro-1-butene and 2-chloro-2-butene, the initial increase in absorption due to nitroxy and nitroperoxy groups were followed by a decrease and by a simultaneous formation of aldehydes and acid chlorides.

The yields of the main reaction products in reactions of aromatics with NO_3 were determined by GC-MS after sampling on a carbon column or a XAD-2 trap [7]: For benzene only nitro-benzene was identified as a product, but benzene was essentially unconverted in the reaction with NO_3. Benzaldehyde and benzyl nitrate were identified in the reaction with toluene. For *ortho-*, *meta-*, *para*-xylene and 4-fluorotoluene the corresponding aldehydes and nitrates were identified as main products, which suggests that the below mechanism for toluene and the xylenes is the main reaction pathway [8]:

(1) $Ar–CH_3 + NO_3$ $\rightarrow$ $Ar–CH_2^{\bullet} + HNO_3$
(2) $Ar–CH_2^{\bullet} + O_2 + M$ $\rightarrow$ $Ar–CH_2O_2^{\bullet} + M$
(3) $2\ Ar–CH_2O_2^{\bullet} + M$ $\rightarrow$ $2\ Ar–CH_2O^{\bullet} + O_2$
(4) $Ar–CH_2O^{\bullet} + O_2$ $\rightarrow$ $Ar–CHO + HO_2$
(5) $Ar–CH_2O^{\bullet} + NO_2 + M$ $\rightarrow$ $Ar–CH_2ONO_2 + M.$

However, minor yields of aromatic nitroderivatives found in these reactions indicated that the nitrate radical may also react by addition to the aromatic ring.

Interconversion of NO_y species

The rapidly established equilibrium between NO_3, NO_2 and N_2O_5

(6) $N_2O_5 + M$ $\rightleftharpoons$ $NO_3 + NO_2 + M$

was investigated by direct measurement of all three components, by FTIR (NO_2 and N_2O_5) and UV/VIS (NO_3 and NO_2). The equilibrium constant was found as $(2.7 \pm 0.8) \times 10^{10}$ molecule cm^{-3} at 298 K [9].

The rate constant of the reaction

(7) $NO_3 + NO_2$ $\rightarrow$ $NO + NO_2 + O_2$

was determined as $(5.1 \pm 1.8) \times 10^{-16}$ cm^3 molecule^{-1} s^{-1} at 296 K in a study where low concentrations of NO was measured by TDL spectroscopy [10].
The OH forming reaction

(8) $HO_2 + NO_3$ $\rightarrow$ $OH + NO_2 + O_2$

was studied by kinetic modelling of a complex chemical system at 750 and 50 Torr total pressure where HO_2 and NO_3 were formed by the thermal dissociation of HO_2NO_2 and N_2O_5, respectively, and [OH] was estimated from the rate of co-oxidation. The best fits to the experimental data were obtained by applying values of k_8 below 1×10^{-12} cm^3 molecule^{-1} s^{-1} [11].

Atmospheric implications

The fast reaction of NO_3 radicals with alkenes was found to give substantial yields of bifunctional carbonyl-nitroxy products and the mechanism of their formation is expected to be relevant also for the conditions of the troposphere. Studies performed by other workers [12] have shown that such organic nitrates may contribute to the atmospheric transport of NO_x by acting as reservoirs, releasing NO_x by their degradation in the atmosphere, but bifunctional nitrates may also be subject to fast removal by wet deposition.

It is of particular relevance to atmospheric photochemistry that isoprene, a main naturally emitted hydrocarbon, was found to form bifunctional organic nitrates as the predominant products of its reaction with the NO_3 radical in air.

Although the OH initiated oxidation of aromatics is generally more important under atmospheric conditions than the reaction with NO_3 the latter type of reaction becomes more important with increasing alkyl-substitution and predominates for phenolic and heterocyclic aromatics [12]. Thus a basic understanding of the mechanism of these reactions is needed to investigate the role of aromatic species in tropospheric photochemistry.

Acknowledgements

The authors gratefully acknowledge Mr. G. Ottobrini for his help with the experimental setup and Prof. C. Lohse, Prof. B. Rindone, Prof. C.J. Nielsen and Dr. I. Wängberg for helpful discussions. This study has been supported by the CNR-ENEL project "Interaction of Energy Systems with Human Health and Environment", Rome, Italy.

References

1. B.J. Finlayson-Pitts, J.N. Pitts, *Atmos. Chemistry*, Wiley, New York 1986.
2. R. Atkinson, S.M. Aschmann, J.N. Pitts, Rate constant for the gas-phase reactions of the NO_3 radical with a series of organic compounds at 296 ± 2 K, *J. Phys. Chem.* **92** (1988) 3454.
3. J. Hjorth, L. Lohse, C.J. Nielsen, H. Skov, G. Restelli: Products and mechanisms of the gas-phase reactions between NO_3 and a series of alkenes, *J. Phys. Chem.* **94** (1990) 7494.
4 H. Skov, J. Hjorth, C. Lohse, N.R. Jensen, G. Restelli: Products and mechanism of the reactions of the nitrate radical (NO_3) with isoprene, 1,3-butadiene and 2,3-dimethyl-1,3-butadiene in air, *Atmos. Environ.* **26A** (1992) 2771.
5. H. Skov, Th. Benter, R.N. Schindler, J. Hjorth, G. Restelli: Epoxide formation in the reactions of the nitrate radical with 2,3-dimethyl-2-butene, *cis-* and *trans*-2-butene and isoprene, *Atmos. Environ.* **28A** (1994) 1583.
6. Th. Benter, M. Liesner, R.N. Schindler, H. Skov, J. Hjorth, G. Restelli, REMPI-MS and FTIR study of NO_2 and oxirane formation in the reactions of unsaturated hydrocarbons with NO_3 radical, *J. Phys. Chem.* **98** (1994) 10492.
7. I. Wängberg, E. Ljungström, J. Hjorth, G. Ottobrini: FTIR studies of reactions between the nitrate radical and chlorinated butenes. *J. Phys. Chem.* **94** (1990) 8036.
8. G. Chiodini, B. Rindone, F. Carlati, S. Polesello, G. Restelli, J. Hjorth: Comparison between the gas-phase and the solution reaction of the nitrate radical and methylarenes, *Environ. Sci. Technol.* **27** (1993) 1659.
9. J. Hjorth, J. Notholt, G. Restelli: A spectroscopic study of the equilibrium $N_2O_5 + M \rightleftharpoons NO_3 + NO_2 + M$ and the kinetics of the N_2O_5 / NO_3 / NO_2 / O_3 /air system, *Int. J. Chem. Kinet.* **24,** (1991) 51.
10. J. Hjorth, F. Cappellani, C.J. Nielsen, Restelli, G. Determination of the $NO_3 + NO_2 \rightarrow NO + NO_2 + O_2$ rate constant by infrared diode laser and Fourier transform spectroscopy, *J. Phys. Chem.* **93** (1989) 5458.
11. J. Hjorth, F. Cappellani, G. Restelli: A TDL and FTIR study of the reaction $HO_2 + NO_3 \rightarrow OH + NO_2 + O_2$, in: *Air Pollution Research Report* **45**, Brussels 1992.
12. I. Barnes, K.H. Becker, T. Zhu: Near UV absorption spectra and photolysis products of bifunctional organic nitrates: possible importance as NO_x reservoirs, *J. Atmos. Chem.* **17** (1993) 353.
13. C. Zetzsch: Atmospheric oxidation processes of aromatics studied within LACTOZ and STEP', in: *Proc. Sixth European Symp. on Physico-Chemical Behaviour of Atmospheric Pollutants*, EC, Luxembourg 1994, 118.

3.7 Laboratory Studies of Peroxy Radical Reactions of Importance for Tropospheric Chemistry

Michael E. Jenkin[1], Garry D. Hayman[1], R. Anthony Cox[2] and Timothy P. Murrells[1]

[1]AEA Technology, National Environmental Technology Centre, E5 Culham Laboratory, Oxfordshire, OX14 3DB, UK

[2]Department of Chemistry, Lensfield Rd., Cambridge, UK

Summary

Studies of the UV absorption spectra as well as the kinetics and products of reactions of the following organic peroxy radicals have been carried out: CH_3O_2, $CH_2=CHCH_2O_2$, $HOCH_2CH_2O_2$, $CH_3CH(OH)CH(O_2)CH_3$, $(CH_3)_2C(OH)CH_2O_2$, $(CH_3)_2CH(OH)CH(O_2)(CH_3)_2$, $CH_3OCH_2O_2$, $CH_3C(O)CH_2O_2$ and $C_2H_5C(O)O_2$. These radicals were chosen to be representative of those formed from the degradation of some classes of volatile organic compound (VOC) likely to be important in tropospheric chemistry (*i.e.* alkanes, alkenes, conjugated dienes, alcohols, ethers, aldehydes and ketones). Kinetic studies of the self reactions and reactions with HO_2 were carried out using the molecular modulation technique and the laser flash photolysis technique, coupled with UV absorption spectroscopy. Product studies were carried out by long pathlength Fourier transform infra-red spectroscopy and UV-visible diode array spectroscopy. This work has also allowed investigation of a variety of laboratory sources of peroxy radicals.

The results have been used, along with those obtained by other LACTOZ participants, to establish trends of reactivity of peroxy radicals, and to assign rate coefficients to complex peroxy radicals, in particular those formed from the OH-initiated degradation of isoprene. A degradation mechanism of isoprene to first generation products has been assembled, incorporating these data, which gives a good description of the product yields measured in other laboratory studies.

Aims of the research

The reactions of organic peroxy radicals and HO_2 are of interest for tropospheric chemistry, since they are chain carriers in VOC oxidation mechanisms, converting NO to NO_2 and consequently producing O_3. The reactions of RO_2 with HO_2 and the permutation reactions of organic peroxy radicals ($RO_2 + RO_2$ and $RO_2 + R'O_2$) lead to radical termination, and may therefore significantly reduce O_3 formation chain lengths, particularly at lower levels of NO_x.

Although some comparatively simple peroxy radicals of tropospheric importance can be readily generated and studied in laboratory systems, the majority are difficult to generate independently from isomeric species. Furthermore, the extreme diversity of emitted organic compounds implies the production of many hundreds of structurally different peroxy radicals which will never realistically be studied, but for which the rates and products of key reactions may need to be estimated. The main aims of the work presented in this report were:

- To study the UV absorption spectra as well as the kinetics and products of the self-reactions and reactions with HO_2 for simple peroxy radicals derived from tropospherically abundant VOCs (*e.g.* CH_4 and C_2H_4).
- To investigate novel methods for generating peroxy radicals in laboratory systems.
- To study the UV absorption spectra, and kinetics and products of the self-reactions and reactions with HO_2 for comparatively simple peroxy radicals containing structural features or functional groups representative of different classes of VOC.
- To identify and, where possible, rationalise the effect of the different functionalities on the UV absorption spectra and reactivity of organic peroxy radicals as a first step to defining structure-reactivity relationships.

Principal scientific results

The work performed within the duration of the LACTOZ subproject has allowed laboratory data to be obtained on the UV absorption spectra, and kinetics and products of the self-reactions and reactions with HO_2 for a range of peroxy radicals. Some additional kinetic and mechanistic information related to other aspects of VOC degradation has also been obtained, particularly concerning the behaviour of some of the "oxy" radicals (RO) formed from reactions of the peroxy radicals. Much of this work has appeared in the open literature [1–12].

Generation of peroxy radicals in laboratory systems

The generation of peroxy radicals may be achieved in a number of ways, as recently discussed in detail [6]. In the present work programme, the photolysis of organic halides (RX, X = Br, I) has been considered as a novel method of generating specific peroxy radicals, independently of the isomeric species which would result in many cases from the general oxidation of a parent VOC:

$$RX + h\nu \quad \rightarrow \quad R + X \tag{1}$$
$$R + O_2 \, (+M) \quad \rightarrow \quad RO_2 \, (+M) \tag{2}$$

Some success was achieved with this method, although certain complications were apparent which appear to arise from both the subsequent behaviour of the halogen atom X in the system, and from the generation of the organic fragment with a significant amount of excess energy. Both Br and I atoms were found to react with

RO_2 radicals to form ROOX adducts with very intense UV absorption spectra, peaking at *ca.* 265 and 295 nm respectively. This can cause significant complications in kinetics studies, owing to interfering UV absorptions and sequestration of the peroxy radical. However, it appears to be less of a problem in steady state systems (*e.g.* product studies) where the adduct acts as a comparatively short-lived intermediate in the RO_2 catalysed recombination of the halogen atoms, and does not accumulate significantly. In some systems, the generation of a "hot" organic fragment may complicate the subsequent chemistry, as was believed to be the case during the study of $HOCH_2CH_2O_2$ formation from iodoethanol photolysis [4, 5].

UV absorption spectra

The maximum UV absorption cross-sections and wavelengths of the maxima for all the peroxy radicals measured by this group are given in Table 1. For those radicals not containing a carbonyl group, the spectra at wavelengths longer than 200 nm have been found to consist of a single, unstructured UV band resulting from the "-OO" chromophore. The maximum absorption cross-sections for most of the species lie in the range *ca.* $4-5 \times 10^{-18}$ cm^2 molecule^{-1}, suggesting that the presence of a range of functionalities on the organic group has only a comparatively minor influence on the intensity of the absorption spectra. The spectrum of the allyl peroxy radical ($CH_2=CHCH_2O_2$) appears to be slightly more intense than that of the corresponding alkyl peroxy radical (n-$C_3H_7O_2$ [6]), and those for the α-oxygenated radicals ($HOCH_2O_2$ and $CH_3OCH_2O_2$) less intense than the corresponding alkyl peroxy radicals (CH_3O_2 and $C_2H_5O_2$). The positions of the absorption maxima are also only influenced slightly by the presence of those functional groups investigated in the present work, although clear trends are emerging which apparently relate to the electron-donating ability of the organic group. All alkyl peroxy radical spectra appear to have maxima very close to 240 nm, which is some 35 nm to longer wavelength than that of HO_2, presumably since organic groups are significantly more electron-donating than a hydrogen atom. The electron-withdrawing influence of the α-oxygenated groups in $HOCH_2O_2$ and $CH_3OCH_2O_2$ leads to a shift to shorter wavelengths by some 5–10 nm compared with the corresponding alkyl peroxy radicals. A similar effect is also observed for other peroxy radicals containing electron-withdrawing functionalities (*e.g.* halogen atoms) a to the peroxy radical centre [6]. In contrast, the spectra of the β-oxygenated peroxy radicals ($HOCH_2CH_2O_2$, $CH_3CH(OH)CH(O_2)CH_3$ and $(CH_3)_2CH(OH)CH(O_2)(CH_3)_2$) are shifted to longer wavelength by some 5 nm compared with alkyl peroxy radicals. This may be interpreted in terms of the effect of possible internal hydrogen bonding in these radicals, together with the reduced electron-withdrawing influence compared with the α-substituted radicals.

Peroxy radicals containing carbonyl groups have been found to contain two maxima (*e.g.* $CH_3C(O)O_2$ [6]). Collaborative studies with Risø [2] and Bordeaux

[11] have established that this is also the case for the acetonyl peroxy radical ($CH_3C(O)CH_2O_2$), although there is some disagreement concerning the precise positions and intensities of the absorption maxima.

Table 1: Maximum UV absorption cross-sections and positions of the maxima for a series organic peroxy radicals and HO_2 measured in this laboratory.

Peroxy radical	λ_{max} / nm	σ / 10^{-20} cm^2 molecule^{-1}
HO_2	205	430
CH_3O_2	235–240	500
$C_2H_5O_2$[a]	235–240	470
$CH_2=CHCH_2O_2$	235	620
$HOCH_2O_2$[a]	230	340
$HOCH_2CH_2O_2$	245	510
$CH_3CH(OH)CH(O_2)CH_3$	245	420
$(CH_3)_2CH(OH)CH(O_2)(CH_3)_2$	245	410
$CH_3OCH_2O_2$	230	400
$CH_3C(O)CH_2O_2$[b]	230	220
	290	250
$CH_3C(O)CH_2O_2$[c]	220	420
	300	150

[a] Measured in this laboratory prior to LACTOZ. Displayed for comparison.
[b] Measured at University of Bordeaux in a collaborative study.
[c] Measured at Risø National Laboratory in a collaborative study.

Kinetics and branching ratios of self reactions

Self-reaction rate coefficients (k_3) and branching ratios (a) obtained in this laboratory for a series of peroxy radicals, are presented in Table 2, where a is defined as the fraction of the reaction proceeding via the propagating channel, i.e. k_{3a}/k_3:

$$RO_2 + RO_2 \quad \rightarrow \quad RO + RO + O_2 \tag{3a}$$
$$\rightarrow \quad R_{-H}O + ROH + O_2 \tag{3b}$$

It is apparent that rate coefficients for peroxy radical self reactions are strongly influenced by the structure of the radical. All peroxy radicals containing substituent functional groups have been found to be significantly more reactive than their unsubstituted alkyl peroxy counterparts: for example both $HOCH_2CH_2O_2$ and $CH_3OCH_2O_2$ are about 30 times more reactive than $C_2H_5O_2$. The work from this laboratory has concentrated particularly on the effect of the allyl group, and the β-hydroxy group, since the peroxy radicals derived from the OH-initiated oxidation of isoprene possess these functionalities. A trend of decreasing reactivity for the primary, secondary and tertiary β-hydroxy peroxy radicals $HOCH_2CH_2O_2$, $CH_3CH(OH)CH(O_2)CH_3$ and $(CH_3)_2CH(OH)CH(O_2)(CH_3)_2$ has been established, which is very similar to that observed for unsubstituted alkyl peroxy radicals [6].

Table 2: Summary of room temperature self-reaction parameters measured in this laboratory.

Peroxy radical	a	k_4 10^{-12} cm^3 s^{-1}	chemical system
CH$_3$O$_2$	0.33[b]	0.36	Cl$_2$ / CH$_4$ / O$_2$
		0.35	CH$_3$I / O$_2$
C$_2$H$_5$O$_2$[c]	0.63[b]	0.080	Cl$_2$ / C$_2$H$_6$ / O$_2$
		0.072	azoethane / O$_2$
CH$_2$=CHCH$_2$O$_2$	0.61	0.68	1,5-hexadiene / O$_2$
HOCH$_2$O$_2$[c]	0.89	6.1	HCHO / O$_2$
HOCH$_2$CH$_2$O$_2$	0.50[d]	2.1	H$_2$O$_2$ / C$_2$H$_4$ / O$_2$
		2.2	HOCH$_2$CH$_2$I / O$_2$
CH$_3$CH(OH)CH(O$_2$)CH$_3$	0.75[e]	0.48	H$_2$O$_2$ / *trans* 2-butene / O$_2$
(CH$_3$)$_2$CH(OH)CH(O$_2$)(CH$_3$)$_2$	1.0[f]	0.0057	H$_2$O$_2$ / 2,3-dimethyl 2-butene / O$_2$
CH$_3$OCH$_2$O$_2$	0.7	2.1	Cl$_2$ / CH$_3$OCH$_3$ / O$_2$
CH$_3$C(O)CH$_2$O$_2$[g]	0.75	8.0	Cl$_2$ / CH$_3$COCH$_3$ / O$_2$

[a] Units 10^{-12} cm^3 molecule^{-1} s^{-1}.
[b] Recommended by Lightfoot *et al.* [6].
[c] Measured in this laboratory prior to LACTOZ. Displayed for comparison.
[d] Measured by Barnes *et al.* [13].
[e] Estimated.
[f] By definition.
[g] Measured at University of Bordeaux in a collaborative study.

The values of a for primary peroxy radicals $\geq$ C$_2$ have been found to be remarkably insensitive to the nature of the organic group (lying in the range 0.5–0.75), despite rate coefficients ranging from 0.08×10^{-12} cm^3 molecule^{-1} s^{-1} for C$_2$H$_5$O$_2$ to 8.0×10^{-12} cm^3 molecule^{-1} s^{-1} for CH$_3$C(O)CH$_2$O$_2$.

Kinetics and branching ratios of reactions with **HO$_2$**

Rate coefficients for the reactions of peroxy radicals with HO$_2$ (k_4) measured in this laboratory are given in Table 3:

$$\text{RO}_2 + \text{HO}_2 \quad \rightarrow \quad \text{ROOH} + \text{O}_2 \qquad (4a)$$
$$\rightarrow \quad \text{R}_{-H}\text{O} + \text{H}_2\text{O} + \text{O}_2 \qquad (4b)$$

Although channel (4a) tends to dominate for most peroxy radicals, strong evidence for a significant contribution (*ca.* 40 %) from channel (4b) has been found for the reaction involving CH$_3$OCH$_2$O$_2$ [9] (collaborative study with Ford Motor Company, Dearborn), which is similar to that observed previously for the closely related radical HOCH$_2$O$_2$ [6]. The kinetics studies indicate that the values of k_4 for primary secondary and tertiary β-hydroxy peroxy radicals are equivalent, within experimental error, suggesting that the rates of such reactions are insensitive to both the size and degree of substitution of the organic group. These data also indicate that the β-hydroxy group itself apparently has little influence on the reactivity of peroxy radicals with HO$_2$. From the available data base, it appears that alkyl and β-hydroxy peroxy radicals $\geq$ C$_3$ have room temperature rate coefficients in the region of 1.5 to 2×10^{-11} cm^3 molecule^{-1} s^{-1}, with no significant trend with increasing carbon number (see also data from University of Bordeaux).

In contrast, available information on the reactions of alkyl and β-hydroxy peroxy radicals with NO [6, 14, 15] indicate a progressive decrease in reactivity as the carbon number increases, suggesting that the reactions with HO$_2$ become progressively more favoured over the reactions with NO, as the size of the organic group becomes larger.

Oxy radical reactions

Some information was also obtained which relates to the behaviour of oxy radicals, formed from reactions of peroxy radicals. The thermal decomposition of acetonoxy radicals (CH$_3$C(O)CH$_2$O) was found to predominate over the reaction with O$_2$ under tropospheric conditions (collaboration with MPI Mainz and University of Bordeaux [10, 11]):

$$CH_3C(O)CH_2O \ (+M) \quad \rightarrow \quad CH_3CO + HCHO \ (+ M) \tag{5}$$

Table 3: Room temperature RO$_2$ + HO$_2$ rate coefficients measured in this laboratory.

Peroxy radical	k_4	pressure	chemical system
	10^{-12} cm^3 s^{-1}	Torr	
CH$_3$O$_2$	5.4	10	Cl$_2$ / CH$_4$ / H$_2$O$_2$ / O$_2$
	6.8	760	Cl$_2$ / CH$_4$ / H$_2$ / O$_2$
C$_2$H$_5$O$_2$[a]	6.3	2	Cl$_2$ / C$_2$H$_6$ / CH$_3$OH / O$_2$
	9.5	760	Cl$_2$ / C$_2$H$_6$ / CH$_3$OH / O$_2$
HOCH$_2$O$_2$[a]	12.0	2	HCHO / O$_2$
HOCH$_2$CH$_2$O$_2$	8.4	10	HOCH$_2$CH$_2$I / O$_2$
	12.0	760	HOCH$_2$CH$_2$Cl / O$_2$
	15.0	760	H$_2$O$_2$ / C$_2$H$_4$ / O$_2$
CH$_3$CH(OH)CH(O$_2$)CH$_3$	15.0	760	H$_2$O$_2$ / *trans*-2-butene / O$_2$
(CH$_3$)$_2$CH(OH)CH(O$_2$)(CH$_3$)$_2$	~ 20	760	H$_2$O$_2$ / 2,3-dimethyl 2-butene / O$_2$
CH$_3$C(O)CH$_2$O$_2$[b]	9.0	760	Cl$_2$ / CH$_3$COCH$_3$ / CH$_3$OH / O$_2$

[a] Measured in this laboratory prior to LACTOZ. Displayed for comparison.
[b] Measured at University of Bordeaux in a collaborative study.

The methoxymethoxy radical (CH$_3$OCH$_2$O) was found to decompose by a mechanism involving H-atom ejection, in competition with its reaction with O$_2$ [8], thereby explaining complications in a previously published study [16] of the kinetics of CH$_3$OCH$_2$O$_2$:

$$CH_3OCH_2O \ (+M) \quad \rightarrow \quad CH_3OCHO + H \ (+ M) \tag{6}$$

Achievements

The work performed on the UV spectra and kinetics of reactions of peroxy radicals at this laboratory, and by collaborating partners within LACTOZ, has significantly expanded the available data base and has consequently improved our understanding of some of the processes describing the degradation of VOCs in the troposphere. Consistent with the primary objective of this work, significant progress has been made in establishing the influence on peroxy radical reactivity, of a variety of functional groups likely to be prevalent in peroxy radicals involved in tropospheric chemistry. This is perhaps best illustrated by the detailed mechanism of the OH initiated degradation of isoprene to first generation products, which was constructed by this group [12] using information mainly obtained by collaborating partners within the LACTOZ subproject. Due to the complexity of the organic compound, the mechanism was defined on the basis of correlations and trends of reactivity of intermediates possessing similar functional groups. The resulting mechanism is able to give a good description of the yields of stable products measured in the laboratory in both the presence and absence of NO_x [12], indicating a sound understanding of the detailed chemistry. The work performed to date provides, therefore, a sound basis for further systematic expansion of the data base on peroxy radical reactivity, and the future development of structure-reactivity rules which can be applied to peroxy radicals derived from many classes of VOC.

Acknowledgements

The support of the European Commission, the Natural Environment Research Council and the UK Department of the Environment is gratefully acknowledged. We also wish to thank Geert Moortgat and John Crowley (MPI Mainz), Robert Lesclaux, Phil Lightfoot and Bernard Veyret (University of Bordeaux), Tim Wallington (Ford Motor Company) and Ole John Nielsen (Risø) for close collaboration on peroxy radicals studies during the past seven years.

References

1.   M.E. Jenkin R.A. Cox, G.D. Hayman, L.J. Whyte, *J. Chem. Soc. Faraday Trans. 2*, **84** (1988) 913–930.
2.   R.A. Cox, J. Munk, O.J. Nielsen, P. Pagsberg, E. Ratajczak, *Chem. Phys. Lett.* **173** (1990) 206–210.
3.   J.N. Crowley, F.G. Simon, J.P. Burrows, G.K. Moortgat, M.E. Jenkin, R.A. Cox, *J. Photochem. Photobiol.* **60**A (1991) 1–10.
4.   M.E. Jenkin, R.A. Cox, *J. Phys. Chem.* **95** (1991) 3229–3237.
5.   T.P. Murrells, M.E. Jenkin, S.J. Shalliker, G.D. Hayman, *J. Chem. Soc. Faraday Trans.* **87** (1991) 2351–2360.
6.   P.D. Lightfoot, R.A. Cox, J.N. Crowley, M. Destriau, G.D. Hayman, M.E. Jenkin, G.K. Moortgat, F. Zabel, *Atmos. Environ.* **26**A (1992) 1805–1964.

7. M.E. Jenkin, T.P. Murrells, S.J. Shalliker, G.D. Hayman, *J. Chem. Soc. Faraday Trans.* **89** (1993) 433–446.
8. M.E. Jenkin, G.D. Hayman, T.J. Wallington, M.D. Hurley, J.C. Ball, O.J. Nielsen, T. Ellermann, *J. Phys. Chem.* **97** (1993) 11712–11723.
9. T.J. Wallington, M.D. Hurley, J.C. Ball, M.E. Jenkin, *Chem. Phys. Lett.* **211** (1993) 41–47.
10. M.E. Jenkin, R.A. Cox, M. Emrich, G.K. Moortgat, *J. Chem. Soc. Faraday Trans.* **89** (1993) 2983–2991.
11. I. Bridier, B. Veyret, R. Lesclaux, M.E. Jenkin, *J. Chem. Soc. Faraday Trans.* **89** (1993) 2993–2997.
12. M.E. Jenkin, G.D. Hayman, *J. Chem. Soc. Faraday Trans.* (1995) in press.
13. I. Barnes, K.H. Becker, L. Ruppert, *Chem. Phys. Lett.* **203** (1992) 295.
14. J. Sehested, O.J. Nielsen, T.J. Wallington, *Chem. Phys. Lett.* **213** (1993) 457.
15. S. Langer, E. Ljungstrom, J. Sehested, O.J. Nielsen, *Chem. Phys. Lett.* **226** (1994) 165.
16. P. Dagaut, T.J. Wallington, M.J. Kurylo, *J. Photochem. Photobiol.* **48** (1989) 187.

3.8   The Rates and Mechanisms for VOC Photo-oxidation Reactions under Simulated Tropospheric Conditions

J. Alistair Kerr, David W. Stocker, Marco Semadeni, Jürg Eberhard and Claudia Müller

Swiss Federal Institute of Environmental Science and Technology, EAWAG/ETH Zürich, CH-8600 Dübendorf, Zürich, Switzerland

Summary

Laboratory kinetic studies of gas-phase elementary reactions of importance to tropospheric chemistry were initiated as part of the LACTOZ programme at the Swiss Federal Institute of Environmental Science and Technology (EAWAG) in October 1989. Two types of experiments have been carried out: (i) studies of the kinetics of competitive OH radical reactions with VOC in an atmospheric flow reactor and (ii) detailed end-product analyses of the OH radical initiated photo-oxidations of VOCs carried out in a static Teflon bag reactor.

The first type of experiment involves the application of a flow reactor designed to operate over the temperature range –25 to 80 °C and over the pressure range 100 to 760 Torr. This reactor has been used to study competitive OH reactions with pairs of aliphatic ethers or with an ether plus 2,3-dimethylbutane as the reference compound. The temperature coefficients of the OH reactions with diethyl ether, methyl n-butyl ether, ethyl n-butyl ether and di-n-butyl ether have been determined. This system has also been used to study competitive OH reactions with pairs of aromatic compounds or with an aromatic compound plus a reference compound. The temperature coefficients of the OH reactions with benzene, toluene, phenol, benzaldehyde, o-cresol, m-cresol and p-cresol have been determined.

In the second series of experiments, the products from the photo-oxidation of diethyl ether, carried out in a Teflon bag reactor at ppm and ppb levels, have been determined by withdrawing vapour samples and monitoring by gas chromatography, HPLC and by chemiluminescence analysis. The major reaction products which have been measured are ethyl formate, ethyl acetate, acetaldehyde, formaldehyde, PAN, methyl nitrate and ethyl nitrate. The products observed arise from the decomposition reactions of the 1-ethoxyethoxy radical and from its reaction with oxygen. The data enable the establishment of a quantitative mechanism for the photo-oxidative reaction. In addition the rate of conversion of NO to NO_2, determined by chemiluminescence analysis, shows that for each molecule of ether reacted only one molecule of NO is converted to NO_2. In further end-product analyses experiments, the OH radical initiated photo-oxidation of n-hexane or the photolyses of 2- or 3-hexyl nitrites were studied to examine the importance of alkoxy radicals undergoing isomerization versus unimolecular decomposition or reaction with oxygen. Based on the first observations of products arising from the isomerization of the hexoxy radicals in these systems, the fractions of alkoxy radicals undergoing isomerization were determined and are in good agreement with theoretical predictions.

Aims of the research

The aims of the research project have been to obtain information on the rates and mechanisms for a range of VOC photo-oxidation reactions under simulated atmospheric conditions. By deriving detailed quantitative kinetic data on the reactions involved in the complex chemistry of the tropospheric photodegradations of VOC, we seek to provide key parameters required in tropospheric modelling. Such mathematical modelling studies will ultimately form the basis for drawing up scientifically sound control strategies for the release of VOC into the atmosphere.

Principal scientific findings

Kinetic studies

The experiments were performed in an irradiated thermostated flow reactor over the temperature range 248–353 K. Assuming loss of the ether (ROR) and of the reference compound (REF) occurs only by OH attack via reactions (1) and (2):

REF + OH	$\rightarrow$	products	(1)
ROR + OH	$\rightarrow$	products	(2)

then integration and combination of the rate expressions for reactions (1) and (2) leads to

$$\ln\frac{[ROR]_0}{[ROR]_i} = \frac{k_2}{k_1}\ln\frac{[REF]_0}{[REF]_i} \qquad (i)$$

where the subscripts o and i refer to the concentrations measured in the dark and under reactor illumination respectively. Plots of the data according to equation (i) are linear with zero intercepts, confirming the general correctness of the treatment of the data. The rate coefficient ratios at a given temperature were then derived from such plots and relative Arrhenius parameters were obtained from measurements of the ratios over a range of temperatures. These were converted to absolute values from a knowledge of the temperature dependent rate coefficients of the reference reactions.

Rate coefficients for the reactions of OH radicals with ethers

The Arrhenius parameters of the reactions of hydroxyl radicals with aliphatic ethers, shown in Table 1, are based on the temperature independent value of

$k_1 = 6.2 \times 10^{-12}$ cm^3 molecule^{-1} s^{-1} for OH + 2,3-dimethylbutane [1]. Previous discrepancies in the room temperature rate coefficients for the OH reactions with ethyl n-butyl ether and di-n-butyl ether, obtained in the flow and static experiments of Bennett and Kerr [2] have been resolved.

Table 1: Arrhenius parameters for OH radical reactions with ethers.

Ether	$(E_a / R) / K$	A / cm^3 molecule^{-1} s^{-1}
Diethyl ether	−262 ± 150	5.2×10^{-12}
Methyl n-butyl ether	−309 ± 150	5.4×10^{-12}
Ethyl n-butyl ether	−335 ± 150	7.3×10^{-12}
Di-n-butyl ether	−502 ± 150	5.5×10^{-12}
Di-n-pentyl ether	−417 ± 150	8.5×10^{-12}

Rate coefficients for the reactions of OH radicals with aromatic compounds

The Arrhenius parameters of the reactions of hydroxyl radicals with aromatic compounds, listed in Table 2, are based on the rate coefficients (cm^3 molecules^{-1} s^{-1}) of the reference compounds taken from recent evaluations of OH radical reactions: k(2,3-dimethylbutane) = 6.2×10^{-12} [1], independent of temperature; k(diethyl ether) = 7.3×10^{-12} exp(158 K / T), 242–440 K [3]; k(1,3-butadiene) = 1.48×10^{-11} exp(448 K / T), 295–483 K [1]; k(phenol) = 6.75×10^{-12} exp(405 K / T), 245–300 K [3]. Our results are in good agreement with existing literature data on OH + aromatic compounds and where possible evaluated rate coefficients have been deduced or updated.

Table 2: Arrhenius parameters for OH radical reactions with aromatic compounds.

Aromatic	$(E_a / R) / K$	A / cm^3 molecule^{-1} s^{-1}
Benzene	231 ± 84	2.6×10^{-12}
Toluene	−614 ± 114	7.9×10^{-13}
o-Cresol	−1170 ± 248	9.8×10^{-13}
m-Cresol	−686 ± 231	5.2×10^{-12}
p-Cresol	−943 ± 449	2.2×10^{-12}
Phenol	−1270 ± 233	3.7×10^{-13}
Benzaldehyde	−243 ± 85	5.3×10^{-12}

Product studies

The experiments were carried out in 200 dm^3 Teflon bag reactors at room temperature (297 ± 3 K) and at atmospheric pressure (725 ± 5 Torr). The decay of the VOC concentration and build up of the product concentrations were monitored during the course of the experiment. The concentration of VOC was measured by GC-FID. The formation of organic nitrates and carbonyls were measured by GC with sample preconcentration using adsorption tubes filled with Tenax TA. Detection was performed with FID and ECD coupled in series. The carbonyl products were determined by derivatisation with 2,4-dinitrophenylhydrazine (2,4-DNPH) followed by HPLC using a three-component gradient solvent programme and u.v. detection.

OH initiated photo-oxidation of diethyl ether

The major products were ethyl formate and formaldehyde and the minor products were ethyl acetate, acetaldehyde, peroxyacetyl nitrate, and methyl and ethyl nitrates. The products arise from the decomposition reactions of the 1-ethoxyethoxy radical and from its reaction with molecular oxygen:

$$C_2H_5OCH_2CH_3 + OH/O_2 \rightarrow CH_3CH(O_2)OC_2H_5 + H_2O \qquad \text{I}$$
$$CH_3CH(O_2)OC_2H_5 + NO \rightarrow CH_3CH(O)OC_2H_5 + NO_2 \qquad \text{II}$$
$$CH_3CH(O)OC_2H_5 \rightarrow HCOOC_2H_5 + CH_3 \qquad 1$$
$$CH_3CH(O)OC_2H_5 \rightarrow CH_3CHO + C_2H_5O \qquad 2$$
$$CH_3CH(O)OC_2H_5 + O_2 \rightarrow CH_3COOC_2H_5 + HO_2 \qquad 3$$
$$CH_3CH(O)OC_2H_5 \rightarrow CH_3(OH)OCH_2CH_2 \qquad 4$$

The fractions of the 1-ethoxyethoxy radicals undergoing reactions 1–4 are shown in Table 3 and compared with previous FTIR studies and with theoretical estimates. Although diethyl ether is rapidly photo-oxidised, its contribution to tropospheric ozone formation is limited as one of the major products is ethyl formate which is relatively unreactive in the troposphere. This is also confirmed from the $NO-NO_2$ oxidation stoichiometry, $ca.$ 1 mole NO is oxidised per mole diethyl ether reacted.

Table 3: Product formation from the alkoxy radical $CH_3CH(O)OC_2H_5$.

Reference	f_1	f_2	f_3	f_4
This work	0.84 ± 0.05	0.06 ± 0.01	0.10 ± 0.03	0
[6]	0.92 ± 0.06	< 0.05	< 0.05	0
[7]	0.95	0	0.04	0.01

Alkoxy radicals from the photolyses of hexyl nitrites and the photo-oxidation of hexane.

The primary products which were quantified include 3-hexyl nitrate, 2-hexyl nitrate, n-butyl nitrate, 3-hexanone, 2-hexanone, formaldehyde, acetaldehyde, propionaldehyde, butyraldehyde and 5-hydroxy-2-hexanone. 2,5-hexanedione was quantified but is thought to be a secondary product. and 5-nitrooxy-2-hexanol was detected but reliable quantification was not possible.

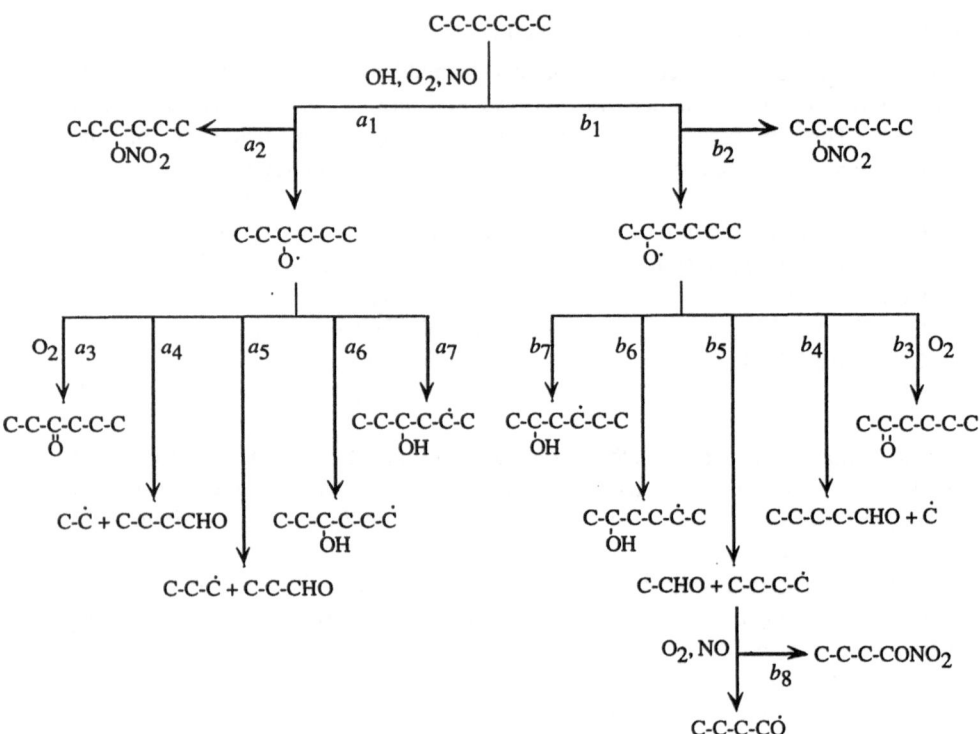

Fig. 1: General reaction scheme for the photo-oxidation of hexane.

A general reaction mechanism for the OH initiated photo-oxidation of hexane is shown in Fig. 1. For a quantitative assessment of the importance of each reaction channel for the reactions of the hexoxy radicals, the observed yields of the primary products were corrected for secondary photo-oxidation, and the fractions were calculated as described by Eberhard *et al.* [4]. As shown in Table 4 the derived fractions for the isomerization reaction channels a_6 and b_6 are in very good agreement with the theoretical predictions. These values indicate that the 2-hexoxy radical undergoes rapid isomerization by abstracting a secondary H-atom. In comparison the isomerization of the 3-hexoxy radical is significantly slower owing to the need to abstract a primary H-atom.

Table 4: Summary of derived fractions for hexoxy radical reactions.

Reaction channel[a]	Process	Fraction (%)	Estimation[b,c] (%)
3-Hexoxy			
a_3	reaction with O_2	6.7 ± 3.5	25.1
a_4	decomposition	10.4 ± 7.3	2.5
a_5	decomposition	9.1 ± 2.9	2.5
a_6	1,5-H shift isomerization	73.8 ± 12.5[d]	68.1
a_7	1,4-H shift isomerization	0.0	1.8
2-Hexoxy			
b_3	reaction with O_2	0.8 ± 0.5	0.3
b_4	decomposition	0.0	0.0
b_5	decomposition	0.2 ± 0.1	0.0
b_6	1,5-H shift isomerization	99.0 ± 0.5	99.6
b_7	1,4-H shift isomerization	0.0	0.0

[a] See Fig. 1;
[b] Based on estimated rate coefficients by the method of Carter and Atkinson [5],
[c] Estimate for $RO+O_2$ reaction rate from Atkinson *et al.* [6],
[d] Calculated by difference.

Achievements of the project

The project has resulted in a considerable extension of the kinetic data base for the reactions of OH radicals with the VOC classes of aliphatic ethers and of aromatic compounds. Such data are essential in the determination of tropospheric lifetimes of the VOC.

The detailed product studies of the atmospheric photo-oxidations of ethers have revealed much finer detail than hitherto reported and at the same time have been carried out under experimental conditions, with respect to mixing ratios of VOC, much closer to the real polluted atmosphere. Here the important conclusion is that data obtained from experiments carried out at VOC mixing ratios of ppm are consistent with those carried out at VOC mixing ratios much closer to troposheric conditions, *i.e.* ~ ppb.

The highlight of the product studies has undoubtedly been the first observations and measurements of products arising from the isomerization reactions of alkoxy radicals in the photo-oxidations of hexane and hexyl nitrites. This is an important finding in directly confirming the occurrence of such isomerization reactions. Furthermore the quantitative agreement of the isomerization kinetic data with thermochemical kinetic estimates confirms that these estimation procedures can be used with confidence in the chemical mechanisms of modelling studies of the troposphere. The achievements of these projects have certainly matched and in many respects exceeded the original aims which were set out in initiating the research programme.

Acknowledgement

This work was supported in part by a research grant from Schweizerischer Nationalfonds zur Förderung der wissenschaftlichen Forschung.

References

1. R. Atkinson, *J. Phys. Chem. Ref. Data, Monograph* No.1. (1989).
2. P.J. Bennett, J.A. Kerr, *J. Atmos. Chem.* **10** (1990) 29.
3. M. Semadeni, D.W. Stocker, J.A. Kerr, *J. Atmos. Chem.* **16** (1993) 79.
4. J. Eberhard, C. Müller, D.W. Stocker, J.A. Kerr, *Int. J. Chem. Kinet.* **25** (1993) 639.
5. W.P.L. Carter, R. Atkinson, *J. Atmos. Chem.* **3** (1985) 377.
6. R. Atkinson, D.L. Baulch, R.A. Cox, R.F. Hampson, J.A. Kerr, J. Troe, *J. Phys. Chem. Ref. Data* **21** (1992) 1125.

3.9   Laboratory Studies of NO₃ and OH Reactions of Tropospheric Relevance

Georges Le Bras, Gilles Poulet, N. Butkovskaya, V. Daële, D. Johnstone, I.T. Lançar, G. Laverdet, A. Mellouki, A. Ray, S. Téton and I. Vassali

Laboratoire de Combustion et Systèmes Réactifs-CNRS, F-45071 Orléans - cedex 2, France

Summary

Kinetic studies have been performed on the individual steps occurring in the NO₃ and OH initiated oxidation of VOCs. The studied reactions include essentially reactions of NO₃ with alkenes, di-alkenes and dimethyl sulfide (DMS), reactions of NO₃ with intermediate peroxy radicals (HO₂, CH₃O₂, C₂H₅O₂) and reactions of OH with methane and oxygenated VOCs (ethers, alcohols). The rate constants for these reactions have been measured, and mechanistic information has been determined. The experimental methods used were discharge-flow reactors coupled with mass spectrometry, electron paramagnetic resonance (EPR), laser-induced fluorescence (LIF) analysis and the laser photolysis associated with LIF analysis. The discharge-flow LIF and laser photolysis LIF experiments have been especially developed for these studies.

Aims of the research

The aims of the research were to provide kinetic and mechanistic data for NO₃ and OH reactions of potential atmospheric relevance. In particular, a large body of the studies aimed at improving the knowledge of the NO₃ initiated oxidation of VOCs during night-time in order to assess the contribution of the night-time processes to the overall atmospheric oxidation of VOCs and photo-oxidant formation.

Principal scientific findings

NO₃ reactions in the night-time oxidation of VOCs

Reactions of NO₃ with alkenes and di-alkenes

The kinetics of the reaction of NO₃ with the following molecules have been investigated using the discharge-flow mass spectrometry method: 2,3 dimethyl-2, butene ((CH_3)₂C=C(CH_3)₂), 1,3 butadiene (CH_2=CH–CH=CH_2), 2 methyl-1,3 butadiene (CH_2=C(CH_3)–CH=CH_2, isoprene) and 2,3 dimethyl-1,3 butadiene (CH_2=C(CH_3)–C(CH_3) = CH_2).

The rate constants measured at 298 K (see Table III of the report) have contributed to extending the data base and to improving its quality. The data followed the same trend as for the similar electrophilic addition reaction of OH: the rate constant increases with the degree of methyl substitution of the butadiene.

Mass spectrometry analysis of the products provided some insight into the reaction mechanisms. Especially, the observations indicate that the NO_3 addition primarily occurs on both π bonds of the isoprene molecule.

Reaction NO_3 + DMS

Although not directly relevant to LACTOZ, the study of the NO_3 + DMS reaction is mentioned since it has led to development of a method to study reaction mechanisms applicable to other VOCs. The method consisted of mass spectrometry chemical titration of the primary radicals produced (R), using Cl_2 or Br_2 as the titrant ($R + Cl_2 \rightarrow RCl + Cl$ and $R + Br_2 \rightarrow RBr + Br$). The results showed that the reaction predominantly ($\geq$ 98 %) proceeds to completion: $NO_3 + CH_3SCH_3 \rightarrow CH_3SCH_2 + HNO_3$, and hence converts NO_x into HNO_3. The peroxy radical, $CH_3SCH_2O_2$, was observed by mass spectrometry upon addition of O_2 to the discharge-flow reactor.

NO_3 reactions with peroxy radicals

Reaction of NO_3 with HO_2

The reaction of $NO_3 + HO_2$ was considered of potential atmospheric relevance since both NO_3 and HO_2 radicals are present during night-time (PAN and PNA decomposition, reactions of NO_3 with H_2CO and some hydrocarbons are sources of HO_2 radicals). The reaction of $NO_3 + HO_2$ was investigated together with the reaction of $NO_3 + OH$ which interacted under the conditions of the flow-tube experiments. OH and HO_2 (after conversion into OH via $NO + HO_2 \rightarrow NO_2 + OH$) were monitored by gas-phase EPR. The kinetic treatment of the OH and HO_2 decays (Fig. 1) yielded a rate constant at 298 K for the following reactions (units of cm^3 $molecule^{-1}$ s^{-1}):

$$
\begin{array}{llll}
NO_3 + OH & \rightarrow & NO_2 + HO_2 & k = (2.6 \pm 0.6) \times 10^{-11} \\
NO_3 + HO_2 & \rightarrow & NO_2 + OH + O_2 & k = (3.6 \pm 0.9) \times 10^{-12} \\
NO_3 + HO_2 & \rightarrow & HNO_3 + O_2 & k = (9.2 \pm 4.8) \times 10^{-13}
\end{array}
$$

Thus, the $NO_3 + HO_2$ reaction was found to be a potentially significant source of OH radicals during night-time.

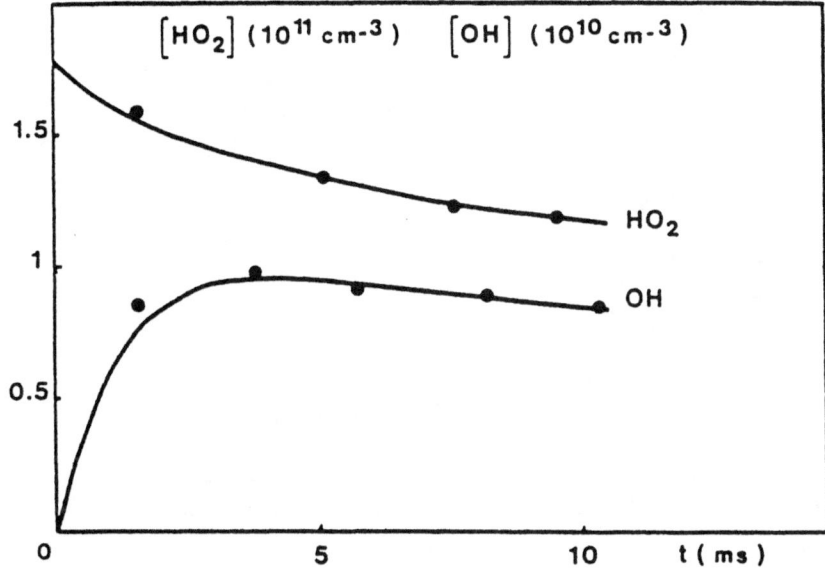

Fig. 1: Reactions of NO_3 with OH and HO_2: typical experimental (points) and calculated (solid lines) profiles of OH and HO_2 ; $[NO_3]_0 = 2.5 \times 10^{13}$ cm^{-3}.

Reactions of NO_3 with CH_3O_2 and $C_2H_5O_2$

The reactions of NO_3 with RO_2 (CH_3O_2, $C_2H_5O_2$) have been investigated as potential conversion steps of RO_2 into RO as the $NO_3 + HO_2$ reaction converts HO_2 into OH. CH_3O_2 and $C_2H_5O_2$ were chosen as models for peroxy radicals produced in the atmosphere as intermediates in the oxidation of hydrocarbons. The reactions of NO_3 with RO_2 (CH_3O_2 and $C_2H_5O_2$) were effectively found to be coupled with the reactions of NO_3 and RO (CH_3O and C_2H_5O) under laboratory conditions:

$$NO_3 + RO_2 \quad \rightarrow \quad NO_2 + RO + O_2$$
$$NO_3 + RO \quad \rightarrow \quad NO_2 + RO_2$$

The rate constants of these reactions were determined in discharge-flow experiments from kinetic analysis of RO concentrations measured by LIF using as initial reactants NO_3 and RO or NO_3 and RO_2 (Figs 2 and 3).

The data obtained at 298 K are summarised in Table 1 together with those for reactions of NO_3 with HO_2 and OH.

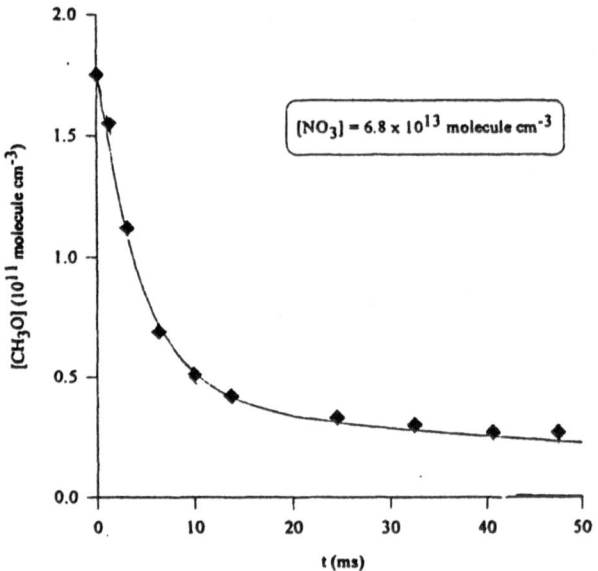

Fig. 2: Reaction of NO_3 with CH_3O and CH_3O_2, with NO_3 and CH_3O as initial reactants: experimental (points) and calculated (solid line) profiles of CH_3O.

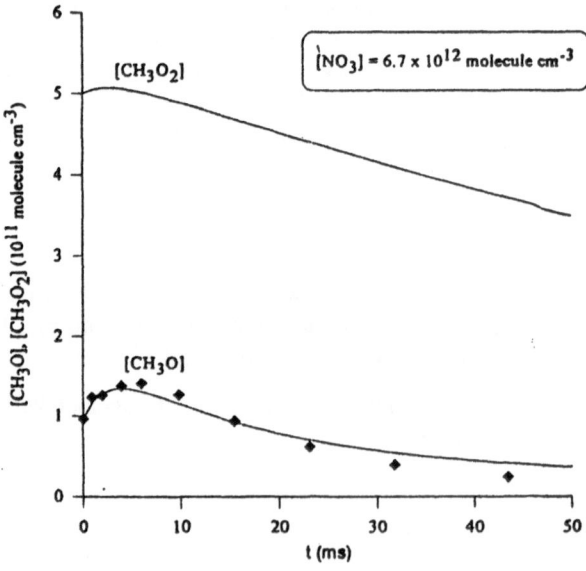

Fig. 3: Reaction of NO_3 with CH_3O and CH_3O_2, with NO_3 and CH_3O_2 as initial reactants: experimental (points) of CH_3O and calculated (solid line) profiles of CH_3O and CH_3O_2.

Table 1: Rate constants in cm^3 molecule^{-1} s^{-1} at 298 K for the reactions of NO_3 with RO and RO_2

R	H	CH₃	C₂H₅
$NO_3 + RO \rightarrow NO_2 + RO_2$	$(2.6 \pm 0.6) \times 10^{-11}$	$(1.8 \pm 0.5) \times 10^{-12}$	$(3.3 \pm 0.8) \times 10^{-12}$
$NO_3 + RO_2 \rightarrow NO_2 + RO + O_2$	$(3.6 \pm 0.9) \times 10^{-12}$	$(1.2 \pm 0.6) \times 10^{-12}$	$(1.3 \pm 1.0) \times 10^{-12}$

Mechanism for the NO_3 initiated oxidation of VOCs in the atmosphere

We suggested the following chain mechanism for the night-time oxidation of VOCs [1]:

$$NO_3 + VOC + O_2 \rightarrow RO_2$$
$$NO_3 + RO_2 \rightarrow RO + NO_2 + O_2$$
$$RO + O_2 \rightarrow HO_2 + products$$
$$NO_3 + HO_2 \rightarrow OH + NO_2 + O_2$$
$$OH + VOC + O_2 \rightarrow RO_2$$

The present results together with those obtained at PCL, Oxford and MPI, Mainz, indicate that reactions of NO_3 with HO_2 and RO_2 (CH_3O_2, $C_2H_5O_2$) are potentially important chain propagation steps in the above mechanism: their rate constants are fairly high, and they convert HO_2 into OH and RO_2 into RO. Such mechanisms used in box modelling of NO_3 – DMS interaction in semi-polluted coastal areas have been shown to produce high levels of peroxy radicals and related so-called photo-oxidants (CH_2O, H_2O_2, CH_3OOH) during night-time [1,2]. Further kinetic and mechanistic data are indeed required for reactions of NO_3 with other peroxy radicals before such a mechanism could be generalised.

NO_3 reactions with inorganic species

Reactions of NO_3 with hydrogen halides, HX (X = Cl, Br, I)

The reactions of NO_3 with HX have been studied by discharge-flow EPR, by monitoring the halogen atom (X) resulting from the sequence: $NO_3 + HX \rightarrow HNO_3 + X$; $NO_3 + X \rightarrow NO_2 + XO$; $NO + XO \rightarrow NO_2 + X$. No reactivity was observed at 298 K between NO_3 and HCl and HBr. In contrast NO_3 was found to react with HI: $k = 1.3 \times 10^{-12} \exp\{(-1830 \pm 300)$ K $/ T\}$ cm^3 molecule^{-1} s^{-1} over the range 298–373 K. This reaction is nevertheless too slow to be of atmospheric significance.

Reactions of NO_3 with Br and BrO

These reactions have been studied in a discharge-flow reactor by monitoring Br and BrO by gas-phase EPR. Similarly to the reactions of NO_3 with OH and HO_2, these two reactions were found to interact under laboratory conditions:

$$NO_3 + Br \rightarrow NO_2 + BrO$$
$$NO_3 + BrO \rightarrow NO_2 + Br + O_2$$

The rate constants obtained at 298 K were: $(1.6 \pm 0.7) \times 10^{-11}$ and $0.3 \times 10^{-12} < k < 3 \times 10^{-12}$ (cm^3 molecule^{-1} s^{-1}), respectively. These reactions have been recently postulated to possibly control both NO_3 and active bromine in the atmosphere. It was suggested that the reaction of NO_3 + BrO could be the missing process responsible for the loss of NO_3 in some field observations [3].

OH *reactions with VOCs*

Reaction of OH with CH_4

The reaction OH + CH_4 has been investigated subsequent to a study yielding a rate constant value 25 % lower than that recommended so far over the atmospheric temperature range [4]. The data obtained using both the discharge-flow EPR and LP-LIF methods are in excellent agreement with this determination. The LP-LIF value is $k = (2.56 \pm 0.53) \times 10^{-12}$ exp$\{-(1765 \pm 146)\ /T\}$ cm^3 molecule^{-1} s^{-1} over the temperature range 233–343 K with $k = (6.34 \pm 0.56) \times 10^{-15}$ cm^3 molecule^{-1} s^{-1} at 298 K. These data illustrate the requirement of high precision measurements for key reactions such as the reaction OH + CH_4 which controls the oxidative capacity of the troposphere and the atmospheric concentration of the greenhouse CH_4 species.

Reactions of OH with oxygenated VOCs (ethers, alcohols)

Oxygenated VOCs such as ethers and alcohols are used as solvents or fuel additives. The rate constants for OH reactions with several ethers and alcohols have been measured over the temperature range 230–370 K, using the LP-LIF method. The studied ethers and alcohols are: dimethyl ether $((CH_3)_2O)$, dipropyl ether $((C_3H_7)_2O)$, ethyl butyl ether $(C_2H_5OC_4H_9)$, methyl *tert*-butyl ether $(C_2H_5OC(CH_3)_3)$, methyl *tert*-amyl ether $(CH_3OC(CH_3)_2C_2H_5)$ and *tert*-butanol $((CH_3)_3COH)$. The Arrhenius plots ln $k = f(1\ /\ T)$ are presented in Fig. 4 for the reactions of OH + ethers. The rate constant data are summarised in Chapter 2, Table 1. These measurements improve the existing data base for these reactions [5], especially by providing additional data at the lowest temperatures. Slight curvatures of the Arrhenius plot are observed for some reactions which were not previously considered.

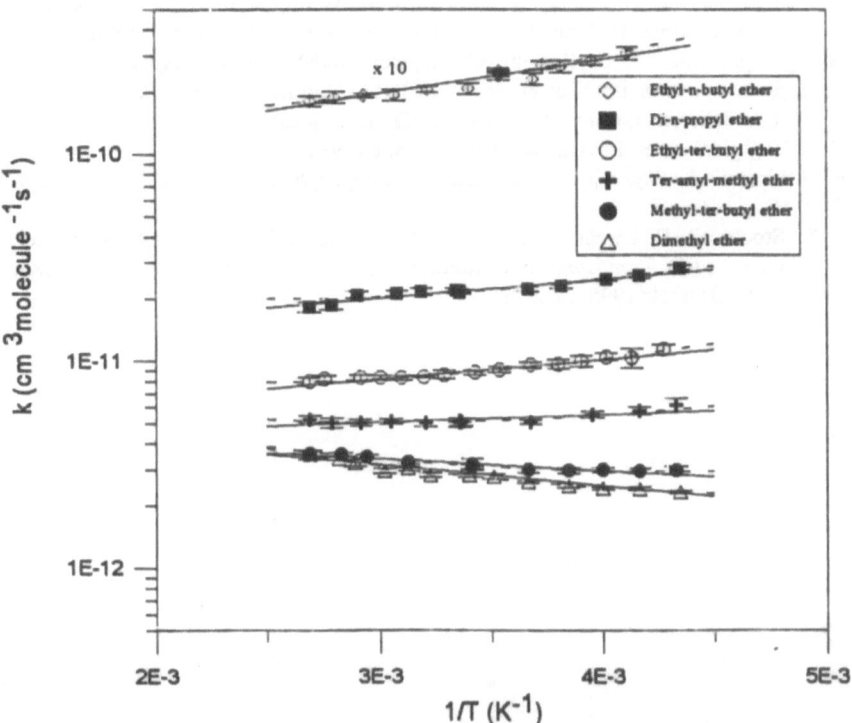

Fig. 4: Arrhenius plot for the reactions of OH with the studied oxygenated VOCs

Achievements

Kinetic and mechanistic data have been provided for reactions involved in the NO_3 and OH initiated oxidation mechanisms of VOCs. These data can to be used, and some of them have already been used (*e.g.* [6]), in models for assessing the role of VOCs and NO_x precursors in ozone and other photo-oxidant formation.

Acknowledgements

- J. P. Burrows, J. N. Crowley, C. Golz, G. Moortgat, U. Platt, R. Schindler, R. Singer and U. Ville for contributing to joint experiments or modelling calculations.
- EC (through Environment Programme), CNRS (Programme Environnement and CNRS / MPG agreement), Ministère de l'Environnement and Institut Français du Pétrole, for support.

References

1.   U. Platt, G; Le Bras, G. Poulet, J.P. Burrows, G.K. Moortgat, *Nature* **348** (1990) 147.
2.   G. Le Bras, C. Golz, U. Platt: Production of peroxy radicals in the DMS oxidation during night-time, in: G. Restelli, G. Angeletti (eds), *Dimethylsulphide, Oceans, Atmosphere and Climate,* Kluwer Academic Publ., Dordrecht 1993, p. 251.
3.   U. Platt, C. Janssen, *J. Chem. Soc. Faraday Trans.* in press.
4.   G. L. Vaghjiani, A. R. Ravishankara, *Nature* **350** (1991) 406.
5.   R. Atkinson, *J. Phys Chem. Ref. Data, Monograph* **1** (1989) and *Monograph* **2** (1994).
6.   W.R. Stockwell, F. Kirchner, J.B. Milford, D. Gao, Y.J. Yang, in: K.H. Becker, P. Wiesen (eds), *Tropospheric Oxidation Mechanisms, Air Pollution Research Report*, EC, Brussels 1995, in press.

3.10 Peroxy Radical Reactions of Tropospheric Interest

R. Lesclaux, A.A. Boyd, I. Bridier, F. Caralp, V. Catoire, F.F. Fenter,
P.D. Lightfoot, B. Nozière, D.M. Rowley, B. Veyret and E. Villenave

Laboratoire de Photophysique et Photochimie Moléculaire, Université Bordeaux I,
F-33405 Talence Cedex, France

Summary

Our work during the LACTOZ project has essentially focused on the investigation
of peroxy radical reactivity under atmospheric conditions, with the principal aim of
providing data important in modelling ozone formation in the troposphere. A set of
typical radicals have been studied in order to build structure-reactivity
relationships for the principal reactions of atmospheric importance. Significant
advances have been achieved for cross-reactions such as RO_2 + HO_2,
RO_2 + CH_3O_2 and RO_2 + $CH_3C(O)O_2$. Concerning self-reactions, a great deal of
information has been collected, particularly about the influence of functional
groups, but more work is necessary to ascertain relationships between structure
and reactivity. It should be emphasised that the latter work has provided valuable
information about the way alkoxy radicals react under tropospheric conditions.

In addition, the important reaction of PAN formation has been investigated in great
detail.

Aims of the scientific research during the project

It is well recognised that peroxy radicals play a key role in the oxidation processes
of hydrocarbons under the conditions of the atmosphere, leading to the formation
of tropospheric ozone and other oxidised molecules, in the presence of nitrogen
oxides. Their relative reaction rate with nitrogen oxides and with other radicals is
one of the principal factors determining the ozone balance in the troposphere.

The principal aim of our work was to establish the general characteristics of the
reactivity of peroxy radicals with the main atmospheric reactive trace species, *i.e.*
nitrogen oxides and other peroxy radicals. Most of our work focused on the
reactions of peroxy radicals with themselves, on reactions with HO_2 and on their
cross-reactions. The objective was to build structure-reactivity relationships by
studying typical peroxy radicals. Collaboration with other groups within LACTOZ
has resulted in significant progress in understanding the reactivity of peroxy
radicals and in assessing their actual role in the chemical processes prevailing in
the troposphere and leading to the formation of ozone. Note that in the course of
our studies, important information has also been obtained on the reactions of
several alkoxy radicals.

All radicals and their particular reactions investigated during the project are listed in Table 1.

Table 1: Peroxy radicals investigated during the project.

Radical	Spectrum	Self-reaction	α^a	Reaction with HO_2	Reaction with CH_3O_2	Reaction with NO_2	Alkoxy[b]
HO_2	+	+		+	+		
CH_3O_2	+	+	+	+	+	+	
$C_2H_5O_2$	+	+		+	+		+
$AllylO_2$		+		+	+		+
t-butylO_2	+	+					+
neopentylO_2	+	+	+	+	+		+
c-$C_5H_9O_2$	+	+		+			+
c-$C_6H_{11}O_2$	+	+	+	+	+		+
BenzylO_2	+	+	+	+	+		+
$CH_3C(O)O_2$	+	+		+	+	+	
$C_2H_5C(O)O_2$				+			
$CH_3C(O)CH_2O_2$	+	+	+	+	+		+
$HOCH_2O_2$	+	+		+			+
$HOC(CH_3)_2CH_2O_2$	+	+	+	+			+
CH_2ClO_2	+	+	+	+	+		+
$CHCl_2O_2$	+	+	+	+			+
CCl_3O_2	+	+		+	+		+
CH_2BrO_2	+	+		+			
$CH_3CHBrCH(O_2)CH_3$	+	+					

[a] Branching ratio for the non-terminating channel in self-reaction.

[b] Indicates information has been obtained on the reaction of the corresponding alkoxy radical.

Principal scientific findings during the project

Experimental methods

Most of our experiments were performed using the flash-photolysis technique with UV detection of peroxy radicals, in the spectral range extending from 190 to 300 nm. The temperature could be varied from 248 up to 600 K and the pressure from 30 to 760 Torr.

In most our studies, radicals were generated by photolysis of molecular chlorine in the presence of an appropriate precursor:

$$Cl_2 + hn \qquad \rightarrow \quad 2\,Cl$$
$$Cl + RH + O_2 \qquad \rightarrow \quad HCl + RO_2$$

Laser photolysis was also utilised when the above method could not be applied, in particular when other precursors had to be used (for the allylperoxy radical, for example) or when two radicals had to be generated simultaneously in order to study their cross-reaction. For example, the following system was used to study the cross-reaction between RO_2 and $R'O_2$:

$$RCl + hn + O_2 \qquad \rightarrow \quad RO_2 + Cl$$
$$Cl + R'H + O_2 \qquad \rightarrow \quad R'O_2 + HCl$$

Rate constants were obtained by numerically simulating decay traces, using the complete reaction mechanism and including all reactive and absorbing species. Thus, in most cases, the primary rate constant was obtained instead of the observed rate constant (which include secondary loss processes).

The reaction mechanism had to be determined in many instances by collaborating with other groups: Moortgat (Mainz), Zabel and Becker (Wuppertal), Cox, Hayman and Jenkin (Harwell), Wallington (Ford).

UV spectra of peroxy radicals

The most appropriate way of detecting peroxy radicals in studies of their reaction kinetics is the UV absorption spectrometry. The UV absorption spectra of 17 radicals have been determined in our group during the LACTOZ project, as indicated in Table 1. The radicals all have the characteristic broad absorption band peaking around 230–250 nm, with a maximum absorption cross-section of generally $(4 \pm 1) \times 10^{-18}$ cm^2 molecule^{-1}. Halogen substitution apparently tends to lower the maximum absorption cross-section. For example, the absorption cross-section of chloromethyl peroxy radicals decreases regularly with chlorine substitution to reach the low value of 1.6×10^{-18} cm^2 molecule^{-1} for CCl_3O_2.

A second absorption band appearing at shorter wavelength (200–210 nm) can be seen when the C–O–O chromophore is conjugated with an unsaturated group. This second band has been well characterised in the case of acylperoxy radicals, bearing the O=C–O–O group. The start of a second band has also been seen for the benzylperoxy radical but not for allylperoxy. However, the spectrum of the latter has not been fully investigated at wavelength shorter than 230 nm. Spectra of peroxy radicals have been discussed in detail in the review on peroxy radicals written by LACTOZ contributors [1].

The acetonyl peroxy radical exhibits a distinctive spectrum having a second absorption band at longer wavelengths (» 300 nm). This band apparently corresponds to the n–π^* transition of the carbonyl group, which has been enhanced. However, the reason for this enhancement is not presently understood and has not been explained by semi-empirical calculations.

Self-reactions of peroxy radicals

Self-reactions of peroxy radicals may occur in the atmosphere under low NO_x concentrations, particularly for the fastest cases, where rate constants may reach values as high as 10^{-11} cm^3 molecule^{-1} s^{-1}. In addition, self-reactions must be well characterised before studying other reactions of peroxy radicals, particularly those with HO_2 and CH_3O_2. The general mechanism for RO_2 self-reactions is the following:

$$RO_2 + RO_2 \qquad \rightarrow \qquad 2\,RO + O_2 \qquad\qquad\qquad (1a)$$
$$\rightarrow \qquad ROH + R_{-H}O + O_2 \qquad\qquad (1b)$$

with $R_{-H}O$ being an aldehyde or a ketone and a the branching ratio of the non-terminating channel (1a): $\alpha = k_{1a}/k_1$. A possible third reaction channel, forming the peroxide ROOR, has never been observed. The non-terminating channel generates an alkoxy radical which is either converted into HO_2 and a carbonyl compound by reaction with O_2:

$$RO + O_2 \qquad\qquad \rightarrow \qquad R_{-H}O + HO_2 \qquad\qquad\qquad (2)$$

or into new peroxy radical after decomposition, for example:

$$CH_3C(O)CH_2O \qquad \rightarrow \qquad CH_3CO + HCHO \qquad\qquad (3)$$
$$CH_3CO + O_2 \qquad \rightarrow \qquad CH_3C(O)O_2 \qquad\qquad\qquad (4)$$

giving rise to new reaction routes for the initial RO_2 radical ($RO_2 + HO_2$ or $RO_2 + R'O_2$). In most of our experiments, the determination of rate constants were performed by using the complete reaction mechanism in which the subsequent reactions of RO_2 were included. This has necessitated the determination of α and of the reaction mechanism when necessary, using the collaborations mentioned above.

All self-reactions investigated during the project are listed in Table 1 and rate constants values are reported in Chapter 2, Table 8. Also indicated in Table 1 are the radicals for which we have participated in the determination of both α and of the reaction mechanism. The detailed analysis of each reaction has been presented in preceding reports and only general considerations regarding structure-reactivity relationships are considered here.

Primary peroxy radicals (RCH_2O_2)

The simplest alkyl peroxy radicals have rate constant values at 298 K of about 10^{-13} cm^3 molecule^{-1} s^{-1}, which may increase in the case of chain branching, *e.g.* 1.2×10^{-12} cm^3 molecule^{-1} s^{-1} for neopentylperoxy. Apparently, these values are roughly the lower and upper limits for primary alkylperoxy radicals and more data would be required to refine the relationship between structure and reactivity of such radicals.

The most striking feature of the results is the dramatic increase of the rate constant under substitution of an H atom in the alkyl peroxy radicals with halogen atom, an

OH radical or a particular functional group such as a carbonyl group, unsaturated bond or aromatic ring. Apparently, this effect is independent of the position of the substituent (α or β) relative to the peroxy function.

In the case of the methyl peroxy radical, for example, the presence of an F atom, Cl atom (s), or OH group, increases the rate constant by more than a factor of 10. The effect of bromine, on the contrary seems smaller. The enhancement is even larger with the presence of a carbonyl group (*e.g.* acetonylperoxy) or an aromatic ring (*e.g.* benzylperoxy).

The acetylperoxy radical, and probably other acylperoxy radicals $RC(O)O_2$ exhibit even higher rate constants: 1.7×10^{-11} cm^3 molecule^{-1} s^{-1}. Acylperoxy radicals are certainly the most reactive peroxy radicals, as seen below for other type of reactions.

A negative temperature dependence of the rate constant seems to be a general characteristic of self-reactions of primary alkylperoxy radicals.

Secondary peroxy radicals ($RCHO_2R'$)

Very few data are available for self-reactions of secondary alkylperoxy radicals. Our main contribution in this case is the study of cycloalkylperoxy radicals c-$C_5H_9O_2$ and c-$C_6H_{11}O_2$ radicals. The rate constants are 40 times larger than the only linear secondary peroxy radical studied so far, i-$C_3H_7O_2$ and exhibit a slight positive temperature dependence. More data would be necessary for a better description of such reactions but, due to their relatively low rate constants ($< 5 \times 10^{-14}$ cm^3 molecule^{-1} s^{-1}) their contribution to atmospheric chemistry is negligible.

It should be emphasised, however, that the presence of a functional group may result in a significant increase of the rate constant, up to nearly 10^{-12} cm^3 molecule^{-1} s^{-1}. We have confirmed this behaviour in the case of bromine substitution and shown that the temperature dependence has become negative. More data are needed to confirm if this spectacular effect occurs with other functional group substitution.

Tertiary peroxy radicals

We have confirmed that the rate constant of the t-butyl peroxy radical self-reaction is very slow, with a fairly high activation energy. However, as above, the effect of functional groups on the rate constant should be investigated.

Reactions of peroxy radicals with HO$_2$

Due to the fairly high concentration of HO$_2$ in the atmosphere, RO$_2$ + HO$_2$ reactions play an important role under moderately low concentrations of NO$_x$. Investigations of such reactions have represented a significant part of our work during the project. The reactions that have been studied are listed in Table 1 and the results are given in the Chapter 2, Table 5. Structure to reactivity rules are much easier to establish than in the case of self-reactions, as all rate constant values are restricted within a factor of about three.

There is no clear systematic effect of functional groups on the rate constants except, perhaps, for halogen substitution which seems to maintain the rate constant at the lower limit. Apparently, the main factor influencing the rate constant is the size of the radical. The lowest rate constants are those of CH_3O_2 and of the corresponding halogen substituted radicals, around 6×10^{-12} cm^3 $molecule^{-1}$ s^{-1}. The rate constant increases rapidly, over 10^{-11} cm^3 $molecule^{-1}$ s^{-1} for bigger radicals and reaches values of 1.8×10^{-11} cm^3 $molecule^{-1}$ s^{-1} for c-$C_5H_9O_2$ and c-$C_6H_{11}O_2$. A recommendation might be as follows (units of cm^3 $molecule^{-1}$ s^{-1}): 6×10^{-12} for CH_3O_2 and halogenated peroxy radicals, 1.0×10^{-11} for C_2–C_3 radicals and 1.5×10^{-11} for bigger radicals (> C_4) and acyl peroxy radicals. All reactions exhibit strong negative temperature dependence with exponential factors varying from 800 K / T to 1600 K / T.

Reactions of peroxy radicals with CH_3O_2

It has been shown that CH_3O_2 is the most abundant alkylperoxy radical in the atmosphere and, under low NO_x level its concentration can be as high as that of HO_2 [2]. Therefore, there is a possibility that reactions of the type $RO_2 + CH_3O_2$ play a role in the chemistry of remote atmospheres, depending on the rate constants of such cross-reactions. We have investigated a series of this type of reaction. They are listed in Table 1 and the values of the rate constants are reported in Chapter 2, Table 5. It is observed that the rate constants for cross-reactions are generally intermediate between the rate constants for self-reactions of RO_2 and CH_3O_2. A geometrical mean such as

$$k(RO_2 + CH_3O_2) = \sqrt{k(RO_2 + RO_2) \cdot k(CH_3O_2 + CH_3O_2)}$$

might be a good estimate for the cross-reaction rate constant. However, it is observed that when the RO_2 self-reactions is fast, $k(RO_2 + CH_3O_2)$ is much closer to $k(RO_2 + RO_2)$ than to $k(CH_3O_2 + CH_3O_2)$. This is true, for example, for acetyl, acetonyl and chloromethylperoxy radicals. If this behaviour is indeed general, a better approximation would be to set $k(RO_2 + CH_3O_2)$ equal to $k(RO_2 + RO_2)$. This would be wrong for slow reactions, but as far as atmospheric chemistry is concerned, slow reactions are unimportant.

Reactions of peroxy radicals with $CH_3C(O)O_2$

A few cross-reactions of RO_2 radicals with the acetylperoxy radical have been investigated: CH_3O_2, $C_2H_5O_2$ and c-$C_6H_{11}O_2$. The reactions are very fast, with rate constants equal to 10^{-11} cm^3 $molecule^{-1}$ s^{-1}. This is even true for the secondary radical c-$C_6H_{11}O_2$, in spite of the much lower rate constant observed for the self-reaction (4×10^{-14} cm^3 $molecule^{-1}$ s^{-1}). This may be important since the acetylperoxy radical is also an abundant radical in the atmosphere and, if all cross-reactions of this type are as fast as those reported above, they should be taken into account in modelling the chemistry of remote atmospheres. Further experiments are in progress.

Reactions of peroxy radicals with NO_2

The reaction of the acetylperoxy radical with NO_2, forming PAN, is one of the most important reactions influencing the formation of ozone in the troposphere. We have performed a very detailed investigation of this reaction, in collaboration with the group from Wuppertal having considered the kinetics of the forward and reverse reaction, the equilibrium constant and thermodynamics, pressure and temperature effects and the interpretation of results using statistical theories [3]. This is obviously one of our most significant results and is now the reference work for this reaction. The equilibrated reaction of CH_3O_2 with NO_2 has also been reinvestigated.

Reactions of alkoxy radicals

The analysis of peroxy radical self-reactions, including the complete reaction mechanism, has provided information on the way alkoxy radicals react either under tropospheric conditions or under the conditions of laboratory studies. CH_3O, C_2H_5O, allylO and benzylO, all react with oxygen to yield HO_2 and an aldehyde. The same reaction occurs in the atmosphere for $HOCH_2O$ and CH_2ClO, but these radicals may decompose at low oxygen partial pressure (by H elimination for $HOCH_2O$ and HCl elimination for CH_2ClO). Competition between reaction with oxygen and decomposition has been observed for c-$C_6H_{11}O$ under atmospheric conditions and for neopentylO at high O_2 partial pressure. Only decomposition was observed for all other radicals: c-C_5H_9O, $CH_3C(O)CH_2O$, $CHCl_2O$, CCl_3O and $HOC(CH_3)_2CH_2O$.

It is certainly difficult, at the present time, to establish relationships between structure and reactivity for reactions of alkoxy radicals. More experimental and theoretical approaches are necessary to build a general picture of alkoxy radical reactivity.

Conclusions

Various types of information have been obtained from our studies of kinetics and mechanisms of peroxy radical reactions. The principal achievements, relevant to tropospheric chemistry, are the following:

- Reactions with NO_2: PAN formation and decomposition reactions have been studied in detail; pressure and temperature dependencies of the kinetics, thermodynamic properties and thermal parameters of the equilibrium constant have been measured.
- Self-reactions: significant information on relationships between structure and reactivity have been obtained, but more data are still necessary; we have shown

that substitution by functional groups in alkyl peroxy radicals may result in a dramatic increase of the rate constant.

- Reactions with HO_2: the data obtained for this important type of reaction allow us to propose a fairly reliable structure-reactivity relationship.
- Cross-reactions of peroxy radicals: original data have been obtained on the kinetics of RO_2 reactions with CH_3O_2 and $CH_3C(O)O_2$; some reactions with CH_3O_2 are fast enough to play a role in the tropospheric chemistry and all reactions with $CH_3C(O)O_2$ are apparently very fast and should be taken into account.

All the results obtained in this work are necessary for modelling the atmospheric oxidation of hydrocarbons and for mechanism reduction in complex oxidation processes.

Acknowledgements

The authors wish to thank the European Commission for funding this work.

References

1.   P.D. Lightfoot, R.A. Cox, J.N. Crowley, M. Destriau, G.D. Hayman, M.E. Jenkin, G.K. Moortgat, F. Zabel: Organic peroxy radicals: kinetics, spectroscopy and tropospheric chemistry, *Atmos. Environ.* **26A** (1992) 1805.
2.   N.M. Donahue, R.G. Prinn: Nonmethane hydrocarbon chemistry in the remote marine boundary layer, *J. Geophys. Res.* **95** (1990) 18337.
3.   I. Bridier, F. Caralp, H. Loirat, R. Lesclaux, B. Veyret, K.H. Becker, A. Reimer, F. Zabel,: Kinetic and theoretical studies of the reactions $CH_3C(O)O_2 + NO_2 + M <=> CH_3C(O)O_2NO_2 + M$ between 248 and 393 K and between 30 and 760 Torr, *J. Phys. Chem.* **95** (1991) 3595.

3.11 Laboratory Studies of Some Thermal Reactions Involving NO_y Species

E. Ljungström, M. Hallqvist, R. Karlsson, S. Langer and I. Wängberg

Department of Inorganic Chemistry, University of Göteborg and Chalmers University of Technology, S-412 96 Göteborg, Sweden

Summary

An investigation of the mechanism of heterogeneous PAN decomposition was made. It was found that the reactions of the acetylperoxy radical were influenced slightly by the presence of surfaces. The temperature dependence of the side-channel of the reaction between nitrate radicals and nitrogen dioxide, producing nitric oxide, nitrogen dioxide and molecular oxygen, was investigated. Monte Carlo simulation calculations of the photolysis of nitrogen dioxide were performed. Reaction products and in some cases also kinetic information were determined for several alkenes reacting with nitrate radicals. Structurally related, simple carbonyl compounds were among the more abundant products together with nitrooxy-substituted ketones and alcohols. A systematic investigation of reaction kinetics and product distributions from nitrate radical reaction with aliphatic esters, ethers and alcohols was made. The rate coefficients for ethers and alcohols were found to be substantially higher than expected from simple correlations, *e.g.* for alkanes. This is interpreted as an effect of the oxygen atom, weakening adjacent hydrogen-carbon bonds. The oxygen atom also has an additional influence, possibly via stabilisation of the transition complex. The nitrate radical is quite specific for abstraction of the weakest bonded hydrogen atom and reaction products can often be deduced from this information combined with common knowledge of atmospheric gas-phase reactions.

Aims of the research

This work was intended to give information about non-photochemical processes affecting species that are involved in hydrocarbon oxidation and ozone formation, such as the hydrocarbons themselves or the active nitrogen compounds. The main part of the work was concerned with rates and products in nitrate radical reactions with various hydrocarbons. Some interest was also taken in purely inorganic reactions.

Principal scientific results

PAN decomposition

The results show that reaction (I) is a gas-phase reaction that is not influenced by the surface / volume (S / V) ratio [1].

$$CH_3C(O)OONO_2 \quad <=> \quad CH_3C(O)OO + NO_2 \quad\quad (I)$$

The disappearance of the acetylperoxy radical is the rate controlling step for PAN destruction in a situation where equilibrium (I) is established. The peroxy radical disappearance is the result of at least two competing processes with different temperature dependencies and where at least one is heterogeneous. A mathematical model of the decomposition was set up and calculations were made for some realistic cases. The results show that the ambient concentration of PAN is unlikely to be influenced by heterogeneous decomposition on aerosol surfaces, nor can the aerosol composition be significantly changed due to deposition of PAN. Laboratory investigations involving PAN and which are made using equipment with a high S/V ratio could be sensitive to heterogeneous PAN decomposition. The stability of gas mixtures containing PAN for calibration of analytical systems and other laboratory work will benefit from containers having a low S/V ratio. The greatest gain in stability can be realised if some NO_2 can be tolerated together with the PAN. The higher the NO_2/PAN ratio is, the better, since acetylperoxy radical formation is suppressed.

NO_2 photolysis

An exception from the studies of non-photochemical processes is the theoretical work on NO_2 photolysis [2]. Stochastic simulation of NO_2 (X^2A_1) ® $O(^3P_j)$ + NO ($X^2\Pi_\Omega$) photodissociation in the wavelength range between 348 and 379 nm lends no support to the hypothesis that vibrationally excited NO_2 molecules should give a distinct contribution to the quantum yield in the vicinity of the nominal dissociation wavelength.

The reaction $NO_3 + NO_2 <=> NO + NO_2 + O_2$

The temperature dependence of the reaction rate was investigated using FTIR and TDL spectroscopy [3]. The rate coefficient can be described by $k = 5.4 \times 10^{-14}$ $\exp(-1488\ K / T)$.

Alkene-NO_3 reactions

Halogenated compounds

The products formed in the reactions between the nitrate radical and 2-chloro-1-butene, 3-chloro-1-butene, 1-chloro-2-butene and 2-chloro-2-butene were investigated by FTIR spectroscopy [4]. The experiments were performed in synthetic air and using N_2O_5 as the NO_3 source. The principal products formed from chloro-butenes, where the chlorine atom was substituted at a carbon participating in the double bond, were acid chlorides, aldehydes and NO_2. The products formed from chlorobutenes with the chlorine atom substituted at a carbon next to the double bond, were chlorinated carbonyl-nitrate compounds and NO_2.

The kinetic behaviour was investigated [5] in a collaboration with PCL, Oxford. Some of the observed rate coefficients are given in Table 1. Later this work was carried on to its full potential, by further experimental work and theoretical calculations by the Oxford group [6].

Butenes

Nitrate radical-initiated oxidation of 2-butenes was investigated using long optical path FTIR-spectroscopy [7]. At 295 K, 1020 mbar and 21 % O_2, major products were 3-nitroxy-2-butanone, 3-nitroxy-2-butanol and acetaldehyde. The oxygen dependence of the product formation was investigated by varying the O_2 concentration. The yield of 3-nitroxy-2-butanone increased and the formation of 3-nitroxy-2-butanol and acetaldehyde decreased with increasing oxygen concentration. At low O_2 concentrations (< 0.01 %) and 1020 mbar pressure, large amounts of 3-nitro-2-butylnitrate were formed together with small amounts of trans-2-butene-oxide.

Cyclic alkenes

Cyclic alkenes have some structural resemblance with certain terpenes. Since compounds with a smaller molecular weight than the terpenes themselves, are easier to handle experimentally, they are interesting as model compounds.

Absolute rate coefficients for the reaction between NO_3 and cyclopentene, cyclohexene and 1-methyl cyclohexene were determined using the fast flow-discharge technique [8] and are listed in Table 1. The simple cyclic alkenes react slightly faster than cis-2-butene which contains a similar active structural element. The rate coefficient for reaction with 1-methyl cyclohexene is close to that for D^3 carene or 2-carene but significantly greater than that for a-pinene.

Cyclohexene, cyclopentene and 1-methylcyclohexene were investigated con-cerning product formation [9]. The main products from these three cycloalkenes are likely to be di-carbonyls formed by ring opening. Thus, glutaric aldehyde was identified from cyclopentene experiments. The corresponding compounds from cyclohexene and 1-methylcyclohexene, i.e. adipic aldehyde and 6-oxo-heptanal were not positively identified due to the lack of reference compounds. However, the appearance of characteristic aldehyde bands in spectra from the experiments indicated the existence of such compounds. In addition, 2-nitrooxycyclopentanone, 2-nitrooxycyclopentanol and 1,2-cyclopentanedinitrate were identified in a molar relation of 4/2/1 from cyclopentene. From cyclohexene, 2-nitrooxycyclohexanone, 2-nitrooxycyclohexanol and 1,2-cyclohexanedinitrate were observed in a molar relation of 4/2/1. In the case of 1-methyl-cyclohexene, 2-methyl-2-nitrooxycyclo-hexanone and 1-methyl-2-nitrooxycyclohexanol were found in a 1/1 relation. A detailed mechanism that accounts for the formation of these compounds is given in the original paper [9].

Table 1: Rate coefficients or rate coefficient expressions for nitrate radical reactions. When no temperature expression is given the value refers to room **temperature.**

compound	rate coefficient expression or k / cm^3 molecule^{-1} s^{-1}		ref
2-chloro-1-butene	7×10^{-14}		5
3-chloro-1-butene	2.4×10^{-12} exp(−1992 K / T)		5
1-chloro-2-butene	6.0×10^{-13} exp(−981 K / T)		5
2-chloro-2-butene	11.0×10^{-14}		5
cyclopentene	$(5.9 \pm 1.1) \times 10^{-13}$		8
cyclohexene	$(6.3 \pm 1.3) \times 10^{-13}$	($E_a = 1 \pm 4$ kJ mol^{-1})	8
1-methyl cyclohexene	$(1.5 \pm 0.5) \times 10^{-11}$	($E_a = 0 \pm 5$ kJ mol^{-1})	8
methyl formate	$(0.36 \pm 0.08) \times 10^{-17}$		10
methyl acetate	$(0.7 \pm 0.2) \times 10^{-17}$		10
methyl propionate	$(3.3 \pm 0.8) \times 10^{-17}$		10
methyl butyrate	$(4.8 \pm 0.5) \times 10^{-17}$		10
ethyl formate	$(1.7 \pm 0.3) \times 10^{-17}$		10
ethyl acetate	$(1.3 \pm 0.3) \times 10^{-17}$	($E_a = 23 \pm 8$ kJ mol^{-1})	10
ethyl propionate	$(3.3 \pm 0.4) \times 10^{-17}$		10
propyl formate	$(5.4 \pm 0.9) \times 10^{-17}$		10
propyl acetate	$(5 \pm 2) \times 10^{-17}$		10
dimethyl ether	$(1.40 \pm 0.42) \times 10^{-12}$ exp[(−2525 ± 600) K / T]		11
diethyl ether	$(3.88 \pm 1.12) \times 10^{-12}$ exp[(−2065 ± 480) K / T]		11
di-n-propyl ether	$(3.33 \pm 0.51) \times 10^{-12}$ exp[(−1870 ± 250) K / T]		11
methyl t-butyl ether	$(2.40 \pm 0.29) \times 10^{-12}$ exp[(−2415 ± 199) K / T]		11
ethyl t-butyl ether	$(2.48 \pm 0.78) \times 10^{-12}$ exp[−(1613 ± 542) K / T]		12
diisopropyl ether	$(2.02 \pm 0.35) \times 10^{-12}$ exp[−(1759 ± 301) K / T]		12
t-amyl methyl ether	$(1.21 \pm 0.22) \times 10^{-12}$ exp[−(1874 ± 304) K / T]		12
methanol	$(1.06 \pm 0.51) \times 10^{-12}$ exp[−(2093 ± 803) K / T]		13
ethanol	$(6.99 \pm 1.21) \times 10^{-13}$ exp[−(1815 ± 419) K / T]		13
isopropanol	$(1.54 \pm 0.75) \times 10^{-12}$ exp[−(1743 ± 1009) K / T]		13

Oxygenated, aliphatic substances

A number of esters [10], ethers [11, 12] and alcohols [13] were investigated with respect to reactivity with nitrate radicals . Both absolute and relative rate methods were employed. Rate coefficients for the reaction of NO$_3$ are given in Table 1. The rate coefficients for aliphatic esters may be predicted from available group reactivity factors for alkanes provided that formate carbonyl hydrogen atoms are treated as primary hydrogen atoms. The rate coefficients with temperature dependence for ethers and alcohols are valid between 268 to 363 K.

Fig. 1, adapted from Wayne *et al.* [14] shows that the rate coefficients for ethers and alcohols are high when the effect of the weakened carbon-hydrogen bonds in the vicinity of the oxygen atom is accounted for. This effect could arise from an increased stability of a transition complex, caused by interaction between the nitrate radical and the oxygen atom.

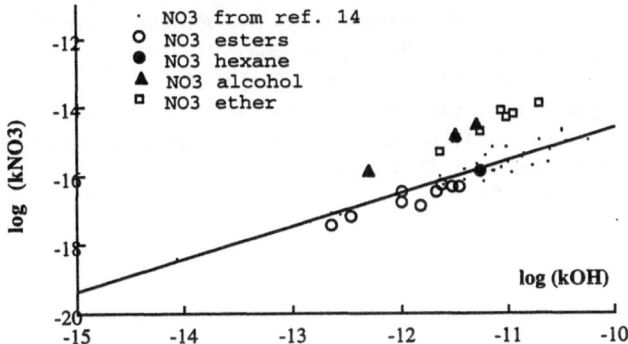

Fig. 1: Graph showing the correlation between the logarithms of rate coefficients for a certain organic substance, reacting by hydrogen abstraction, for reaction with nitrate and hydroxyl radicals. The straight line is the best fit to a number of compounds (dots), taken from [14]. The other symbols represent data from [10–13].

The product studies [11-13] show that the nitrate radical is quite selective for the weakest bonded hydrogen atom in these compounds. This knowledge, combined with common knowledge of atmospheric oxidation reactions makes possible a prediction of the products. For a detailed account of the branching of some competitive processes, the reader is referred to the original work.

Acknowledgements

The work included in this report was supported financially by the Swedish Environmental Protection Agency (SNV), AB VOLVO Technological Development and the Swedish National Science Research Council (NFR).

References

1.   S. Langer, I. Wängberg, E. Ljungström, *Atmos. Environ.* **26**A (1992) 3089.
2.   P.-A. Elofsson, E. Ljungström, *Chem. Phys.* **165** (1992) 323.
3.   I. Wängberg, E. Ljungström, B.E.R. Olsson, J. Davidsson, *J. Phys. Chem.* **96** (1992) 7640.
4.   I. Wängberg, E. Ljungström, J. Hjorth, G. Ottobrini, *J. Phys. Chem.* **94** (1990) 8036.
5.   R.W.S. Aird, C.E. Canosa-Mas, D.J. Cook, G. Marston, P.S. Monks, R.P. Wayne, E. Ljungström, *J. Chem. Soc. Faraday Trans.* **88** (1992) 1093.
6.   G. Marston, P.S. Monks, C.E. Canosa-Mas, R.P. Wayne, *J. Chem. Soc. Faraday Trans.* **89** (1993) 3899.

7. M. Hallquist, I. Wängberg, E. Ljungström, in: G. Angeletti, G. Restelli (eds), *Physico-Chemical Behaviour of Atmospheric Pollutants, Air Pollution Research Report* **50**, EC, Luxembourg 1994, p. 79.

8. E. Ljungström, I. Wängberg, S. Langer, *J. Chem. Soc. Faraday Trans.* **89** (1993) 2977.

9. I. Wängberg, *J. Atmos. Chem.* **17** (1993) 229.

10. S. Langer, E. Ljungström, I. Wängberg, *J. Chem. Soc. Faraday Trans.* **89** (1993) 425.

11. S. Langer, E. Ljungström, *Int. J. Chem. Kinet.*, **26** (1994) 367.

12. S. Langer, E. Ljungström, *J. Phys. Chem.* **98** (1994) 5906.

13. S. Langer, E. Ljungström, *J. Chem. Soc. Faraday Trans* **91** (1995) 405.

14. R.P. Wayne , I. Barnes, P. Biggs, J.P. Burrows, C.E. Canosa-Mas, J. Hjorth, G. Le Bras, G.K. Moortgat, D. Perner, G. Poulet, G. Restelli, H. Sidebottom, *Atmos. Environ.* **25**A (1991) 1.

3.12 High-Resolution Measurements of the Absorption Cross-Sections for O_3 and NO_2

Brion Jean, A. Chakir, B. Coquart, D. Daumont, A. Jenouvrier, J. Malicet and M.F. Merienne

U.A. D.1434 Spectrométrie Moléculaire et Atmosphérique, U.F.R. Sciences, B.P. 347, F-51062 Reims cedex, France

Summary

Two objectives have been achieved during these last years:

- the determination of the absorption cross-sections of O_3 and NO_2 in the UV and visible regions at different temperatures.
- the construction of a modulated photolysis apparatus for the study of atmospheric reactions initiated by photolysis.

Aims of the scientific research

The main work in this study is the determination of the absolute absorption cross-sections of the two important molecules, O_3 and NO_2, involved in atmospheric chemistry.

These parameters are used in the UV-visible region by the ground-based and satellite networks to measure the total amount and the concentration profile of the two species in the atmosphere. They are also involved in the calculation of the photodissociation coefficients related to the processes:

$$O_3 + h\nu \quad \rightarrow \quad O_2 + O\,(^1D)$$
$$NO_2 + h\nu \quad \rightarrow \quad NO + O\,(^3P)$$

These reactions are particularly important because the $O\,(^3P)$ and $O\,(^1D)$ atoms interact with molecules such as H_2O, CH , hydrocarbons and O_2. Moreover, the absorption cross-sections of O_3 are of great importance to calculate the solar flux penetration especially in the UV wavelength range where ozone absorbs in the Huggins and Hartley bands.

In order to provide sets of data directly available for the modelling or for optical *in situ* concentration measurements, the laboratory experiments need to be made in pressure and temperature conditions similar to those in the atmosphere.

Thus ozone absorption spectrum has been studied in the UV (200–500 nm) and visible (450–700 nm) regions at ambient and low temperatures down to 218 K.

For NO_2, absorption data have been obtained with very low (NO_2 - N_2O_4) pressures in the spectral region (300–500 nm) under similar temperature conditions.

Principal scientific findings during the project

O_3 spectroscopy

i) In a first step, the Hartley and Huggins bands which play a very important role in the atmosphere since they are located in the critical UV region, 200–350 nm, have been studied under high resolution (0.01 nm). The first work was to test our apparatus and our calculation method for the determination of the ozone concentration during the experiments. From this work, cross-section values were obtained at ambient temperature [1]. Measurements were extended to low temperatures (273, 243, 228 and 218 K) corresponding to tropospheric and stratospheric conditions [2, 3].

The absorption spectrum consists of a large continuum from 200 to 350 nm with a maximum near 250 nm ($\sigma(O_3)$ » 10^{-17} cm^2) which overlaps with small diffuse structures spaced by approximately 200 cm^{-1}. At higher wavelengths (310–350 nm) the spectrum is dominated by diffuse vibrational bands corresponding to the Huggins bands.

We have shown that the effect of temperature on the cross-section values is weak in the Hartley band (λ < 310 nm); at the mercury wavelength (λ < 253.6 nm), commonly used as reference, the cross-section value decreases from 1.0 % for a temperature variation from 218 to 295 K. Our results are in very good agreement ($\Delta\sigma / \sigma = 0.5$ %) with those determined by Barnes and Mauersberger [4] at the same temperatures.

On the other hand, in the Huggins bands, the temperature effect is important and increases progressively at increasing wavelengths where variations higher than 50 % can be observed.
Comparison with the most recent data [5–7] have been made, showing that our results are reliable.

ii) In the continuation of this work, the study of the Chappuis band has been undertaken in the visible region (515–650 nm) at ambient and low (218 K) temperatures, since differences up to 10 % were observed between the results of previous data [8–13].

The comparison at ambient temperature shows that our results are very close to those of Hearn [11], Anderson and Mauersberger [13] determined at some fixed wavelengths, while the data of Amoruso et al. [12] and Inn and Tanaka [9] are 5 % and 10 % below, respectively. The temperature dependence has been found to be very small (< 1 %) in the whole spectral range, in agreement with the conclusions of Vigroux [8] and Amoruso et al. [12].

iii) In order to provide data required by the *in situ* O_3 concentration measurements (SANOA/SAOZ experiments) we have extended our cross-section determinations in the whole UV-visible region (200–820 nm) with the same conditions of resolution.

In the region located between the Huggins and the Chappuis bands, the cross-sections are very weak (10^{-22}–10^{-23} cm^2), and the data are either missing (390 nm < λ < 420 nm) or obtained under insufficient resolution (Johnston for λ > 420 nm and Wahner for λ > 390 nm). To measure such a weak absorption we have built a 1 metre multipass absorption cell, coolable down to –60 °C. The body of the cell, made of stainless steel, is well suited to being filled with corrosive gases such as pure ozone. Measurements are in progress; they concern the 440–515 nm and 650–720 nm ranges at ambient temperature.

NO_2 *spectroscopy*

Before the beginning of our work, there were data for the NO_2 absorption in the UV visible region, but these data, obtained under various experimental conditions, exhibit differences up to 20 %.

In the 300–500 nm region, a new analysis was performed at three temperatures (293, 240 and 220 K) with long absorption paths (60 m) coupled with very low gas pressures. Under these conditions, the partial pressure of the dimer N_2O_4 is negligible at ambient temperature. Therefore reliable NO_2 cross-sections have been obtained in the 300–500 nm region at high resolution (0.01 nm) [14].

On the other hand, at lower temperatures, the determinations below 400 nm are complicated by the contribution of the N_2O_4 dimer that absorbs in this region which is also the photodissociation region of NO_2. Nevertheless for wavelengths higher than 400 nm, owing to the low pressures used, we were able to obtain the NO_2 cross-sections which are now available at 0.01 nm intervals [15]. A temperature effect is clearly shown for this structured region (400–500 nm).

Below 400 nm, even with low pressures, N_2O_4 shows a significant absorption compared to the total absorption, and it is necessary to make finer measurements at different pressures to separate the contribution of the two species. New experiments are necessary to improve the accuracy of our results.

Our values, acquired at ambient temperature in the 300–500 nm and at lower temperatures in the 400–500 nm, are an over previous ones [16–19]. With our experimental conditions, accuracies better than 3 %, 4 % and 6 % were obtained at ambient temperature, at 240 and 220 K. Our values provide a good set of data for the atmospheric purposes such as the modelling and the NO_2 concentration measurements by differential absorption techniques.

Kinetic studies

Recently a modulated photolysis system was constructed in our laboratory in order to study the kinetics of some atmospheric reactions. It consists of a coolable multipass triple jacket cell (1 m long, 7 cm in diameter) in which the reactants are photolysed by 12 surrounding fluorescent "sun lamps". The absorption of the reactants can be controlled by means of a classical mounting with an absorption source, a monochromator and a photomultiplier. Absorption paths up to 40 m allow to detect the presence of low reactant concentrations, and measurements can be carried out in a wide range of experimental conditions: from 1 to 760 Torr and from -30 °C to 100 °C, with frequencies varying up to 10 Hz.

Currently, the setup has been tested with simple peroxy radicals such as $C_2H_5O_2$ and HO_2. The first results are in agreement with the published data.

Achievements of the project

Our objective initially based on the determination of the absorption cross-sections of O_3 in the Hartley and Huggins bands and of NO_2 in the 300–500 nm region has been achieved. Additional regions towards the higher wavelengths have been studied to provide new data required by the extension of new *in situ* measurements.

Moreover a modulated photolysis apparatus has been set up for kinetic studies of atmospheric compounds.

Acknowledgements

This project was supported by the French EUROTRAC committee. We gratefully acknowledge this help.

References

1. D. Daumont, J. Brion, J. Charbonnier, J. Malicet, *J. Atmos. Chem.* **15** (1992) 145.
2. J. Brion, A. Chakir, D. Daumont, J. Malicet, C. Parisse, Chem. Phys. Letters **213** (1993) 610.
3. J. Malicet, D. Daumont, J. Charbonnier, C. Parisse, A. Chakir, J. Brion, *J. Atmos. Chem.* (1995) in press.
4. J. Barnes, K. Mauersberger, *J. Geophys. Res.* **92** (1987) 14 861.
5. K. Yoshino, J.R. Esmond, D.E. Freeman, W.H. Parkinson, *J. Geophys. Res.* **98** (1993) 5205.
6. A.M. Bass, R.J. Paur, in: *Proc. Quadrennial Ozone Symp.*, D. Reidel, Dordrecht 1984, pp. 606–616.
7. L.T. Molina, M.J. Molina, *J. Geophys. Res.* **91** (1986) 14501.
8. E. Vigroux, *Ann. Phys.* **8** (1953) 709.
9. E.C.Y. Inn, Y. Tanaka, *J. Opt. Soc. Amer.* **43** (1953) 870.
10. M. Griggs, *J. Chem. Phys.* **49** (1968) 857.
11. A.G. Hearn, *Proc. Phys. Soc.* **78** (1961) 932.
12. A. Amoruso, M. Cacciani, A. Di Sarra, G. Fiocco, *J. Geophys. Res.* **95** (1990) 20565.
13. S. Anderson, K. Mauersberger, *Geophys. Res. Lett.* **19** (1992) 933.
14. M.F. Merienne, A. Jenouvrier, B. Coquart, *J. Atmos. Chem.* (1995) in press.
15. B. Coquart, A. Jenouvrier, M.F. Merienne, *J. Atmos. Chem.* (1995) in press.
16. W. Schneider, G.T. Moortgat, G.S. Tyndall, J.P. Burrows, *J. Photochem. Photobiol., A: Chemistry* **40** (1987) 195.
17. J.A. Davidson, C.A. Cantrell, A.H. Mc Daniel, R.E. Shetter, S. Madronich, J.G. Calvert, *J. Geophys. Res.* **93** (1988) 7105.
18. A.M. Bass, A.E. Ledford, H.A. Laufer, *J. Res. Natl. Bur. Stand.* A**80** (1976) 143.
19. T.C. Corcoran, E.J. Beiting, M.O. Mitchell, *J. Mol. Spectros.* **154** (1992) 119.

3.13 Laboratory Studies of Peroxy Radicals, Carbonyl Compounds and Ozonolysis Reactions of Tropospheric Importance

G. K. Moortgat, D. Bauer, J.P. Burrows, J. Crowley, F. Helleis, O. Horie, S. Koch, S. Limbach, P. Neeb, W. Raber, F. Sauer, C. Schäfer, W. Schneider, F. Simon and C. Zahn

Max-Planck-Institut für Chemie, Atmospheric Chemistry Division, Postfach 3060, D-55020 Mainz, Germany

Summary

Spectroscopic, kinetic, photolytic and mechanistic studies have been carried out on simple peroxy radicals such as HO_2, CH_3O_2, $C_2H_5O_2$ and $CH_3C(O)O_2$ using a variety of laboratory techniques. The photo-oxidation studies of selected carbonyl compounds was performed and quantum yields established in order to assess their photolytic lifetime in the troposphere. The product distribution of the ozonolysis of selected alkenes was determined in the presence of water vapour.

Aims of the research

The objective of present research was to provide a better understanding of the chemical processes involved in production and loss of ozone in the troposphere. This was achieved by providing kinetic and mechanistic data for several reactions of peroxy radicals involved in the photo-oxidation of volatile organic compounds (VOC). Additional aims were to determine the product quantum yields in the photolysis of carbonyl compounds, and to investigate the mechanism in the ozonolysis of alkenes, especially in the presence of water vapour.

Principal scientific findings

Studies of peroxy radical reactions of tropospheric importance

UV Absorption spectra of HO_2, CH_3O_2, $C_2H_5O_2$, $CH_3C(O)O_2$ radicals

A vast majority of the kinetic measurements involving RO_2 radicals has made use of the relatively intense UV absorption spectrum of these species in order to obtain time-concentration profiles. Moreover, for the study of peroxy radical self-reactions, absolute concentrations must be known in order to analyse second order decay profiles, indicating the importance of accurately measured absolute absorption cross sections. The cited peroxy radicals were generated by photolysis of mixtures of molecular chorine in the presence of appropriate precursors in air, and monitored by absorption spectroscopy. Specifically, the techniques of modulated photolysis, and laser flash photolysis was used for the determination of absorption cross sections. The spectra of HO_2 [1], CH_3O_2 [2], $C_2H_5O_2$ [3] and $CH_3C(O)O_2$ [4] are displayed in Fig. 1. The absorption cross sections of $CH_3C(O)O_2$ were recently measured and found to be 30 % smaller than a previous determination [5].

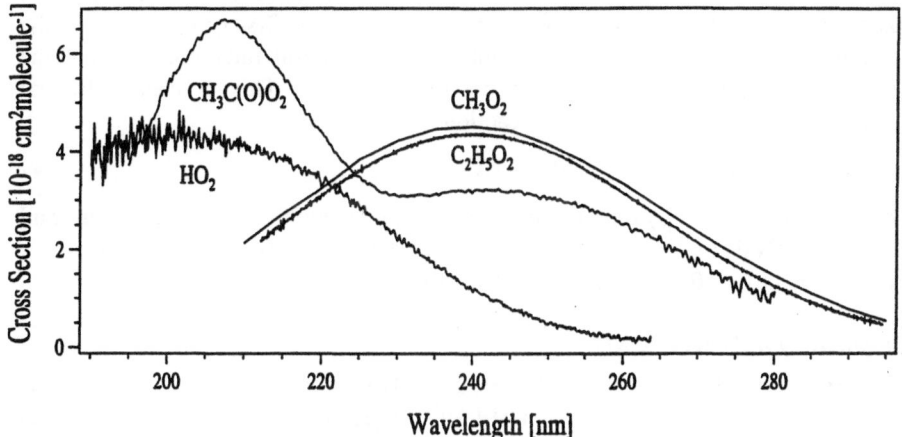

Fig. 1: Absorption spectra of HO_2, CH_3O_2, $C_2H_5O_2$ and $CH_3C(O)O_2$ radicals.

Kinetic studies of peroxy radicals

The self-reaction $CH_3O_2 + CH_3O_2$ $\rightarrow$ $CH_3O + CH_3O + O_2$ (1a)

$\rightarrow$ $HCHO + CH_3OH + O_2$ (1b)

$\rightarrow$ $CH_3OOCH_3 + O_2$ (1c)

The rate constant of the self-reaction was obtained by the modulated photolysis technique combined with long path UV absorption spectroscopy. Absorption-time profiles, obtained at various wavelengths, were analysed with the help of computer simulation. The presence of HO_2 radicals must be taken into account in the kinetics analysis, since they are generated by reaction (2) of CH_3O radicals with O_2, emphasising the importance of the knowledge about channel (1a):

$CH_3O_2 + O_2$ $\rightarrow$ $HO_2 + HCHO$ (2)

The observed rate constant was measured as $k_1(obs) = (1 + \alpha) k_1 = 4.8 \times 10^{-13}$ cm^3 molecule^{-1} s^{-1} at 298 K, where $k_1 = k_{1a} + k_{1b} + k_{1c}$ and $\alpha = k_1/k_{1a} = 0.34$ [2].

The temperature dependence of the products HCHO and CH_3OH, generated in reactions (1b) and (2), was measured by making use of a matrix isolation technique combined with FTIR spectroscopy [6]. Here, in a novel reactor, aliquots of the gaseous reaction mixture were deposited on a cold finger maintained at 7 K by liquid helium to form a cold matrix, which is then analysed by FTIR spectroscopy. In the temperature range 223 and 333 K, the branching ratio a is given by $\alpha = 1 / \{1 + [\exp((1535 \pm 90) \text{ K} / T)] / (51 \pm 15)\}$.

The reaction $CH_3O_2 + NO_3 \rightarrow CH_3O + NO_2 + O_2$ (3)

The kinetics of the reaction $CH_3O_2 + NO_3$ was studied by modulated photolysis spectroscopy [7] and later in a discharge flow reactor combined with molecular beam mass spectrometry [8]. In the latter experiment, the CD_3O_2 radicals were used instead of CH_3O_2, and their first order decay monitored in the presence of excess NO_3 radicals. It was however observed that the first order decay of the methylperoxy radicals did not extrapolate to a common intercept and that the second order plot showed a large positive intercept. This is caused by the regeneration of CH_3O_2 radicals via reaction

$CH_3O + NO_3 \quad \rightarrow \quad CH_3O_2 + NO_2$ (4)

which was confirmed by direct kinetic measurements by the groups in Orléans and Oxford. Analysis of the data resulted in $k_3 = (1.4 \pm 0.3) \times 10^{-12}$ and $k_4 = (3.3 \pm 0.8) \times 10^{-12}$ cm^3 molecule^{-1} s^{-1} [9].

Ethylperoxy radicals ($C_2H_5O_2$)

$C_2H_5O_2 + C_2H_5O_2 \quad \rightarrow \quad C_2H_5O + C_2H_5O + O_2$ (5a)
$\rightarrow \quad CH_3CHO + C_2H_5OH + O_2$ (5b)

The kinetics of the self-reaction was investigated by modulated photolysis spectroscopy in the temperature range 220–330 K [3]. In this temperature range a curvature in the Arrhenius plot is observed. The rate constant $k_5 = k_{5a} + k_{5b}$ is described by $k_5 = 2.49 \times 10^{-13} \exp(-518 / T) + 9.39 \times 10^{-16} \exp(960 \text{ K} / T)$ cm^3 molecule^{-1} s^{-1}, and the branching ratio by $k_{5a}/k_{5b} = 265 \exp(-1478 \text{ K} / T)$. At 298 K $k_5 = 6.7 \times 10^{-14}$ cm^3 molecule^{-1} s^{-1}, with $k_{5a}/k_5 = 0.65$. Substitution of an H atom by a Br atom at the ß-carbon was also studied and found to enhance the rate constant for the self-reaction by a factor 60 [10].

Acetylperoxy radicals $CH_3C(O)O_2$

$CH_3C(O)O_2 \;+\; CH_3C(O)O_2 \rightarrow CH_3C(O)O + CH_3C(O)O + O_2$ (6)
$CH_3C(O)O_2 \;+\; HO_2 \quad\;\; \rightarrow CH_3C(O)OOH + O_2$ (7a)
$\rightarrow CH_3C(O)OH \;+\; O_3$ (7b)
$CH_3C(O)O_2 \;+\; CH_3O_2 \quad \rightarrow CH_3C(O)O \;+\; CH_3O + O_2$ (8a)
$\rightarrow CH_3C(O)OH \;+\; HCHO + O_2$ (8b)

The kinetics of the self-reaction (6) and the cross reactions (7) and (8) were studied by conventional flash photolysis [5, 11] and later by laser flash photolysis [4] combined with UV long path absorption spectroscopy. Rate constants for reaction (6) to (8) were obtained:

$k_6 = (1.4 \pm 0.2) \times 10^{-11}$, $k_7 = (1.3 \pm 0.2) \times 10^{-11}$,
$k_{8a} = (8.4 \pm 0.3) \times 10^{-12}$ cm^3
and $k_{8b} = (8.2 \pm 1.5) \times 10^{-13}$ cm^3 molecule^{-1} s^{-1}.

The temperature dependence of the branching ratio of the cross reactions (7) and (8) was obtained in a separate study [12], whereby biacetyl was photolysed in the presence of O_2, as source of $CH_3C(O)O_2$ radicals. The reaction products, arising from the set of reactions (6) to (8), were analysed by matrix isolation FTIR spectroscopy. The branching ratios were obtained in the temperature range 233–333 K are k_{7a}/k_{7b} = 330 exp(–1430 K/ T) and k_{8a}/k_{8b} = 2.2 × 10^6 exp (–3870 K / T).

Photochemical studies of carbonyl compounds

Photo-oxidation of HCHO

The photo-oxidation of both HCHO–O_2 and Cl_2–HCHO–O_2 mixtures were investigated by modulated and flash photolysis [13, 14] and the spectrum and kinetics of the intermediate radical $HOCH_2O_2$ determined, which is formed in the equilibrium of HO_2 and HCHO. Significant production of HCOOH was observed.

Photo-oxidation of CH_3CHO

The kinetics of the radicals CH_3O_2 and HO_2, which are produced in the photo-oxidation of CH_3CHO, were studied under variable O_2 concentrations [15]. Direct observation of both CH_3O_2 and HO_2 radicals allowed the determination of $k(CH_3O_2 + HO_2)$ = (6.7 ± 2.2) × 10^{-12} cm^3 molecule^{-1} s^{-1} in agreement with previous studies. From the products CH_3OH, CO_2, CH_3OOH, $CH_3C(O)OH$ and $CH_3C(O)OOH$, the kinetic parameters k_8 = (1.5 ± 0.5) × 10^{-11} and $k(CH_3O + CH_3CHO)$ = (1.5 ± 1) × 10^{-13} cm^3 molecule^{-1} s^{-1} were determined.

Photo-oxidation of MEK, MVK, MACR and MGLY

The photo-oxidation of four carbonyl compounds, methylethyl ketone (MEK), methyl-vinyl ketone (MVK), methacrolein (MACR) and methyl glyoxal (MGLY), was studied by long path FTIR spectroscopy at 298 K [16, 17]. The absorption spectra were obtained using diode array spectroscopy [16–18]. Broad-band photolysis was performed using sun lamps (280–360 nm), and overall quantum yields of removal of reactants and formation of products were measured in the total pressure range 50 to 760 Torr. The main photolysis processes in the cited wavelength range are:

$CH_3COC_2H_5$ (MEK) + $h\nu$	→	$CH_3CO + C_2H_5$	(80–90 %)	(9a)
$\varphi(760) = 0.34$	→	other products	(10–20 %)	(9b)
$CH_3COCH=CH_2$ (MVK) + $h\nu$	→	$CH=CH–CH_3 + CO$	(70–80 %)	(10a)
$\varphi(760) < 0.05$	→	other products	(20–30 %)	(10b)
$CH_2=C(CH_3)CHO$ (MACR) + $h\nu$	→	Products	($j(760) < 0.03$)	(11)

$$CH_3COCHO \ (MGLY) + h\nu \quad \rightarrow \quad CH_3CO + CHO \quad (90\ \%) \quad (12a)$$
$$\varphi(760) = 0.64 \quad\quad\quad\quad\quad \rightarrow \quad \text{other products} \quad (10\ \%) \quad (12b)$$

All photolytic systems showed a Stern-Volmer type pressure dependent behaviour. The photolytic lifetimes were calculated on the basis of the cited overall quantum yields: 4 day for MEK, 14 h for MVK, 22 h for MACR and 35 min for MGLY.

In a separate experiment the photolysis of MGLY was investigated and CO quantum yield measured as a function of wavelength at pressures ranging between 50 and 760 Torr [19].

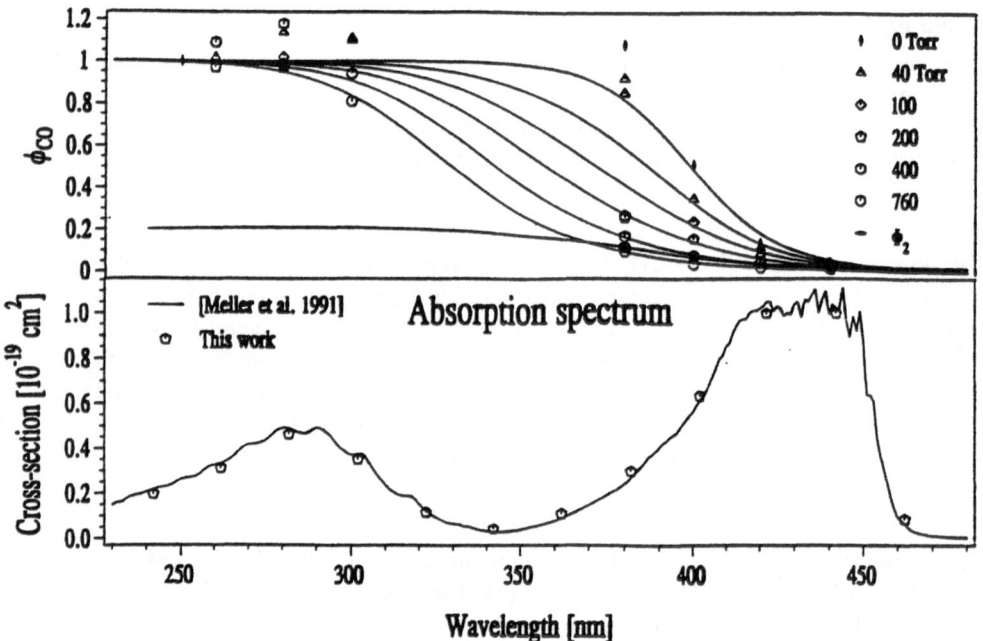

Fig. 2: Wavelength-dependent CO quantum yields of MGLY as a function of total pressure.

The pressure-dependent CO quantum yields are shown in Fig. 2 and can be represented by $\varphi(12a) = 1/(1 + a\ e^{-b/k} + c\ e^{-d/k}\ P)$ with $a = 8.1 \times 10^{12}$, $b = 1.2 \times 10^4$ nm, $c = 2.1 \times 10^4$ Torr^{-1} and $d = 5.5 \times 10^3$ nm. The yield of the other minor channel $\varphi(12b)$ is pressure independent and contributes to about 20 %.

Ozonolysis of alkenes. Influence of water vapour

The ozonolysis of simple alkenes was studied in two different apparatus: a 2 litre stirred tank reactor (for ethene, propene, *trans*-2-butene, butadiene, and isoprene) coupled *via* molecular beam sampling to a matrix isolation FTIR set-up [20, 21], and a 570 L spherical glass vessel "big sphere" (for ethene, 2-butene isomers, isobutene, and isoprene) where products were identified by FTIR spectroscopy, GC and a scrubber sampling unit for analysis with HPLC (for peroxides) and IC (for organic acids). In the latter system, two extreme humidity conditions, one with 0.5 ppm and the other with 2×10^4 ppm (corresponding to *ca.* 60 % relative humidity at 298 K) were used, which are referred to as "dry" and "wet" conditions, respectively. Results of the studies performed in the "big sphere" are summarised here.

According to currently accepted Criegee mechanism [20], the formation of the reaction products in the ozonolysis of ethene is explained in terms of the formation of the excited Criegee intermediate CH_2OO^* and its subsequent decomposition and collisional stabilisation:

$C_2H_4 + O_3$	$\rightarrow$	primary ozonide	$\rightarrow CH_2OO^* + HCHO$	(13)
CH_2OO^*	$\rightarrow$	decomposition products (CO, CO_2, H_2, H, OH, H_2O)		(14)
$CH_2OO^* + M$	$\rightarrow$	stabilisation (CH_2OO)		(15)
$CH_2OO + HCHO$	$\rightarrow$	$CH_2(OH)-O-CHO$ (HMF) $\rightarrow$ $(CHO)_2O$ (FAN) + H_2		(16) (17)

The following observations were made:

a) Both compounds HMF and FAN were identified and their total yield D(HMF) + D(FAN) (relative to O_3 converted) was estimated to *ca.* 0.2. Initially, the rate of HMF formation was found to increase linearly for low O_3 conversions. However at larger O_3 conversions, the rate of the HMF formation tended to decrease, while the FAN formation was accelerated. These results suggest that HMF is the precursor of FAN. Subsequent experiments with the added HCHO yielded D(HMF) + D(FAN) which was about 40 % lower than in the absence of HCHO addition. These observations imply either that HMF has different precursors or that the stabilised Criegee intermediate CH_2OO may not be formed [22]. The formation of trace amounts (< 0.05) of HMF and FAN was also noted in the ozonolysis of isobutene. Analogous products to HMF were assigned in the ozonolysis of *cis*- and *trans*-2-butene isomers [23].

b) Under wet conditions, HCOOH and $HOCH_2OOH$ (HMHP) are formed in the ozonolysis of ethene [24] . However, the rate of the HCOOH formation increased with O_3 conversion, indicating the secondary nature of the HCOOH formation [19]. This was also observed in the ozonolysis of propene, isobutene and isoprene under wet conditions. CH_3COOH was observed in the ozonolysis of *cis*- and *trans*-2-butene isomers. Noteworthy also is the observation an $CH_3CH(OH)OOH$ in the ozonolysis of *cis*- and *trans*-2-butene isomers, and the formation of $HOCH_2OOH$ and higher (not identified) hydroperoxides in ozonolysis of isobutene in the presence of the water vapour.

The observation that these hydroperoxides decompose rapidly under the experimental conditions may be related to the increasing formation rates of HCOOH (and probably of CH_3COOH) toward high O_3 conversion ranges, although further study is required. In view of these observations, both HCOOH and HMHP may be considered formed from the reaction of CH_2OO^* intermediate with H_2O vapour:

$$CH_2OO^* + H_2O \rightarrow HOCH_2OOH \rightarrow HCOOH + H_2O \tag{18}$$

c) The OH yield in the ozonolysis can be directly determined from the amount alkene converted per O_3 reacted. While in the C_2H_4 ozonolysis, the OH yield is low, $Y(OH) \approx 0.05$, $Y(OH)$ values for 2-butenes and isobutene were high; $Y(OH) \approx 0.4$ for cis-2-butene, ca. 0.6 for trans-2-butene, and ca. 0.6 for isobutene. For isoprene, $Y(OH)$ was roughly in the range of 0.1 to 0.15.

d) In the ozonolysis of isoprene, the main products were MVK, MACR, HCHO, CO, CO_2, and FAN. Minor products included HMF, ketene, $CH_2=CO$, and MGLY. The effect of water vapour addition on the relative yield of the two main carbonyl products, R = MACR/MVK, was investigated. Under dry conditions, R = 2.5 and was not affected by the addition of the OH radical scavengers such as HCHO, cyclohexane, or CO. Under the wet conditions, D(MVK) increased while D(MACR) remained unchanged, resulting in the decrease in R to 1.8. The yield of HCOOH increased up to ca. 0.3 under the wet conditions.

Assessment of achievements

The goals set with respect to the spectroscopy and kinetics of the simple peroxy radicals was mainly achieved. A detailed review on the chemistry and role of the peroxy radicals in the photo-oxidation of VOC was written within the LACTOZ project [25]. Further studies on quantum yields determination of photolabile carbonyl compounds and on the mechanism of the ozonolysis of alkenes are required.

Acknowledgements

The work performed within LACTOZ was supported by the Max-Planck-Gesellschaft, the Sonderforschungsbereich 233 and various grants by the CEC. We acknowledge the close collaboration with Robert Lesclaux, Phil Lightfoot and Bernard Veyret (University of Bordeaux), Garry Hayman and Mike Jenkin (AEA, Culham), Tony Cox (Cambridge) and Georges Le Bras and Gilles Poulet (CNRS Orléans).

References

1. J.N. Crowley, F.G. Simon, J.P. Burrows, G.K. Moortgat, M.E. Jenkin, R.A. Cox, *J. Photochem. Photobiol. A: Chem.* **60** (1991) 1.
2. F.G. Simon, W. Schneider, G.K. Moortgat, *Int. J. Chem. Kinet.* **22** (1990) 791.
3. D. Bauer, J.N. Crowley, G.K. Moortgat, *J. Photochem. Photobiol. A: Chem.* **65** (1992) 329.
4. C. Roehl, D. Bauer, G.K. Moortgat, *LACTOZ Meeting*, Leipzig 1994.
5. G.K. Moortgat, B. Veyret, R. Lesclaux, *J. Phys. Chem.* **93** (1989) 2362.
6. O. Horie, J.N. Crowley, G.K. Moortgat, *J. Phys. Chem.* **94** (1990) 8198.
7. J.N. Crowley, J.P. Burrows, G.K. Moortgat, G. Poulet, G. Le Bras, *Int. J. Chem. Kinet.* **22** (1990) 673.
8. F. Helleis, J.N. Crowley, G.K. Moortgat, in: G. Angeletti, G. Restelli (eds), *Physico-Chemical Behaviour of Atmospheric Pollutants, Air Pollution Research Report* **50**, EC, Luxembourg 1994, p. 90, (ISBN 92-826-7922-5).
9. G.K. Moortgat, in: G. Angeletti, G. Restelli (eds), *Physico-Chemical Behaviour of Atmospheric Pollutants, Air Pollution Research Report* **50**, EC, Luxembourg 1994, p. 66, (ISBN 92-826-7922-5).
10. J.N. Crowley, G.K. Moortgat, *J. Chem. Soc., Faraday Trans.* **88** (1992) 2437.
11. G.K. Moortgat, B. Veyret, R. Lesclaux, *Chem. Phys. Lett.* **160** (1989) 443.
12. O. Horie, G.K. Moortgat, *J. Chem. Soc., Faraday Trans.* **88** (1992) 3305.
13. B. Veyret, R. Lesclaux, M.T. Rayez, J.-C. Rayez, R.A. Cox, G.K. Moortgat, *J. Phys, Chem.* **26** (1989) 2368.
14. J.P. Burrows, G.K. Moortgat, G.S. Tyndall, R.A. Cox, M.E. Jenkin, G.D. Hayman, B. Veyret, *J. Phys, Chem.* **26** (1989) 2375.
15. G.K. Moortgat, R.A. Cox, G. Schuster, J.P. Burrows, G.S. Tyndall, R.A., *J. Chem. Soc. Faraday Trans.* **85** (1989) 809.
16. W. Raber, R. Meller, G. K. Moortgat, *Air Pollution Research Report* **33** (1991) 91.
17. W. Raber, G.K. Moortgat, *Advan. Phys. Chem.* (1995) in press .
18. R. Meller, W. Raber, J.N. Crowley, M.E. Jenkin, G.K. Moortgat, *J. Photochem. Photobiol. A: Chem.* **62** (1991) 163.
19. S. Koch, G.K. Moortgat, *LACTOZ Meeting*, Leipzig 1994.
20. O. Horie, G.K. Moortgat, *Atmos. Environ.* **25**A (1991) 1881.
21. O. Horie, G.K. Moortgat, *Fresenius J. Anal. Chem.* **340** (1991) 641.
22. P. Neeb, O. Horie, C. Schäfer, F. Sauer, G. K. Moortgat, *LACTOZ Meeting*, Leipzig, 1994.
23. O. Horie, P. Neeb, G. K. Moortgat, *Int. J. Chem. Kinet.* **26** (1994) 913.
24. O. Horie, P. Neeb, S. Limbach, G. K. Moortgat, *Geophys. Res. Lett.* **21** (1994) 1523.
25. P.D. Lightfoot, R.A. Cox, J.N. Crowley, M. Destriau, G.D. Hayman, M.E. Jenkin, G.K. Moortgat, F. Zabel, *Atmos. Environ.* **26**A (1991) 1805.

3.14 Atmospheric Chemistry of Nitrogen-Containing Species

Ole John Nielsen[1] and Jens Sehested[2]

[1]Ford Forschungszentrum Aachen, Dennewartstraße 25, D-52068 Aachen, Germany
[2]Section for Chemical Reactivity, Risø National Laboratory, DK-4000 Roskilde, Denmark

Summary

The rate constants for the reactions of OH with a series of organic nitroalkanes, nitrites and nitrates and of NO with a series of peroxy radicals were measured at 298 K and a total pressure of 1 atm. The rate constants were obtained using the absolute technique of pulse radiolysis combined with time-resolved UV-VIS spectroscopy. The results are discussed in terms of reactivity trends and the atmospheric chemistry.

Aims of the research

The aim of this research is to provide new information on the kinetics and mechanisms for reactions of nitrogen species for use in atmospheric modelling.

Principal scientific findings during the project

Volatile organic nitrogen-containing compounds may act more or less as long-lived reservoirs for NO_x in the troposphere. The reactions of OH radicals with three series of compounds have been studied: n-nitroalkanes, n-alkyl nitrates and n-alkyl nitrites. Nitroalkanes are employed as propellants and as industrial solvents leading to their potential release into the atmosphere. Alkyl nitrates can be formed in the oxidation of volatile organic compounds in the troposphere when peroxy radicals react with NO. Small amounts of alkyl nitrites may be formed in the urban atmosphere by reaction of alkoxy radicals with NO. The photodissociative lifetime of alkyl nitrites has been calculated to be around 2 min under typical atmospheric conditions. Thus alkyl nitrites can only be a temporary reservoir for NO_x during night-time hours. Reaction of these organonitrogen compounds with hydroxyl radicals is an important factor in determining their atmospheric residence times. However, previously very little information concerning the kinetics and mechanisms for these reactions have been available. Results summarised in Table 1.

Peroxy radicals, RO_2, are important intermediates in the atmospheric degradation of organic compounds emitted to the atmosphere. The alkyl radical, R, can be produced from emitted organic compounds in several ways, *e.g.*, by reactions with OH, or NO_3, radicals, or by photolysis. Alkyl radicals will react rapidly with atmospheric O_2 to form peroxy radicals, RO_2:

$$R + O_2 + M \qquad \rightarrow \qquad RO_2 + M \qquad\qquad (1)$$

where M is a third body, N_2 and O_2, in the atmosphere. The fate of RO_2 radicals in the atmosphere is determined by reaction with NO, NO_2, HO_2, or other RO_2 radicals:

$$RO_2 + NO \qquad \rightarrow \qquad products \qquad\qquad (2)$$
$$RO_2 + NO_2 + M \qquad \rightarrow \qquad RO_2NO_2 + M \qquad\qquad (3)$$
$$RO_2 + HO_2 \qquad \rightarrow \qquad products \qquad\qquad (4)$$
$$RO_2 + R'O_2 \qquad \rightarrow \qquad products \qquad\qquad (5)$$

The relative importance of reactions (2–5) depends on the NO, NO_2, HO_2 and RO_2 concentrations and the values of the respective rate constants. In remote areas NO_x concentrations as low as 1 ppt have been measured, decreasing the importance of reactions (2) and (3). However, in urban areas with higher NO_x concentrations reaction (2) and (3) will be the major sink for RO_2 radicals.

The reaction of RO_2 radicals with NO proceeds through two channels:

$$RO_2 + NO \qquad \rightarrow \qquad RO + NO_2 \qquad\qquad (2a)$$
$$RO_2 + NO + M \qquad \rightarrow \qquad RONO_2 + M \qquad\qquad (2b)$$

In this work we have used the pulse radiolysis technique to investigate reaction 2 for a number of alkyl peroxy and halogenated alkyl peroxy radicals: CH_3O_2, $C_2H_5O_2$, $(CH_3)_3CCH_2O_2$, $(CH_3)_3CC(CH_3)_2CH_2O_2$, CH_2FO_2, CH_2ClO_2, CH_2BrO_2, CHF_2O_2, CF_2ClO_2, $CHF_2CF_2O_2$, $CF_3CF_2O_2$, $CFCl_2CH_2O_2$, and $CF_2ClCH_2O_2$.

Experimental

The experimental set-up used, pulse radiolysis combined with kinetic spectroscopy in the visible wavelength region, has been described in detail previously [1] and will only be discussed briefly here. The experiments were carried out in a 1 litre stainless steel gas cell equipped with an internal White type mirror system adjusted to give an optical path-length of 120 cm. A pulsed xenon lamp delivers the analytical light, which is guided through the reaction cell to a detection system consisting of a 1 m McPherson grating UV-VIS monochromator and a Hamamatsu photomultiplier. The instrumental spectral resolution was 0.8 nm. A Biomation digitiser recorded the transmittance as function of time, and a PDP11 minicomputer handled and stored the data. A chromel/alumel thermocouple measured the temperature inside the reaction cell close to the centre. All experiments were carried out at 298 ± 2 K. The partial pressures of the different gases were measured with a Baratron absolute membrane manometer with a resolution down to 10^{-5} bar.

Table 1: Rate constants for the reaction of OH radicals and Cl atoms with RNO$_x$ at 298 K.

Compound	$10^{13} k_{Cl}$(rel)	$10^{13} k_{OH}$(rel)	$10^{13} k_{OH}$(abs)	Reference
CD$_3$NO$_2$	0.02 ± 0.0	0.9 ± 0.1	1.0 ± 0.2	This work
			0.09 ± 0.004[a]	[1]
			0.04[b]	[6]
CH$_3$NO$_2$	0.06 ± 0.02	1.1 ± 0.1	1.6 ± 0.5	This work
			0.14 ± 0.01[a]	[1]
			0.11[b]	[6]
C$_2$H$_5$NO$_2$	1.7 ± 0.1	1.5 ± 0.1	1.5 ± 0.5	This work
			0.72 ± 0.08[a]	[1]
n-C$_3$H$_7$NO$_2$	99 ± 6	5.5 ± 0.8	3.4 ± 0.8	This work
			3.6 ± 0.2[a]	[1]
n-C$_4$H$_9$NO$_2$	619 ± 20	17.3 ± 1.1	15.5 ± 0.9	This work
			6.6 ± 0.4[a]	[1]
			13.5 ± 1.8	[7]
n-C$_5$H$_{11}$NO$_2$	1377 ± 58	32.7 ± 1.5	33.3 ± 0.5	This work
			10.0 ± 0.6[a]	[1]
CH$_3$ONO$_2$	2.62 ± 0.02	3.4 ± 0.7	3.2 ± 0.5	This work
		3.7 ± 0.9		[8]
		0.34 ± 0.04[c]		[9]
C$_2$H$_5$ONO$_2$	42.7 ± 1.6	4.6 ± 0.3	5.3 ± 0.6	This work
		4.8 ± 2.0		[8]
n-C$_3$H$_7$ONO$_2$	247 ± 15	7.7 ± 0.8	8.2 ± 0.8	This work
		7.0 ± 2.2		[8]
		6.2 ± 1.0		[7]
n-C$_4$H$_9$ONO$_2$	924 ± 2	16.2 ± 0.8	17.4 ± 1.9	This work
		14.0 ± 1.9		[10]
		17.8 ± 1.9		[7]
n-C$_5$H$_{11}$ONO$_2$	1571 ± 33	29.6 ± 0.9	33.2 ± 3.0	This work
CH$_3$ONO	94.4 ± 7.4	3.0 ± 1.0	2.6 ± 0.5	This work
		14.1 ± 1.9[d]		[11]
		10.9 ± 1.7[d]		[12]
		2.1 ± 0.4[f]		[13]
		1.2 ± 0.3[g]		[13]
		10.0 ± 1.5[e]		[14]
C$_2$H$_5$ONO	295 ± 13	7.0 ± 1.5	7.0 ± 1.1	This work
		17.7 ± 2.8[d]		[12]
n-C$_3$H$_7$ONO	646 ± 58	11.0 ± 1.5	12.0 ± 0.5	This work
		24.0 ± 4.5[d]		[12]
		23.1 ± 3.4[e]		[14]
n-C$_4$H$_9$ONO	1370 ± 58	22.7 ± 0.8	27.2 ± 6.0	This work
		52.0 ± 18[d]		[12]
		48.0 ± 7.2[e]		[14]
n-C$_5$H$_{11}$ONO	2464 ± 444	37.4 ± 5.0	42.5 ± 8.0	This work

[a] FP-RF at 25–50 Torr;　[b] LP-LIF at 100 Torr;　[c] total pressure of 2–3 Torr;
[d] at 100 Torr,　[e] at 1–2 Torr;　[f] relative to n-hexane;　[g] relative to dimethyl ether.

The radical reactions were initiated by irradiation of the gas mixtures in the cell with a 30 ns pulse of 2 MeV electrons from a Febetron 705B field emission accelerator. The irradiation dose was varied by inserting stainless steel attenuators between the accelerator and the cell. The doses are given relative to the maximum dose, which is set to unity. SF_6 was used as diluent gas and the experiments were carried out at a total pressure of 1 bar. The pulsed irradiation is used to rapidly (<1 ms) produce a high concentration ($10^{14} - 3 \times 10^{15}$ cm^{-3}) of radical species $e.g.$ OH and F atoms.

For the RO_2 + NO investigation the first chemical step is reaction with fluorine atoms which produce the alkyl radical, R, by hydrogen abstraction from the compounds in question, RH,

$SF_6 + e^-$ (2 MeV)	$\rightarrow$	F + products	(6)
F + RH	$\rightarrow$	HF + R	(7)

The alkyl radical, R, reacts with O_2 to form RO_2, which subsequently reacts with NO:

$R + O_2 + M$	$\rightarrow$	$RO_2 + M$	(1)
$RO_2 + NO$	$\rightarrow$	products	(2a)

The formation of NO_2 was followed by its absorption of light at both 400 and 450 nm in order to assure that there was no interference from spectral features from other transient species in the system. All kinetic traces could be fitted using simple first-order formation kinetics.

Results

The observed formation of NO_2 was always first order. No significant difference between experiments performed at 400 nm and 450 nm was detected. Hence the data from 400 nm and 450 nm have been treated together.

Before a value for k_2 can be extracted from the observed kinetics of the NO_2 formation the impact of potential secondary reactions needs to be considered. Peroxy radicals, RO_2, are known to react with NO_2 to form peroxy nitrates [2, 3]. Hence, following the formation of NO_2 from reaction (2), there is a competition between reactions (2) and (3) for the available RO_2 radicals:

$RO_2 + NO$	$\rightarrow$	products	(2)
$RO_2 + NO_2 + M$	$\rightarrow$	$RO_2NO_2 + M$	(3)

Similarly, alkoxy radicals produced in reaction (2) also react with both NO and NO_2:

$RO + NO + M$	$\rightarrow$	$RONO + M$	(8)
$RO + NO_2 + M$	$\rightarrow$	$RONO_2$	(9)

Reactions (3) and (9) remove NO_2 from the system with an efficiency: which increases at long reaction times. The removal decreases the time taken for the NO_2 concentration to reach a maximum. Hence, reaction (3) and (9) lead to an increase in the apparent pseudo first order rate constant for NO_2 formation. This effect is most pronounced under low $[NO]_0$ concentrations leading to positive intercepts in plots of k^{1st} versus $[NO]$. Positive intercepts were observed for all the alkyl peroxy radicals studied here. Interestingly, similar plots for the halogenated peroxy radicals have intercepts which are zero, within the experimental uncertainty. To assess the impact of reactions (3), (8) and (9) and, hence, to compute corrections for such, detailed modelling of the experimental data was performed using the Acuchem chemical kinetic modelling program [4] with a mechanism consisting of reaction (2), (3), (8), and (9) (see publication for details).

Table 2: Summary of rate constants for RO_2 + NO products reactions.

Species	Technique	$10^{12} k_2$	Reference
CH_3O_2	FP-UV	7.1 ± 1.4	[15]
	DF-MS	6.5 ± 2.0	[16]
	FP-UV	7.7 ± 0.9	[17]
	DF-MS	8.6 ± 2.0	[18]
	LP-LIF	7.8 ± 1.2	[19]
	LP-LA	7.0 ± 1.5	[20]
	PR-UV	8.8 ± 1.4	This work
$C_2H_5O_2$	DF-MS	8.9 ± 3.0	[21]
	PR-UV	8.5 ± 1.2	This work
$i\text{-}C_3H_7O_2$	FP-UV	3.5 ± 0.2	[22]
	DF-MS	5.0 ± 1.2	[23]
$t\text{-}C_4H_9O_2$	DF-MS	4.0 ± 1.1	[14]
$(CH_3)_3CCH_2O_2$	PR-UV	4.7 ± 0.4	This work
$(CH_3)_3C(CH_3)_2CH_2O_2$	PR-UV	1.8 ± 0.2	This work
CF_3CHFO_2	PR-UV	12.8 ± 3.6	[24]
$CHF_2CF_2O_2$	PR-UV	$>9.7 \pm 1.3$	This work
$CF_3CF_2O_2$	PR-UV	$>10.7 \pm 1.5$	This work
$CFCl_2CH_2O_2$	PR-UV	12.8 ± 1.1	This work
$CF_2ClCH_2O_2$	PR-UV	11.8 ± 1.0	This work
CH_2FO_2	PR-UV	12.5 ± 1.3	This work
CH_2ClO_2	PR-UV	18.7 ± 2.0	This work
CH_2BrO_2	PR-UV	10.7 ± 1.1	This work
CHF_2O_2	PR-UV	12.6 ± 1.6	This work
CF_3O_2	DF-MS	17.8 ± 3.6	[25]
	LP-MS	14.5 ± 2.0	[26]
	PR-UV	16.9 ± 2.6	[27]
CF_2ClO_2	LP-MS	16.0 ± 3	[17]
	PR-UV	13.1 ± 1.2	This work
$CFCl_2O_2$	LP-MS	16.0 ± 2.0	[28]
	LP-MS	14.5 ± 2.0	[17]
CCl_3O_2	DF-MS	18.6 ± 2.8	[29]
	LP-MS	17.0 ± 2.0	[17]

In view of the corrections applied to the data for CH_3O_2, $C_2H_5O_2$, $(CH_3)_3CCH_2O_2$, and $(CH_3)_3CC(CH_3)_2CH_2O_2$ radicals, we choose to add an additional 10 % and 20 % uncertainty range to the measured rate constants of the reactions of CH_3O_2 and $C_2H_5O_2$ radicals, and $(CH_3)_3CCH_2O_2$ and $(CH_3)_3CC(CH_3)_2CH_2O_2$ radicals with NO, respectively. Values cited in Table 2 reflect these additional uncertainties. The absolute accuracy of the rate constants depends on the precision of reactant concentrations and possible unidentified systematic errors. The overall accuracy of the measured rate constants is estimated to be of the order ±25 %. In the case of experiments including neopentyl peroxy radicals two different total pressures were used; 1 and 0.5 bar. No effect of total pressure on k_2 was evident.

Discussion

In Table 2 we compare our results to those available in the literature. Also all data has been presented graphically in Fig. 1. For the reactions of $CF_3CF_2O_2$ and $CHF_2CF_2O_2$ radicals with NO, the alkoxy radicals formed decompose to give another fluorinated alkyl radical which can then react with NO to give more NO_2. The presence of additional processes forming NO_2 subsequent to reaction (2) increases the time taken for the NO_2 to reach a maximum. Hence, only lower limits can be established for the reaction of $CF_3CF_2O_2$ and $CHF_2CF_2O_2$ radicals with NO. In all cases the halogenated C_2 peroxy radicals are significantly more reactive towards NO than ethyl peroxy radicals.

As shown in Fig. 1, the rate constants for RO_2 + NO reactions vary by one order of magnitude. Comparison of all the available data demonstrates two different trends. Firstly, the rate constants decrease with increasing alkyl chain length and branching going from C_1 to C_8. Secondly, halogen substituents, especially chlorine, increase the RO_2 + NO rate constant.

It is possible to rationalise the first trend in terms of steric effects associated with the increasing alkyl chain length and branching. The same trend is observed in the peroxy radical self-reaction kinetics. The second trend is more difficult to explain but may reflect a decrease in the RO–O bond strength caused by the electron withdrawing halogen atom. *Ab initio* theoretical studies are needed to shed further light on the reactivity trends.

Finally we need to consider the implication of the results from the present work on our understanding of atmospheric chemistry of halogenated organic compounds. In all cases the peroxy radicals derived from haloalkanes were observed to react rapidly with NO with rate constants in the range of $(9.7–18.7) \times 10^{-12}$ cm^3 molecule^{-1} s^{-1}. A reasonable estimate for the global average tropospheric NO concentration is 2.5×10^8 cm^{-3} [5]. Hence the lifetimes of halogenated peroxy radicals with respect to reaction with NO will be 4–7 minutes. In the atmosphere reaction (2) competes with reactions (3), (4) and (5) for available RO_2 radicals. Kinetic data for reactions (4) and (5) are not available for the halogenated RO_2 radicals studied here. Kinetic data are available for the reaction of CF_2ClO_2, $CFCl_2O_2$, CCl_3O_2 and CF_3O_2 with NO_2.

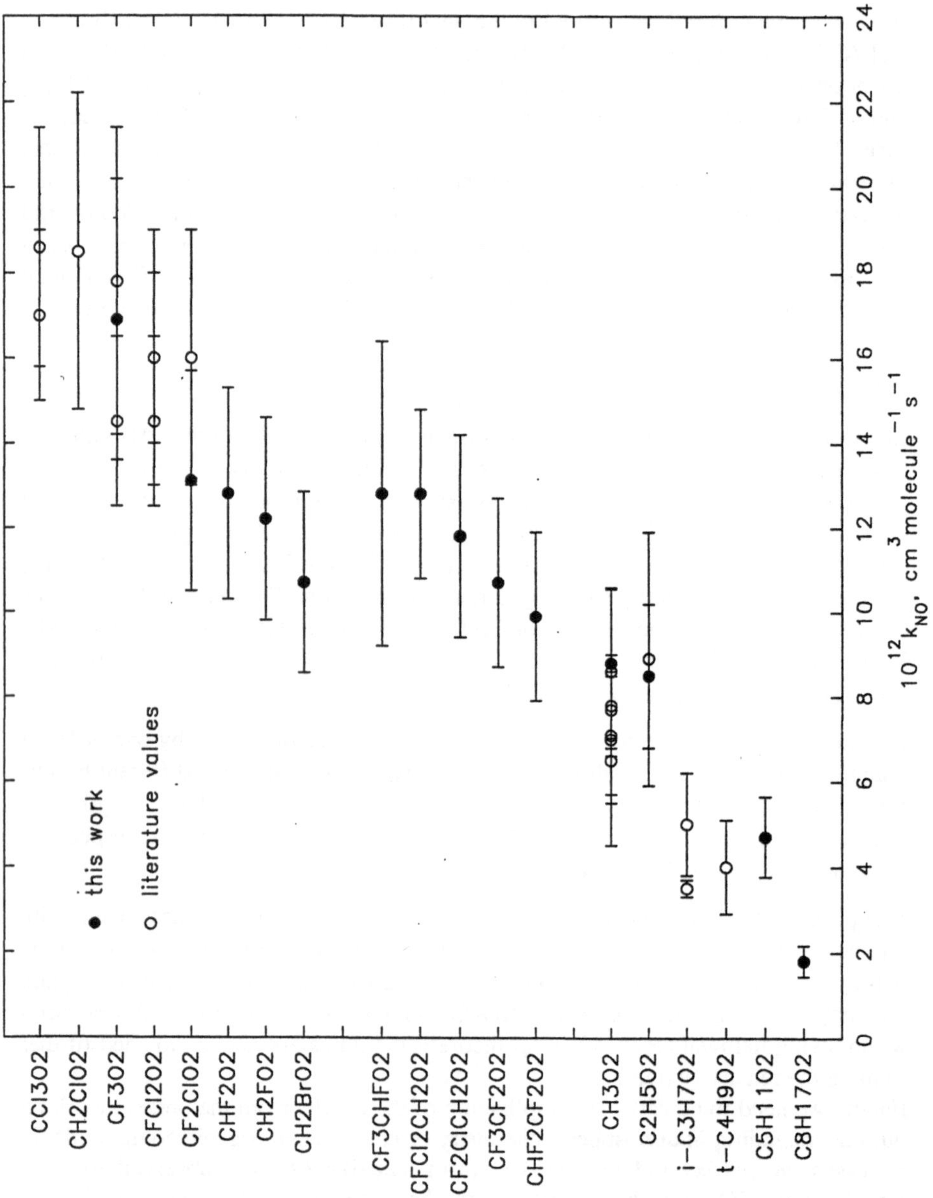

Fig 1: Rate constants for the reactions of NO with RO_2.

At one atmosphere total pressure these reactions are at or close to the high pressure limit with rate constants of 6.0×10^{-12} cm^3 molecule^{-1} s^{-1}. Reasonable estimates of the global average tropospheric NO_2, HO_2, and RO_2 (represented by CH_3O_2) are 2.5×10^8, 10^9, and 1.3×10^8 cm^{-3} [5]. Based upon the data base for other peroxy radicals we estimate that the halogenated peroxy radicals studies have $k_3 = 6 \times 10^{-12}$, $k_4 = 6 \times 10^{-12}$, and $k_5 = 4 \times 10^{-12}$ cm^3 molecule^{-1} s^{-1}. The atmospheric lifetimes of the halogenated RO_2 radicals with respect to reactions (3), (4) and (5) are likely to be of the order of 11, 3, and 32 minutes, respectively.

Clearly reactions (2), (3), (4) and (5) could all play significant roles in the atmospheric degradation of halogenated organic compounds. Kinetic data for reactions (3), (4) and (5) are needed for a more complete understanding of the atmospheric chemistry for halogenated organic compounds.

Acknowledgements

Thanks are due to all our collaborators during this 7 year project.

References

1. O.J. Nielsen, *Risø Report* 480 (1984).
2. T.J. Wallington, P. Dagaut, M.J. Kurylo, *Chem. Rev.* **92** (1992) 667.
3. P.D. Lightfoot, R.A. Cox, J.N. Crowley, M. Destriau, G.D. Hayman, M.E. Jenkin, G.K. Moortgat, F. Zabel, *Atmos. Environ.* **26A** (1992) 1805.
4. W. Braun, J.T. Herron, D.K. Kahaner, *Int. J. Chem. Kinet.* **20** (1988) 51.
5. R. Atkinson, Alternative fluorocarbon environmental acceptability study, in: *WMO Global Ozone Research and Monitoring Project, Report* No. 20, *Scientific assessment of stratospheric ozone*, Vol. 2, WMO, Geneva 1989, p. 167.
6. S. Zabarnick, J.W. Flemming, M.C. Lin, *Chem. Phys.* **120** (1988) 319.
7. R. Atkinson, S.M. Aschmann, *Int. J. Chem. Kinet.* **21** (1989) 1123.
8. J.A. Kerr, D.W. Stocker, *J. Atmos. Chem.* **4** (1986) 253.
9. J.S. Gaffney, R. Fagre, G.I. Senum, J.H. Lee, *Int. J. Chem. Kinet.* **18** (1986) 399.
10. R. Atkinson, S.M. Aschmann, W.P.L. Carter, A.M. Winer, *Int. J. Chem. Kinet.* **14** (1982) 919.
11. I.M. Campbell, K. Goodman, *Chem. Phys. Lett.* **36** (1975) 382.
12. G.J. Audley, D.L. Baulch, I.M. Campbell, D.J. Waters, G. Watling, *J. Chem. Soc. Faraday Trans. 1* **78** (1982) 611.
13. E.C. Tuazon, W.P.L. Carter, R. Atkinson, J.N. Pitts, Jr., *Int. J. Chem. Kinet.* **15** (1983) 619.
14. D.L. Baulch, I.M. Campbell, S.M. Saunders, *Int. J. Chem. Kinet.* **17** (1985) 355.
15. S.P. Sander, R.T. Watson, *J. Phys. Chem.* **85** (1980) 2960.
16. R.A. Cox, G.S. Tyndall, *J. Chem. Soc. Faraday Trans. 2* **76** (1980) 153.
17. R. Simonaitis, J. Heicklen, *J. Phys. Chem.* **95** (1981) 2946.
18. I.C. Plumb, K.R. Ryan, J.R. Steven, M.F.R. Mulcahy, *J. Phys. Chem.* **85** (1981) 3136.
19. A.R. Ravishankara, F.L. Eisele, N.M. Kreutter, P.H. Wine, *J. Chem. Phys.* **74** (1981) 2267.

20. R. Zellner, B. Fritz, K. Lorentz, *J. Atmos. Chem.* **4** (1986) 241.

21. I.C. Plumb, K.R. Ryan, J.R. Steven, M.F.R. Mulcahy, *Int. J. Chem. Kinet.* **14** (1982) 183.

22. H. Adachi, N. Basco, *Int. J. Chem. Kinet.* **14** (1982) 1243.

23. J. Peeters, J. Vertommen, I. Langhans, *Ber. Bunsenges. Phys. Chem.* **96** (1992) 431.

24. T.J. Wallington, O.J. Nielsen, *Chem. Phys. Lett.* **187** (1991) 33.

25. I.C. Ryan, K.R. Ryan, *Chem. Phys. Lett.* **92** (1982) 236.

26. A.M. Dognon, F. Caralp, R. Lesclaux, *J. Chim. Phys.* **82** (1985) 349.

27. J. Sehested, O.J. Nielsen, *Chem. Phys. Lett.* **206** (1993) 369.

28. R. Lesclaux, F. Caralp, *Int. J. Chem. Kinet.* **16** (1984) 1117.

29. K.R. Ryan, I.C. Plumb, *Int. J. Chem. Kinet.* **16** (1984) 591.

3.15 Kinetic Studies of Reactions of Alkylperoxy and Haloalkylperoxy Radicals with NO. A Structure-Activity Relationship for Reactions of OH with Alkenes and Polyalkenes

Jozef Peeters, J. Vertommen, I. Langhans, W. Boullart, J. Van Hoeymissen and V. Pultau

Department of Chemistry, University of Leuven, Celestijnenlaan 200F, B-3001 Leuven, Belgium

Summary

The kinetics of the reactions of higher alkylperoxy radicals and of H(C)FC-derived haloalkylperoxy radicals with NO have been studied at T = 290 K and p = 2 Torr He using the fast-flow technique combined with molecular-beam sampling mass spectrometry. The total rate constant of RO_2 + NO $\rightarrow$ RO + NO_2 / $RONO_2$ was determined from the shape of the NO_2-growth profile. The method was validated by measurements of $k(CF_3O_2 + NO)$ = (1.51 ± 0.4) 10^{-11} cm^3 s^{-1} and of $k(C_2H_5O_2 + NO)$ = (8.9 ± 3.5) 10^{-12} cm^3 s^{-1}. Thus, the first direct, validated measurements were made of $k(RO_2 + NO)$ for higher alkylperoxy radicals: $k(i\text{-}C_3H_7O_2 + NO)$ = (5.0 ± 1.2) 10^{-12}, $k(s\text{-}C_4H_9O_2 + NO)$ = (4.1 ± 1.0) 10^{-12} and $k(t\text{-}C_4H_9O_2 + NO)$ = (4.0 ± 1.1) 10^{-12} cm^3 s^{-1}, showing a marked decrease of the rate coefficient with increasing CH_3-substitution, contrary to earlier assumptions. The $k(RO_2 + NO)$ rate constants for the peroxy radicals from H(C)FC's 134a, 143a, 142b and 123 were all found to be in the range 1.2 to 1.6 × 10^{-11} cm^3 s^{-1}.

In a second part of the work, a Structure-Activity Relationship has been developed for the addition of OH to (poly)alkenes. It was shown that the total rate constants can be expressed in very good approximation by a sum of partial, site-specific rate constants for addition to a given (double-bonded) carbon atom, the values of these partial rate constants depending solely on the stability-type of the ensuing β-hydroxy radical. The SAR is particularly useful in that it also allows the prediction of the detailed primary product distributions of (poly)alkene + OH reactions. Therefore, the SAR is a powerful tool in the modeling of the tropospheric OH-initiated oxidation of biogenic VOC.

Aims of the project

The major aim of the research during the project was to gather kinetic and mechanistic information on reactions of VOC-derived peroxy radicals, in particular reactions with NO, in order to provide data relevant to the formation of tropospheric ozone. A secondary objective was to determine $k(RO_2 + NO)$

coefficients for H(C)FC-derived peroxy radicals and to gain information on the subsequent fate of the resulting oxy-radicals.

Finally, it was also aimed to obtain quantitative information on the primary-product distributions of reactions of OH with alkenes and polyalkenes.

Principal scientific findings during the project

A. Reactions of alkylperoxy- and haloalkylperoxy radicals with NO

Experimental and data analysis

The apparatus is a multi-stage fast-flow reactor, coupled to a molecular-beam sampling mass spectrometer (MBMS). The reactor consists of a 2.8 cm inner diameter quartz tube, with a side-arm equipped with a microwave cavity, and a set of two (or three) coaxial, movable injector tubes. To generate the specific alkyl radicals i-C_3H_7, s-C_4H_9 and t-C_4H_9, hydrogen atoms created upstream in the microwave discharge side-arm were admixed to a large excess of an appropriate alkene admitted through the outer injector tubes; for the above alkyl radicals, the alkenes were respectively propene, 2-butene and isobutene, ensuring yields of the specific radical of 94 % [1], 99 % [2] and 100 % respectively. The radicals C_2H_5, CF_3 and the H(C)FC-derived haloalkyl radicals were produced by H-abstraction from C_2H_6, CF_3H or the pertaining H(C)FC's (in large excess) by F-atoms, created from F_2 in the upstream microwave discharge. The concentration of the (halo)alkyl radicals R, produced in the constant-length "preparation region" of the flow reactor, was between 2 and 6×10^{12} cm^{-3}. A high amount of O_2 ($\sim 5 \times 10^{15}$ cm^{-3}) together with a variable amount of NO (1 to 10×10^{13} cm^{-3}) was admixed to the (halo)alkyl radicals via the inner injector tube, resulting in their very fast conversion to RO_2 radicals, followed by the slow reaction thereof with NO (in at least a five-fold excess), yielding NO_2 and alkyl nitrates. The $i(NO_2^+)$-versus-time growth profiles were recorded by moving the two coaxial injectors together, away from the sampling point. The NO_2^+ signals were monitored at an ionizing electron energy of 14 or 40 eV (depending on the presence or absence of $m/e = 46$ fragment ions from the R-precursor molecule). The NO_2-growth method is prefered over the RO_2-decay method since (for most of the cases at hand) there is no RO_2^+-parent ion nor any interference-proof fragment ion. The experiments were carried out at total pressures of 2 and 0.7 torr (He bath gas) and at 290 K. Flow velocities were about 1800 cm s^{-1}.

The determination of the *total* rate constant k of the process $RO_2 + NO \rightarrow RO + NO_2$/nitrate from the *shape* of $i(NO_2^+)$ is essentially based on the first-order law $[NO_2](t) = [NO_2]_\infty [1 - \exp(-k [NO] t)]$. Corrections were made for the (slight) [NO]-decrease and for axial diffusion effects. Moreover, side-reactions such as $RO_2 + NO_2 \rightarrow$ pernitrate, $RO + NO \rightarrow$ nitrite, $RO + NO_2 \rightarrow$ nitrate, decomposition of RO and subsequent reactions, as well as contributions of nitrates and pernitrates to the total NO_2^+ signal, were duly taken into account by kinetic modeling (resulting in corrections of 5 to 30 %). The k-values were obtained from plots of the corrected pseudo-first-order growth constants versus the [NO]-concentration.

Results

The NO_2-growth method was validated by measurements of the rate constants of CF_3O_2 + NO and of $C_2H_5O_2$ + NO, for which we obtained respectively $k = (1.51 \pm 0.4) \times 10^{-11}$ and $k = (8.9 \pm 3.5) \times 10^{-12}$ cm^3 s^{-1}, in good agreement with the literature values of these known reactions [3, 4, 6, 7, 12]. The validity of the method is further supported by our own measurement $k(CF_3O_2$ + NO) $= 1.54 \times 10^{-11}$ cm^3 s^{-1} from the CF_3O_2 decay as monitored by means of the specific $CF_2O_2^+$ fragment ion.

Table 1 displays our results for alkylperoxy radicals together with literature values, when available.

Table 1: Rate constants of reactions between alkylperoxy radicals and NO; T = 290 K. Data of this work obtained at p(He) = 2 Torr for C_2H_5, i-C_3H_7 and s-C_4H_9, and at both 0.7 and 2 Torr for t-C_4H_9.

RO_2	$k(RO_2 + NO) / 10^{-12}$ cm^3 s^{-1}			
	this work		literature data	
$C_2H_5O_2$	8.9 ± 3.5			
			8.9 ± 3.0	[6]
			7.2 ± 1.3	[7]
			8.5 ± 1.2	[12]
			2.7 ± 0.2	[5]
i-$C_3H_7O_2$	5.0 ± 1.2	[8, 9]		
			3.5 ± 0.2	[10]
s-$C_4H_9O_2$	4.1 ± 1.0	[9]		
t-$C_4H_9O_2$	4.0 ± 1.1	[8, 9]		

Our data, taken in combination with the well-known $k(CH_3O_2$ + NO) $= (7.8 \pm 2) \times 10^{-12}$ and $k(HO_2$ + NO) = $(8.5 \pm 2.0) \times 10^{-12}$ cm^3 s^{-1} [11] show a marked decrease of the kinetic coefficient for RO_2 + NO reactions with increasing CH_3-substitution on the α-carbon atom. This pronounced trend is contrary to earlier assumptions and is observed for the first time. Adachi and Basco had already measured a low value for i-$C_3H_7O_2$ [10] , of 3.5×10^{-12}, but their results for CH_3O_2 and $C_2H_5O_2$, using a similar method, are also only 3.0 and 2.7×10^{-12}. In later work, Nielsen et al. [12] confirmed the downward trend of $k(RO_2$ + NO) with increasing size of the alkyl radical: for $(CH_3)_3CCH_2O_2$ they obtained $k = (4.7 \pm 0.4) \times 10^{-12}$ and for $(CH_3)_3CC(CH_3)_2CH_2O_2$: $k = (1.8 \pm 0.2) \times 10^{-12}$, while confirming also $k = (8.5 \pm 1.2) \times 10^{-12}$ for $C_2H_5O_2$.

The following table lists our results for CF_3O_2 and for several H(C)FC-derived haloethylperoxy radicals.

Table 2: Rate constants of reactions of CF_3O_2 and of haloethylperoxy radicals with NO; T = 290 K (our results at a helium pressure p = 2 Torr).

RO_2	$k(RO_2 + NO) / 10^{-12}$ cm^3 s^{-1}			
	this work		literature data	
CF_3O_2	15.1 ± 4.0	[8]		
	15.3 ± 3.0^a	[8]		
			15.1 ± 2.0	[3]
			17.8 ± 3.6	[4]
			16.9 ± 2.6	[14]
CF_3CHFO_2	15.6 ± 4.1	[13]		
(from HFC 134a)			12.8 ± 3.6	[15]
			13.0 ± 3.0	[16]
$CF_3CH_2O_2$	12.3 ± 3.2	[13]		
(from HFC 143a)			13.0 ± 3.0	[16]
$CF_2ClCH_2O_2$	11.8 ± 3.0	[13]		
(from HCFC 142b)			11.8 ± 1.0	[17]
$CF_3CCl_2O_2$	14.5 ± 3.9	[13]		
(from HCFC 123)			17.5 ± 2.5	[18]

^{a} from CF_3O_2 decays

Our results agree well with other recent data. For all the investigated haloethyl-peroxy radicals, our $k(RO_2 + NO)$ is in the range of 11 to 16×10^{-12} cm^3 s^{-1}. Our data indicate that the kinetic coefficient is highest for the radicals with one or more halogen atoms on the α-carbon atom; yet, a CX_3 substituent on the α-carbon clearly also promotes the reaction. A combination of all the available data on reactions of peroxy radicals with NO confirms the view that e$^-$ acceptors on the α-carbon enhance the reactivity toward NO, whereas e$^-$ donors (such as alkyl groups) lower the reactivity [8].

B.   Product distributions of reactions of OH with (poly)alkenes.

A structure-activity relationship for OH-addition

Existing SARs for OH + polyalkene reactions merely attribute a reactivity to a specific complete C=C structure or to a complete C=C–C=C structure for conjugated dienes. These SARs provide no information with relevance to the detailed product distribution of the various possible OH-adducts.

In the frame of the present project, a detailed analysis of known total rate constants k_{OH} for OH + alkene reactions was carried out, revealing that for monoalkenes and non-conjugated polyalkenes, the total rate constant can be expressed as a sum of independent, partial rate constants k_i for addition to a given double-bonded C-atom. These site-specific k_i depend only on the stability type of the ensuing hydroxyalkyl radical, which can be either primary, secondary or tertiary. Thus, there are only three such k_i, denoted here as k_{prim}, k_{sec} and k_{tert} [19]. It should be understood that for addition of OH to Carbon atom C_a of a $C_a=C_b$ structure, it is the environment of Carbon atom C_b that determines the adduct stability type, and therefore the k_i. The values of the three k_i were determined from the total rate constants k_{OH} for the three symmetric alkenes ethene, 2-butene, and tetramethylethene:

$$k_{prim} = 1/2 \, k_{OH}^{\infty} \text{ (ethene)} = 0.45 \times 10^{-11}$$
$$k_{sec} = 1/2 \, k_{OH} \text{ (2-butene)} = 3.0 \times 10^{-11}$$
$$k_{tert} = 1/2 \, k_{OH} \text{ (tetramethylethene)} = 5.5 \times 10^{-11} \text{ cm}^3 \text{ s}^{-1}$$

Using the above k_i, the additivity hypothesis $k_{OH} = \Sigma \, k_i$ (the sum being made over all double-bonded C-atoms) was verified for the 28 alkenes and nonconjugated dialkenes for which k_{OH} data are available. All predicted k_{OH} are within 20 % of the experimental values, and the standard deviation is only 12.6 % (excepting α-pinene, for which the deviation is 30 %, probably due to steric hindrance caused by the bridged 6-cycle) [19].

For conjugated dienes, one must take into account the resonance stabilisation for some of the possible adduct radicals. This is the case for OH addition to one of the terminal C-atoms of a $C=C-C=C$ structure, because of the resulting allyl-like resonance between the "original" β-hydroxy radical and the resonant δ-hydroxy structure. As an example, when the former is a secondary radical and the latter a tertiary one, the corresponding partial site-specific rate constant is denoted as $k_{sec/tert}$. Generally, for conjugated dienes there are six such additional partial rate constants (beside k_{prim}, k_{sec} and k_{tert}). They were evaluated from the known total k_{OH} for six conjugated dienes; they are listed below, in units of 10^{-11} cm^3 s^{-1}.

$$k_{sec/prim} = 3.0 \qquad k_{sec/sec} = 3.75 \qquad k_{sec/tert} = 5.05$$
$$k_{tert/prim} = 5.65 \qquad k_{tert/sec} = 8.35 \qquad k_{tert/tert} = 9.85$$

One clearly sees the increase of the partial rate constant with increasing stability of the resonant structure. Again, the predicted total rate constants for other conjugated dienes, $k_{OH} = \Sigma \, k_i$, agree very well with the experimental values, with an average deviation of only 10 % (excepting for two cases, with exceptionally large measured k_{OH}, α-phallandrene and α-terpinene) [19].

The present SAR is especially valuable for the prediction of the detailed product distribution of the primary OH-adducts. Since the total rate constant is a sum of independent partial rate constants for addition to specific sites, the yield of a given primary adduct arising from process i should indeed be: yield$_i$ = $k_i \, / \, \Sigma \, k_i$. As an example, application to isoprene $H_2C_a=C_b(Me)-C_cH=C_dH_2$ results in the following prediction: addition to C_a: yield 60 %; addition to both C_b and C_c: yield 5 % each; addition to C_d: yield 30 %.

Measurement of the OH-adduct product distribution

In order to verify and validate the above SAR with respect to its product-distribution implications, measurements have been carried out aiming to quantify the yields of the various possible primary products of the reactions of OH with various alkenes and dienes: propene; 1-butene; isobutene; 1,3-butadiene and isoprene. In discharge/flow-MBMS experiments, a small amount of OH (generated upstream by H + NO$_2$) was reacted with a large excess of the (poly)alkene, and the resulting primary radicals were converted subsequently to stable bromide molecules by reaction with an excess Br$_2$. Quantitative analysis of the various R$_i$Br was undertaken by MBMS, on the basis of the fragment ion mass spectra. At present, the final quantitative results are not yet available; in order to determine the fragmentation spectra of the pure reference compounds, these bromoalcohols have to be synthesised, a process which is now in progress. Only an approximate result for 1,3-butadiene can be given now. The preliminary value for the yield ratio CH$_2$=CH–CH–CH$_2$OH / CH$_2$=CH–CHOH–CH$_2$ is 4.5 with error margins of +2.5 and −1.5, whereas the above SAR predicts a ratio of 5. Generally, as far as can be assessed from estimated fragmentation spectra, the experimental resutls are broadly in line with the SAR predictions.

Assessment in the light of the aims of LACTOZ

Our new finding that the kinetic coefficient of alkylperoxy + NO reactions decreases with increasing size (or CH$_3$-substitution) of the alkyl radical is of considerable importance with respect to tropospheric ozone formation from higher hydrocarbons, the more so since the rate constants of the competing alkylperoxy + HO$_2$ reactions [20] increase significantly for larger R.

Our k(RO$_2$+NO) data, in combination with these provided by other LACTOZ groups are now serving as basis for Structure-Activity Relationships regarding peroxy radical + NO reactions [21]. Moreover, our result for s-C$_4$H$_9$O$_2$ [9] is currently incorporated in tropospheric chemistry models (where butane is a model NMHC alkane).

The major value of the new Structure-Activity Relationship presented and developed by us [19] for (poly)alkene + OH reactions resides in its inherent predictive potential regarding the detailed primary-product distributions of such reactions. This is especially useful for quantitative assessments of the various possible OH-initiated degradation pathways of biogenic VOC, of which there is such a diversity and multitude that one can realistically hope to perform detailed experimental studies on only a few of them.

Acknowledgments

This research was carried out in the frame of the Belgian participation in the EUROTRAC programme, supported by the Belgian State – Prime Minister's Services – Science Policy Office, contract E7/01/01P. Support by the CEC is also acknowledged.

References

1. M.J. Lexton, R.M. Marshall, J.H. Purnell, *Proc. R. Soc. London* A**324** (1971) 433.
2. M.J. Lexton, R.M. Marshall, J.H. Purnell, *ibid.* p. 440.
3. A.M. Dognon, R. Caralp, R. Lesclaux, *J. Chim. Phys.* **82** (1985) 349.
4. I.C. Plumb, K.R. Ryan, *Chem. Phys. Lett.* **92** (1982) 236.
5. H. Adachi, N. Basco, *Chem. Phys. Lett.* **67** (1979) 324.
6. I.C. Plumb, K.R. Ryan, J.R. Steven, M.F.R. Mulcahy, *Int. J. Chem. Kinet.* **14** (1982) 183.
7. R. Zellner, in: *EUROTRAC Annual Report* 1989, Part 8, EUROTRAC ISS, Garmisch-Partenkirchen 1990, p. 152.
8. J. Peeters, J. Vertommen, I. Langhans, *Ber. Bunsenges. Phys. Chem.* **96** (1992) 431.
9. J. Peeters, J. Vertommen, I. Langhans, in: P.M. Borrell, P. Borrell, T. Cvitaš, W. Seiler (eds), *Proc. EUROTRAC Symp. '92*, SPB Academic Publishing, The Hague 1993, p. 399.
10. H. Adachi, N. Basco, *Int. J. Chem. Kinet.* **14** (1982) 1243.
11. R. Atkinson, D.L. Baulch, R.A. Cox, R.F. Hampson, J.A. Kerr, J. Troe, *J. Phys. Chem. Ref. Data* **18** (1989) 881.
12. O.J. Nielsen, J. Sehested, T.J. Wallington, in: G. Angeletti, G. Restelli (eds), *Physico-Chemical Behaviour of Atmospheric Pollutants, Air Pollution Research Report* **50**, EC, Luxembourg 1994, p. 175.
13. J. Peeters, V. Pultau, *ibid.*, p. 372.
14. J. Sehested, O.J. Nielsen, *Chem. Phys. Lett.* **206** (1993) 369.
15. T.J. Wallington, O.J. Nielsen, *Chem. Phys. Lett.* **187** (1991) 33.
16. T.J. Bevilacque, D.R. Hanson, M.R. Jensen, C.J. Howard, in: *Proc. Step-Halocside/AFEAS Workshop*, Dublin 1993, p. 163.
17. J. Sehested, O.J. Nielsen, T.J. Wallington, *Chem. Phys. Lett.* **213** (1993) 457.
18. G.D. Hayman, M.E. Jenkin, T.P. Murrells, C.E. Johnson, *Atmos. Environ.* **28** (1994) 421.
19. J. Peeters, W. Boullart, J. Van Hoeymissen, in: P.M. Borrell, P. Borrell, T. Cvitaš, W. Seiler (eds), *Proc. EUROTRAC Symp. '94*, SPB Academic Publishing, The Hague 1994, p. 110.
20. P.D. Lightfoot *et al.*, *Atmos. Environ.* **26** (1992) 1805.
21. M.E. Jenkin, G.D. Hayman, to be published.

3.16 Mechanism and Kinetics of the Reaction of OH with Volatile Organic Compounds

M. J. Pilling, A. E. Heard, S. M. Saunders and P. I. Smurthwaite

School of Chemistry, University of Leeds, Leeds LS2 9JT, UK

Summary

Laser flash photolysis and discharge flow measurements of the reaction rates of OH with several VOC have been made in a project designed to provide data for the determination of the photochemical ozone creation potentials of the VOC. In related studies, the mechanisms of the atmospheric oxidation of a wide range of compounds have been developed. Finally, objective techniques for extensive mechanism reduction, first applied in combustion chemistry, have been applied to tropospheric mechanisms.

Aims of the research

The major aims of the research are the establishment of quantitative mechanisms of atmospheric oxidation of volatile organic compounds.

Principal scientific results

In order to assess the impact of volatile organic compounds (VOC) on ozone formation in the troposphere it is necessary to know:

(i) The rate constant for the reaction between OH and the VOC. This reaction initiates the oxidation chain and determines the time (and distance) scales of ozone formation.
(ii) The mechanism for the oxidation of the VOC. This aspect draws heavily on the work of the LACTOZ subproject.

In this component of the subproject, OH + VOC rate constants have been measured and oxidation mechanisms developed. In addition, work is continuing on the reduction of the mechanisms, so that they can be more feasibly incorporated in atmospheric transport models.

Experimental studies of OH + VOC

We have employed laser flash photolysis coupled with laser-induced fluorescence (LP) and discharge flow (DF) coupled with LIF to study the rates of reaction of OH with a range of VOC. The compounds studied have significant emission rates and data are needed to determine their photochemical ozone creation potentials.

The systems were tested on the reactions with methane (LP) and cyclohexane (DF) showing good agreement with literature data. The principal results are shown in Table 1. The rate constants were measured at pressures of 760 Torr (LP) and 2 Torr (DF) in a helium diluent.

Table 1: Results of rate constant measurements for reaction of OH with VOC at 295K.

VOC	k / cm^3 molecule^{-1} s^{-1}	Method
methane	5.4×10^{-15}	LP
cyclohexane	6.7×10^{-12}	DF
i-butanol	9.0×10^{-12}	LP
t-butanol	8.1×10^{-13}	DF
n-butanol	8.4×10^{-12}	LP
i-amyl-alcohol	1.3×10^{-11}	DF
1,2-butadiene	1.9×10^{-11}	DF
t-amyl methyl ether	5.0×10^{-12}	LP
n-propyl acetate	2.9×10^{-12}	LP
methyl methcrylate	2.6×10^{-11}	DF
cyclohexane	4.3×10^{-12}	LP
vinyl acetate	2.5×10^{-11}	DF
acrylonitrile	3.2×10^{-12}	DF
cyclohexylamine	3.0×10^{-11}	DF
1,2 dichloroethane	2.0×10^{-13}	LP
$trans$-1,2 dichloroethane	1.9×10^{-12}	DF

LP - laser flash photolysis/laser induced fluorescence.
DF - discharge flow/laser induced fluorescence.

Mechanism development

A programme is in progress to develop, in collaboration with ME Jenkin (AEA Technology), detailed mechanisms for the atmospheric oxidation of 100 VOC of primary interest in the troposphere. The mechanisms draw on recent work, especially in the LACTOZ programme, and incorporate the following features in addition to the central OH initiated radical → peroxy radical → alkoxy radical chain:

- Reaction of alkenes with NO_3
- Reaction of carbonyls, alkenes and aromatics with NO_3.
- Reaction of RO_2 with HO_2 and R/O_2.
- Reactions of alkoxy radicals, including isomerisation, decomposition and reaction with O_2.
- Photochemistry of carbonyls and hydroperoxides.
- PAN chemistry.

The mechanisms are fully documented and are also provided in a reduced format suitable for incorporation in a computer code.

Mechanism reduction techniques

The mechanisms generated by the process described above are very complex. For butane, for example, the overall mechanism describing the total oxidation process contains nearly six hundred reactions and over two hundred species. It is difficult to incorporate such mechanisms in atmospheric transport models, especially those based on Eulerian codes. This difficulty is discussed in a recent review of air pollution modelling by Zlatev *et al.* [1] Thus it is desirable to simplify the representation in some way, and since the bulk of the computing time is spent in solving the equations describing the chemical interactions of the components of the air mass, it is common to minimise the number of chemical species. Two of the chemical schemes most widely used for modelling ground-level ozone formation employ a lumped species (LCC - Lurmann *et al.* [2])or a lumped structure (CBM4 - Gery *et al.* [3]) approach. In this way the number VOCs in the scheme is reduced by a factor of more than ten. Typically the chemical reactions of this reduced set of species are then derived by use of laboratory data on individual reactions and by comparison of the resulting scheme with a large number of environmental chamber experiments. In this work, we apply a suite of systematic reduction techniques, from the program package KINAL [4] developed to study combustion systems [5] to an extended lumped structure mechanism (CBMEX). We then compare the resulting smaller mechanism with a reduced version derived from considerations of typical ambient air conditions and environmental chamber results, CBM4.

Method

The reduction procedure used here has the following steps:

(i)   the important predictions from the model (*e.g.* ozone concentration) and the range of conditions (*e.g.* concentrations, temperature, humidity) of the air mass are chosen;

(ii)   a simulation using the full mechanism is performed and time points for further investigation are selected;

(iii)   the chemical species necessary for accurate prediction of the important variables are found by calculating the change in the rate of production of the important species as a function of a change in the concentration of each species in the mechanism at the selected time points, this is an iterative process;

(iv)   the necessary reactions are found by a rate coefficient sensitivity calculation, in which the resulting changes in rate production of necessary species as a function of change in each rate coefficient, are sorted using a principal component analysis;

(v)   fast reversible reactions are identified from the rates of production and loss, and removed from the scheme in pairs;

(vi) quasi-steady-state species, are found from a comparison of the lifetimes of the remaining species and the calculated error in assuming the quasi-steady-state approximation for the group of species. This procedure replaces the ordinary differential rate equation by an algebraic equation by setting the rate to zero. The algebraic equation can then be solved to obtain explicit relationships between the fast species and the slower species.

The reduction procedure was applied to three sets of initial conditions which were taken from environmental chamber experiments, 134P, 136P and 137P of Hess *et al.* [6]. Without isoprene this gave initial VOC: NO_x ratios of 3.5, 8.7 and 15.6 $\{N(C) / N(NO_x)\}$. The concentrations of the species were simulated for a 48 hour period in which the temperature and solar flux were varied as for a typical diurnal cycle in July at a latitude of 50° N [7]. No further emission or deposition of species were included in the simulation. At each stage of the reduction the ozone profile calculated from the smaller mechanism was compared with that for the CBMEX profile.

Results and discussion

The CBMEX mechanism consists of 204 reactions and 90 species, CBM4 has 81 reactions and 33 species, whereas CBMKINAL, the mechanism derived in this work, has 55 reactions and 30 species. The ozone profiles for the three mechanisms at each of the VOC: NO_x ratios are shown in Fig. 1. For each ratio there is less deviation from the CBMEX profile with CBMKINAL than with CBM4. The performance of the mechanism under more realistic atmospheric conditions has also been investigated. A trajectory model [8] was used in which a box of air is moved over a grid of emissions of VOC and NO_x representative of those for Europe. The chemistry of the air parcel was simulated for a 5 day period in July, with the same diurnal variation in temperature, solar flux and humidity as previously. Dry deposition of the chemical species was also included. Two trajectories have been considered with a background VOC: NO_x value in each case of 2.7. Trajectory (a) passes mostly over the sea and goes from East of Sweden to West of Scotland, whereas trajectory (b) travels overland and goes from Yugoslavia to Northern France. There was a large discrepancy in the ozone concentration for CBMKINAL compared with the full mechanism, most of which could be attributed to the elimination of the reaction between OH and methane, CH_4. The concentrations of all organic surrogates (except FORM and KET) were about 10 times lower in the trajectory runs than in the environmental chamber simulations making CH_4 relatively more important. However, with CH_4 included in KINAL, for case (a) CBM4 deviates further from CBMEX than does CBMKINAL and in case (b) CBMKINAL agrees very well with CBMEX up until 83 hours after which time CBM4 has a better agreement.

From these results we conclude that the systematic reduction techniques discussed can be usefully applied to tropospheric chemical reaction schemes. The process is totally objective and easily accomplished. It is also clear from the results that the reduction must be carried out under conditions closely related to those pertaining in the specific tropospheric scenario under investigation. This conclusion illustrates the need for an approach involving construction of a comprehensive mechanism incorporating all of the appropriate chemistry, followed by systematic reduction, of the type described here, under realistic conditions.

Work is in progress on the reduction of the comprehensive butane scheme. Substantial reductions have already been achieved. This methodology will then be applied to a range of compounds for a variety of conditions.

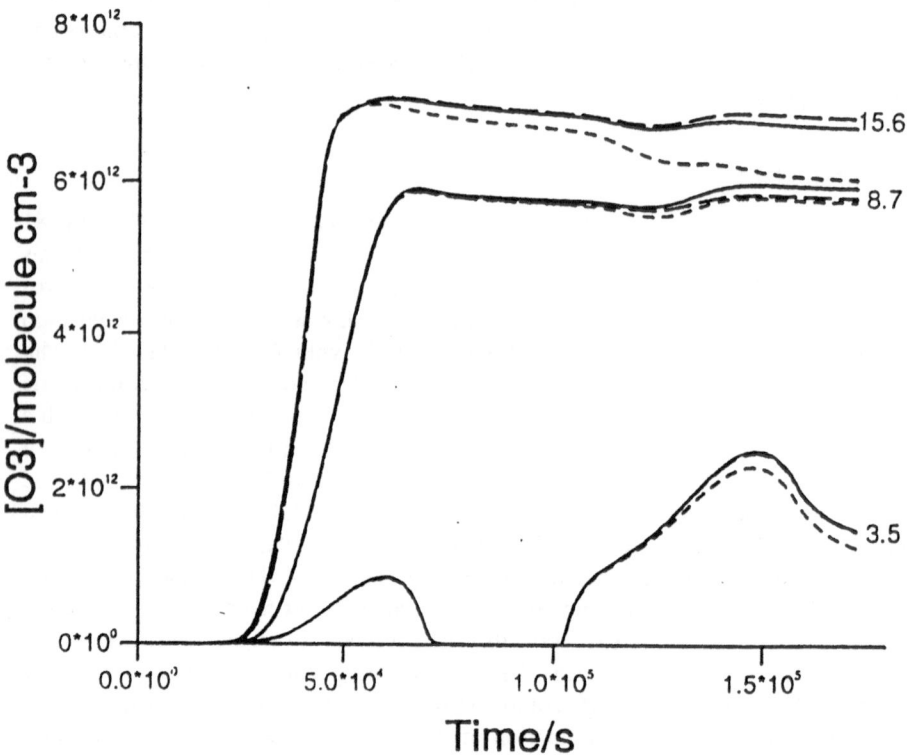

Fig. 1: Ozone concentrations *versus* time for three VOC : NO$_x$ ratios.
(----------- CBMEX;------- CBMIV, - - - - - - - CBMKINAL)

Acknowledgements

Funding from the Department of the Environment and NERC is gratefully acknowledged.

References

1. Z. Zlatev, I. Dimov, K. Georgiev, *IEEE Comp. Sci. Eng.* (1994) (Fall) 45.
2. F.W. Lurmann, W.P.L. Carter, L.A. Coyner, *Final Report*, EPA Contract No. 68-02-4104, Atmospheric Sciences Research Laboratory, Research Triangle Park, NC. 1987.
3. M.W. Gery, G.Z. Whitten, J.P. Killus, M.C. Didge, *J. Geophys. Res.* **94**D (1990) 12925-12956.
4. T. Turanyi, *Combust. Chem.* **14** (1990) 253.
5. A.S. Tomlin, M.J. Pilling, T. Turanyi, J.H. Merkin, J. Brindley, *Combust. Flame* **91** (1992) 107.
6. G.D. Hess, M.E. Carnovale, M.E. Cope, G.M. Johnson, *Atmos. Environ.* **26**A (1992) 625.
7. R.G. Derwent, Ø. Hov, *AERE Rep*-R9434, Her Majesty's Stationary Office, London 1979.
8. R.G. Derwent, M.E. Jenkin, *Atmos. Environ.* **25**A (1991) 1661.

3.17 Reactions of *n*- and *s*-Butoxy Radicals in Oxygen. Importance of Isomerisation and Formation of Pollutants

Krikor Sahetchian and Adolphe Heiss

Laboratoire de Mécanique Physique, CNRS URA 879, Université P. & M. Curie (Paris 6), 2, Place de la Gare de Ceinture, 78210 Saint-Cyr l'Ecole, France

Summary

Reactions of C_4H_9O radicals have been investigated in mixtures of O_2/N_2 under atmospheric pressure in the temperature range 313–503 K, for s-C_4H_9O, and 343–503 K, for n-C_4H_9O radicals. Flow and static experiments were performed in various vessels (quartz or Pyrex, different diameters, walls passivated or not towards reactions of radicals), and products were analysed by HPLC and GC/MS.

The main products formed by reactions of s-butoxy radicals $CH_3CH(O)CH_2CH_3$ are: hydrogen peroxide and ethyl-hydroperoxide as peroxides, acetaldehyde, methylethylketone and traces of propionaldehyde as carbonyl compounds. The reaction of the s-butoxy radical with oxygen, (2) s-C_4H_9O + O_2 → HO_2 + $CH_3COC_2H_5$, yielding HO_2 radicals and methylethylketone, is by far the main channel from room temperature to about 393 K, whereas the decomposition reaction (3) s-C_4H_9O → C_2H_5 + CH_3CHO plays the main role when the temperature is raised above 393 K. The rate constant ratio $k_3/k_2[O_2]$ was obtained from the experimental values of $[CH_3CHO]/[CH_3COC_2H_5]$, and the decomposition rate constant k_3 determined:

$$k_3 = 10^{13.8 \pm 0.2} \exp\{-15\,000 \pm 900 \text{ kcal mol}^{-1})/RT\}.$$

The main products formed by reactions of n-butoxy radicals $CH_3CH_2CH_2CH_2O$ are butyraldehyde, hydroperoxide $C_4H_8O_3$ of MW 104, 1-butanol, butyrolactone, n-propyl hydroperoxide. It is shown that transformation of these RO radicals occurs through two reaction pathways, H shift isomerization (forming C_4H_8OH radicals) and decomposition. A difference in activation energy $\Delta E = (7.7 \pm 0.1$ (σ)) kcal/mol between these reactions is found, leading to an isomerization rate constant k_{isom} (n-C_4H_9O) = 4.4×10^{11} $\exp(-9\,700/RT)$ or 4.4×10^{11} $\exp(-10\,700/RT)$. Oxidation, producing butyraldehyde, would occur after isomerization, in parallel with an association reaction of C_4H_8OH radicals producing $O_2C_4H_8OH$ radicals which, after isomerization lead to a main product, an hydroperoxide of molecular weight 104. The ratio of butyraldehyde/ (butyraldehyde + isomerization products) = 0.290 ± 0.035 (σ), studied from 448 to 496 K, is independent of oxygen concentration. Butyraldehyde would be mainly formed from the isomerised radical $H–O–C–C–C–C^{\bullet}$ + O_2 → $H–O–C–C–C–CO_2$ → $O{=}C–C–C–C$ + HO_2, because the addition of small quantities of NO has no influence on butyraldehyde formation but decreases the concentration of the hydroperoxides (that of MW 104 and C_3H_7OOH). By measuring the decay of [MW 104] as a function of [NO] added (0–22.5 ppm) at 487 K, an estimate of the isomerization rate constant $O_2C_4H_8OH \rightarrow HO_2C_4H_7OH$ (5), $k_5 \approx 10^{11} \exp(-17\,600/RT)$ is made. Implications of these results for atmospheric chemistry are discussed.

Aims of the scientific research

The reactions of RO radicals play a key role in the degradation of hydrocarbons in atmospheric chemistry. These reactions are numerous, depending upon the structure of the hydrocarbon, the temperature and the environment in which they occur. A lot of products are formed through their intermediate, and many of these products are atmospheric pollutants. For alkane hydrocarbons with a carbon number $\geq C_4$, estimations from smog chamber data made by Baldwin et al. [1] have shown, for instance, that the isomerization reaction of the n-butoxy radical is by far the main pathway under atmospheric conditions, the reactions with oxygen or that of decomposition contributing for less than 1 % to the global transformation. Experimental investigations [2–4] found contribution of the oxidation reaction amounting to 25–30 % at room temperature. From calculations about product distributions for more abundant alkanes representative of US emissions into the troposphere [5] it can be concluded that the 1,5-H shift isomerization of the n-butoxy radical accounts for about 73 % of the transformation of this radical at 300 K, the remainder being due to the reaction with oxygen (the decomposition reaction is negligible under these conditions). A recent study of Kerr et al. [6] involving 2- and 3- hexoxy radicals, showed that the 1,5-H shift isomerization could represent 68 ± 48 % and even 96 ± 30 % of the overall reaction. Studies carried out until now consisted mainly of making analyses of end-products present after combustion or after their transformation in the atmosphere. With regard to the formation mechanism of all these products, certainties are hard to achieve.

We have considered that n-butoxy radical is the smallest alkoxy radical we can experimentally study, for which isomerization occurs, generating mechanistic as well as kinetic complexities. Isomerization can play an important role from room to high temperatures, and analysis of the generated products enables us to understand what kind of chemical transformations occurs after this channel. For instance, the role of isomerization is especially interesting to analyse in the formation of pollutants, in combustion and in atmospheric chemistry, especially for tropospheric ozone. Experiments have been made with a variety of reactors (different treatments and diameters), with a wide range of oxygen concentrations and also by adding traces of NO in order to confirm certain aspects of the chemical mechanism; experimental results enable us to estimate not only the isomerization rate constants of n-C_4H_9O radicals but also that of $O_2C_4H_8OH$ radicals.

Principal scientific findings

Reactions of s-butoxy radicals $CH_3CH(O)CH_2CH_3$

The products analysed by HPLC chromatography were: *peroxides* H_2O_2, $C_2H_5O–OH$ and the remaining $s\text{-}C_4H_9O–OC_4H_9$, *carbonyl compounds* CH_3CHO, $CH_3COC_2H_5$ and traces of C_2H_5CHO.

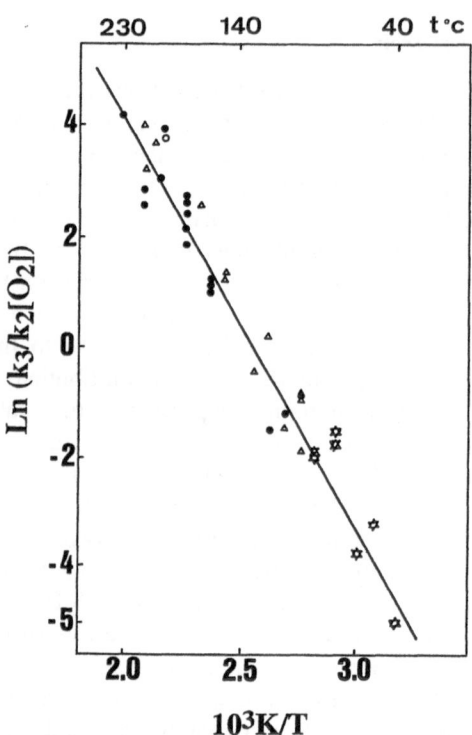

Fig. 1: Arrhenius plot, ln $(k_3/k_2[O_2])$ *versus* $1/T$, decomposition of $s\text{-}C_4H_9O–OC_4H_9$, ● Pyrex vessel ϕ_i 10 mm, △ quartz vessel ϕ_i 20 mm, ✿ quartz vessel ϕ_i 30 mm (stopped-flow experiments in the temperature range 40–80 °C), the straight line is the least-squares fit [7].

The reaction of the *s*-butoxy radical with oxygen, (2) $s\text{-}C_4H_9O + O_2 \rightarrow HO_2 + CH_3COC_2H_5$, yielding HO_2 radicals and methylethylketone, is by far the main channel from room temperature to about 393 K, whereas the decomposition reaction (3) $s\text{-}C_4H_9O \rightarrow C_2H_5 + CH_3CHO$ plays the main role when the temperature is raised above this last temperature. From the study of experimental ratios $CH_3CHO/CH_3COC_2H_5$ the rate constant ratio $k_3/k_2[O_2]$ versus $1/T$ has been determined (Arrhenius plot of Fig. 1) by computer simulation according to a mechanism of 14 reactions [7a]. A difference in the activation energy $E_3 - E_2$ = 14.8 ± 0.9 kcal/mol has been found, and for $s\text{-}C_4H_9O + O_2$, $E_2 = 0.2$ kcal/mol, the decomposition rate constant k_3 was determined:

$$k_3 = 10^{13.8 \pm 0.2} \exp\{(-15\,000 \pm 900) / RT\}$$

Static experiments were made in a treated quartz vessel [7b], at 353, 343, 333, 323 and 313 K, in full agreement with our previous extrapolation to atmospheric conditions [7a].

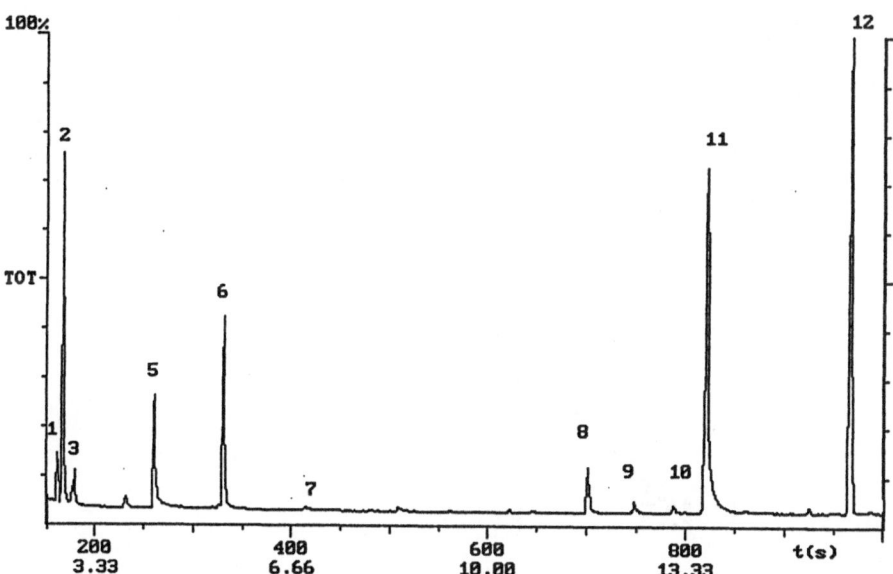

Fig. 2: GC/MS chromatogram, total ion current TIC *vs* elution time in s; capillary column HP-1, 25 m length, id 0.20 mm, film 0.50 μm, carrier gas He 99.9999 %. 1) 2,3 dihydrofurane C_4H_6O, 2) butyraldehyde, 3) propene-methoxy C_4H_8O, 4) tetrahydrofurane C_4H_8O, 5) 1-butanol, 6) *n*-propyl hydroperoxide $CH_3CH_2CH_2O{-}OH$ of MW 76, 7) C_3H_6O ?, 8) butyrolactone $C_4H_6O_2$, 9) tetrahydro-2,5-dimethoxy-furan $C_6H_{12}O_3$, 10) hydroperoxide $C_4H_8O_3$ of MW 104, 11) hydroperoxide $C_4H_8O_3$ of MW 104, 12) remaining di-*n*-butylperoxide $CH_3(CH_2)_3O{-}O(CH_2)_3CH_3$ of MW 146.

Reactions of n-butoxy radicals $CH_3CH_2CH_2CH_2O$

The GC/MS chromatogram of Fig. 2 shows the products obtained in an experiment performed at 498 K, in a (boric acid coated and H_2/O_2 treated) quartz reactor id. 10 mm, O_2/N_2 1:1, residence time 1.6 s.

In the literature, it is commonly admitted that alkoxy radicals RO are involved in three types of reactions, oxidation, isomerization, decomposition, which occur according to parallel pathways. It is the reason for which, the products generated by the reactions of the *n*-butoxy radical have been analysed and separated in three different groups corresponding to each pathway (1-butanol apart, because it is likely produced by RO + RH → ROH + R): butyraldehyde (oxidation), other compounds in C_4 resulting from the isomerization, and species in C_3 formed by the decomposition reaction. We tried to determine the logarithm of the following concentration ratios (butyraldehyde / decomposition products) and (isomerization products / decomposition products) as a function of $1/T$, but we observed a difference of activation energy of about 7–10 kcal/mol.

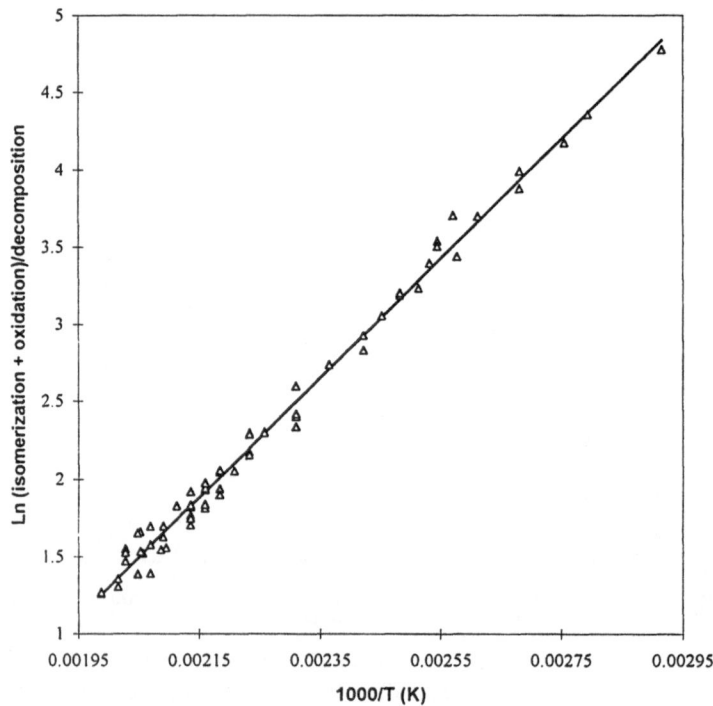

Fig. 3: Arrhenius plot of the logarithm of the ratio (isomerization products + butyraldehyde) / (decomposition products) *versus* reciprocal temperature.

If the ratio R = (butyraldehyde + isomerization products) / (decomposition products) were taken into account, the plot of its logarithm as a function of $1/T$ led (with 66 points for temperatures ranging from 343 to 503 K and a least-mean-squares analysis) to the straight line of Fig. 3, correlation factor $r^2 = 0.9953$. For this ratio R we find a temperature dependence of 7.7 kcal/mol. This value would correspond to the difference of activation energies between the reactions of decomposition and isomerization of the ratio k_{isom}/k_{decomp}:

$$\Delta E = (7.7 \pm 0.1 \ (\sigma)) \ \text{kcal/mol}.$$

If, for the decomposition rate constant of the n-butoxy radical, the values recommended by Carter *et al.* [3] and Altshuller [5] $8 \times 10^{14} \ e^{-18 \ 400/RT}$ or $8 \times 10^{14} \ e^{-17 \ 400/RT}$, are used an activation energy for the isomerization reaction of 9.7 or 10.7 kcal/mol is found, in agreement with values for isomerisation [1, 3, 5, 8]. For 1 H atom, the pre-exponential factor $A_{isom} = 4.4 \times 10^{11}$. The isomerisation rate constant can then be written:

k_{isom} (n-C_4H_9O) = 4.4×10^{11} exp($-9\ 700/RT$) or 4.4×10^{11} exp($-10\ 700/RT$).

To sum up our results for the n-butoxy radical in presence of O_2, they are in favour of two reaction pathways, isomerization and decomposition, according to the following mechanism [9]:

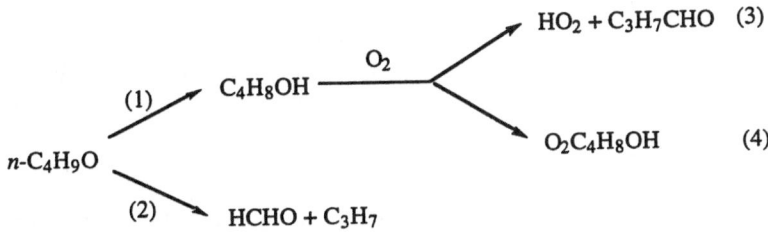

After isomerisation occurs either oxidation (with formation of C_3H_7CHO) or an addition reaction with oxygen, forming the radical $O_2C_4H_8OH$ which in turn can isomerise. It is obvious that this mechanism has to be checked for other alkoxy radicals.

The independence of the ratio butyraldehyde / (butyraldehyde + isomerization products) = 0.290 ± 0.035 (σ) towards O_2 (percentage of O_2 varied from 5 to 80 %), temperature range 448–496 K, agrees with the above mechanism (1–4). By measuring, at 487 K, the decay of hydroperoxides of MW 104 and 76 $vs.$ [NO] added (0–22.5 ppm), a first estimation could be made for the isomerization rate constant $O_2C_4H_8OH \rightarrow HO_2C_4H_7OH$ (5), for which we found

$k_5 \approx 10^{11}$ exp($-17\ 600/RT$).

We observed, in flow experiments with untreated vessels of small internal diameter 10 mm (Pyrex or quartz for which the boric acid coating has been removed), that (i) hydroperoxides MW 104 and 76 totally disappear, (ii) when radicals are destroyed at the walls, almost all compounds were transformed into butyraldehyde. Particularly, it makes us wonder when the major product (reported in the literature) is an aldehyde or a ketone that its formation could be mainly *heterogeneous*.

For tropospheric ozone, the isomerisation of the RO radical, followed by the association with O_2, can be important if it is followed by the isomerization of the $O_2R_{-H}OH$ radical, leading to a complex hydroperoxide bearing an aldehyde or epoxy function. The role played in atmosphere by the isomerisation of the RO_2 radical has been analysed by Carter $et\ al.$ [3a]. They estimated the ratios of 1,5 H shift alkylperoxy isomerisation rate to rate of major competing processes under ambient conditions. Using a NO concentration of 0.1 ppm, they obtained the following values:

reaction		$w_{isom}/w_{compet.\ proc}$

$$\underset{\displaystyle C-C-\underset{\displaystyle C}{|}-C-C}{\overset{\displaystyle \overset{\displaystyle OO^{\bullet}}{|}}{}} \longrightarrow \underset{\displaystyle \overset{\bullet}{C}-C-\underset{\displaystyle C}{|}-C-C}{\overset{\displaystyle \overset{\displaystyle OOH}{|}}{}} \qquad \approx 10^{-4}\text{--}10^{-5}$$

$$\underset{\displaystyle C-C-C-\underset{\displaystyle C}{|}-C}{\overset{\displaystyle \overset{\displaystyle OO^{\bullet}}{|}}{}} \longrightarrow \underset{\displaystyle C-\overset{\bullet}{C}-C-\underset{\displaystyle C}{|}-C}{\overset{\displaystyle \overset{\displaystyle OOH}{|}}{}} \qquad \approx 5 \times 10^{-3} - 5 \times 10^{-4}$$

Measurements made by AIRPARIF [10] show that in rural areas the NO concentration is *ca.* 3 $\mu g/m^3$, whereas it reaches 30–60 $\mu g/m^3$ in city areas; on the contrary, O_3 concentration is about 50–60 $\mu g/m^3$ in rural areas and decreases to 30–40 $\mu g/m^3$ in cities. [NO] is *ca.* 30–40 favours the isomerization of the $O_2R_{-H}OH$ radical. Ratios estimated for these reactions are therefore *ca.* 30–40 times higher and this has to be taken into account. Moreover, we have to add the case, for the $O_2R_{-H}OH$ radical, for which the H atom abstraction is carried out from the carbon atom bearing the OH group. In this case, ΔH can be 4–5 kcal/mol lower than that without O atoms [11]. This fact lowers the activation energy, and consequently increases the isomerization rate of the $O_2R_{-H}OH$ radical and the formation rate of the peroxide in the atmosphere. This peroxide can constitute a complementary source of radicals, which at the end leads to an increase of tropospheric ozone. The reaction sequence, isomerization of the RO radical followed by that of the $O_2R_{-H}OH$ radical, would lead to an increase of concentration of radicals because, one radical (n-C_4H_9O) is replaced by an other (C_4H_8OH), and also an hydroperoxide is generated which can be decomposed by light or thermally and produce two radicals RO and OH; in short, one radical is replaced by three. The rate constant that we have estimated $k_5 \approx 10^{11}$ exp($-17\ 600/RT$) gives, indeed, a lifetime $\tau = (1/k_1) \approx 50$ s at 300 K, and in the absence of termination reactions, a concentration rise (of radicals) of several orders of magnitude in about 10 min, could be reached. A realistic mechanism should be analysed in order to estimate the actual tropospheric impact. Particularly the competition between isomerization of the $O_2R_{-H}OH$ radical and reactions with NO, NO_2, HO_2, and the global RO_2 radicals (such as CH_3O_2 etc.), has to be taken into account, their probability ratio being given by the following expression {reaction (5) being in competition with reaction (9) $O_2C_4H_8OH \rightarrow HO_2C_4H_7OH$ (5), $O_2C_4H_8OH + NO \rightarrow OC_4H_8OH + NO_2$ (9)}:

$$(k_5 \cdot RO_2/k_9 \cdot NO \cdot RO_2) \quad = \quad k_5/k_9 \cdot NO$$

This short analysis points out how important it is to know the reactions in which different alkoxy radicals are involved, particularly their isomerization reactions and reactions ensuing from them, in order to better understand the formation and transformation processes of many pollutants in the troposphere. In the field of combustion, the role of these reactions can be more complex, due to the large inhomogeneity of the medium, and still deeper investigations are therefore needed.

Achievements of the project

In the framework of laboratory studies of elementary reactions relevant to tropospheric chemistry, we think that very interesting results have been achieved in the field of alkoxy radicals, in which exist a lot of theoretical speculations and estimations but only a few reliable experimental results. It seems to us that investigations on reactions of primary and secondary butoxy radicals can serve as models to understand the behaviour of higher RO radicals. Based on this assumption, investigations are in progress on reactions of n-$C_5H_{11}O$ radicals in oxygen, which will soon, as we are convinced, lead to interesting conclusions.

Acknowledgements

All this work has been carried out without the least financial support by either LACTOZ or by the French PROGRAMME ENVIRONNEMENT. In addition, thanks to the French CNRS support, a GC/MS Saturn (Varian) could be bought.

References

1. A.C. Baldwin, J.R. Barker, D.M. Golden, D.G.Hendry, *J. Phys. Chem.* **81** (1977) 2483.
2. H. Niki, P.D. Maker, C.M. Savage, L.P. Breitenbach, *J. Phys. Chem.* **81** (1977) 2483.
3. a) W.P.L. Carter, K.R. Darnall, A.C. Lloyd, A.M. Winer, J.N. Pitts Jr., *Chem. Phys. Lett.* **42** (1976) 22.
    b) W.P.L. Carter, A.C. Lloyd, J.L. Sprung, J.N. Pitts Jr., *Int. J. Chem. Kinet.* **11** (1979) 45.
    c) W.P.L. Carter, R. Atkinson, *J. Atmos. Chem.* **3** (1985) 377.
4. R.A. Cox, K.F. Patrick, S.A. Chant, *Environ. Sci. Technol.* **15** (1981) 587.
5. A.P. Altshuller, *J. Atmos. Chem.* **12** (1991) 19.
6. J.E. Eberhard, C. Muller, D.W. Stocker, J.A. Kerr, 13th Int. Symp. on Gas Kinetics, D14, Dublin 1994, pp. 269–271.
7. a) A. Heiss, J. Tardieu de Maleissye, V. Viossat, K.A. Sahetchian, I.G. Pitt, *Int. J. Chem. Kinet.* **23** (1991) 607.
    b) K. Sahetchian, A. Heiss, R. Rigny, J. Tardieu de Maleissye, in: P.M. Borrell, P. Borrell, T. Cvitaš, W. Seiler (eds), *Proc. EUROTRAC Symp. '92*, SPB Acad. Publ., The Hague 1993, p. 389.
8. S. Dobe, T. Berces, and F. Marta, *Int. J. Chem. Kinet.* **18** (1986) 329.
9. A. Heiss, K. Sahetchian: Reactions of the *n*-butoxy radical in oxygen. Importance of isomerization and formation of pollutants, to be published.
10. AIRPARIF *Edition Annuelle des Mesures de Qualité de l'Air en Ile de France*, Année 1993, 10 rue Crillon, 75004 Paris 1994.
11. J.W. Bozzelli, W.J. Pitz, 25th Int. Symp. on Combustion, Irvine 1994.
    C.K. Westbrook, W.J. Pitz, 1993 Spring Meeting of the Western States Section/The Combustion Institute, Salt Lake City, Utah, March 22-23, 1993.

3.18 Gas-Phase Ozonolysis of Alkenes and the Role of the Criegee Intermediate

W. Sander, M. Träubel and P. Komnick

Lehrstuhl für Organische Chemie II der Ruhr-Universität, D-44780 Bochum, Germany

Summary

The mechanism of the ozonolysis reaction of alkenes has been investigated in the gas phase and solid state using matrix isolation spectroscopy. While alkene/ozone π-complexes and the primary ozonides are readily observed by IR und UV/vis spectroscopy, there is no direct spectroscopic evidence for the *Criegee* intermediate (carbonyl O oxide) in these reactions. However, these elusive species can be synthesized and characterized via the carbene/oxygen route. Comparison of experimental and calculated spectroscopic data allows for the prediction of the spectroscopic properties of carbonyl oxides which are not accessible by this method.

Aims of the scientific research

Ozone plays a major role in the degradation of unsaturated VOCs in the troposphere, especially during night-time. The rate constants of the ozonolysis of a variety of alkenes have been reported [1]. However, in most instances the fate of the primary products of the ozonolysis is unknown, although the secondary reaction products are of crucial importance for the overall understanding of the alkene/ozone chemistry. The classical *Criegee* mechanism of the ozonolysis reaction involves the primary ozonide (POZ, 1,2,3-trioxolane), which cleaves to the *Criegee* intermediate (carbonyl O oxide) and a carbonyl compound [2, 3]. The secondary ozonide (SOZ, 1,2,4-trioxolane) is formed from these components in a [1,3]-dipolar cycloaddition reaction.

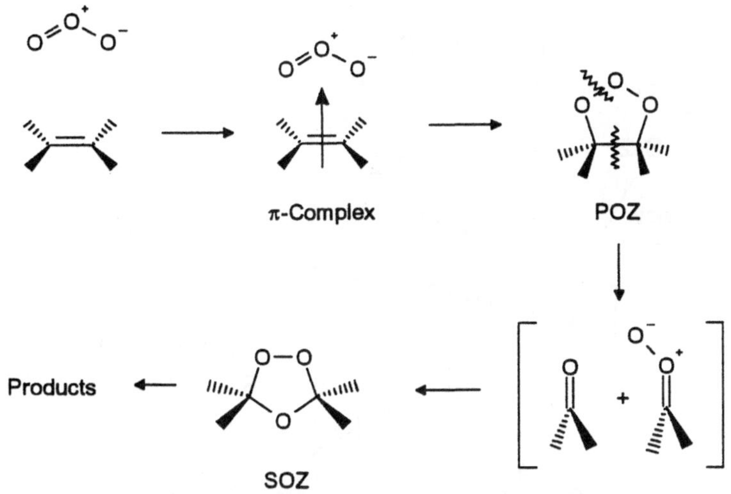

π-Complex   POZ

Products   SOZ

The *Criegee* intermediate has been claimed to be of importance in tropospheric chemistry [4] but never been observed by direct spectroscopic methods in the ozonolysis reaction. The aims of our research were therefore: (i) to provide spectroscopic (IR, UV/Vis) data of a variety of substituted carbonyl O oxides, (ii) to develop a theoretical model which allows the prediction of the spectra of carbonyl O oxides which are not accessible by laboratory studies, but might be of importance to tropospheric chemistry, and (iii) to elucidate the mechanism of the ozonolysis of alkenes and investigate the role of carbonyl O oxides in these reactions.

Principal scientific findings

Spectroscopic properties of carbonyl oxides

The only route to carbonyl O oxides **1** which is suitable for matrix isolation is the oxidation of free carbenes **2** with molecular oxygen $^3O_2$ [5–7]. The carbenes are generated by photolysis of the corresponding diazo compounds or diazirines. Prerequisite for this method is that the carbenes are stable under the conditions of matrix isolation (solid rare gas at 10 K). This excludes simple alkyl substituents, because these carbenes rapidly undergo [1,2]-H shifts to yield alkenes in essentially quantitative yields. During the last years we were able to study the spectroscopy and chemistry of a number of carbonyl oxides (R and R' were H, Ph, HC≡C, CF_3, cyclic systems *etc.*) [8–12] by IR and UV/vis spectroscopy.

All carbonyl oxides proved to be highly photolabile, and on photolysis yield dioxiranes **3** or split off oxygen atoms to produce ketones. Oxygen atoms are also formed thermally from vibrationally excited **1**. Thus, if the large exothermicity of the ozonolysis reaction is taken into account, **1** might be a source of O atoms and OH radicals in the troposphere. The role of dioxiranes has not yet been discussed in context with atmospheric chemistry, although the formation of these species – in contrast to the isomeric carbonyl oxides – in ozone/alkene reactions has been unequivocally demonstrated [13].

Carbonyl oxides **1** exhibit two characteristic spectroscopic features: in the IR an intense absorption in the range 900–1000 cm^{-1} is observed which is assigned to The O–O stretching mode. This assignment has been confirmed by isotopic labeling and *ab initio* calculations. The carbonyl oxides **1** are intensely colored yellow to red, and in the visible spectrum a strong and broad absorption near 400–500 nm is found. This absorption is nicely reproduced by semiempirical CNDO/S calculations [14]. The influence of electron-withdrawing or -donating substituents can now be confidently estimated. These data should allow for the identification of carbonyl oxides in the troposphere.

Ozonolysis of olefins

The gas-phase reaction of ozone and alkenes is very complex, particularly in the presence of excess oxygen, and even the simplest alkenes result in the formation of a large number of partially oxidized products. Some of these products are short-lived and require special techniques to be identified. Our approach was to combine gas-phase reactions at ambient temperature and matrix isolation at low temperature to gain insight into the mechanistic details. Two types of experiments have been performed: (i) Ozone and the alkene were mixed in a flow reactor, and the reaction products either analyzed by mass spectroscopy or trapped in argon at 10 K and analyzed by IR and UV/vis spectroscopy. This allows us to observe many of the secondary products of the ozonolysis. However, highly labile compounds, such as most free radicals, carbonyl oxides *etc.*, are not trapped under these conditions. (ii) Ozone and the alkene are co-deposited with a large excess of argon at 10–30 K. Due to the very low thermal energy accessible, only reactions with low activation barriers (< 5 kcal/mol) can proceed at these temperatures. Labile compounds, *e.g.* π-complexes, primary ozonides, are stable enough to be observed spectroscopically.

The olefins investigated were 1,1-diphenylethylene, stilbene, styrene, 2-methyl-propene, 2,3-dimethyl-2-butene, butadiene, pentadiene, and isoprene. The phenyl-substituted alkenes were chosen because the spectra of the expected carbonyl oxides are well known, the alkyl-substituted because of their relevance in tropospheric chemistry.

6   π-Complex **7**   POZ **8**

The main results of these experiments can be demonstrated with 2-methylpropene **6**. Codeposition of **6** and O_3 with excess Xe at 10 K produces a colorless matrix which turns yellow on warming to 35 K (Fig. 1). The yellow compound (λ_{max} = 378 nm) is assigned the structure of π-complex **7**.

Warming to 50–70 K results in the formation of POZ **8** and small amounts of acetone. The POZ **8** has been identified by comparison with other primary ozonides, and several bands could be assigned by [18]O labeling. Acetone O oxide or the SOZ are not formed under these conditions.

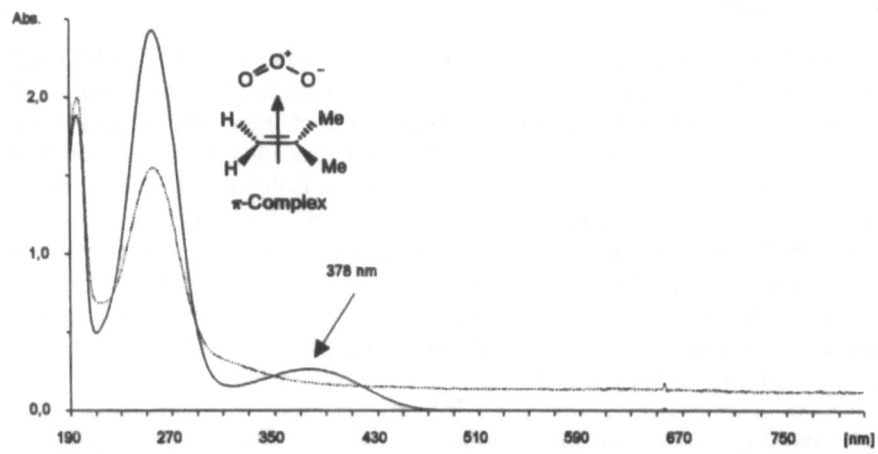

Fig. 1: UV/vis spectrum of 6 (dotted line) and the CT-complex of 6 and O_3 (solid line) in solid Xe.

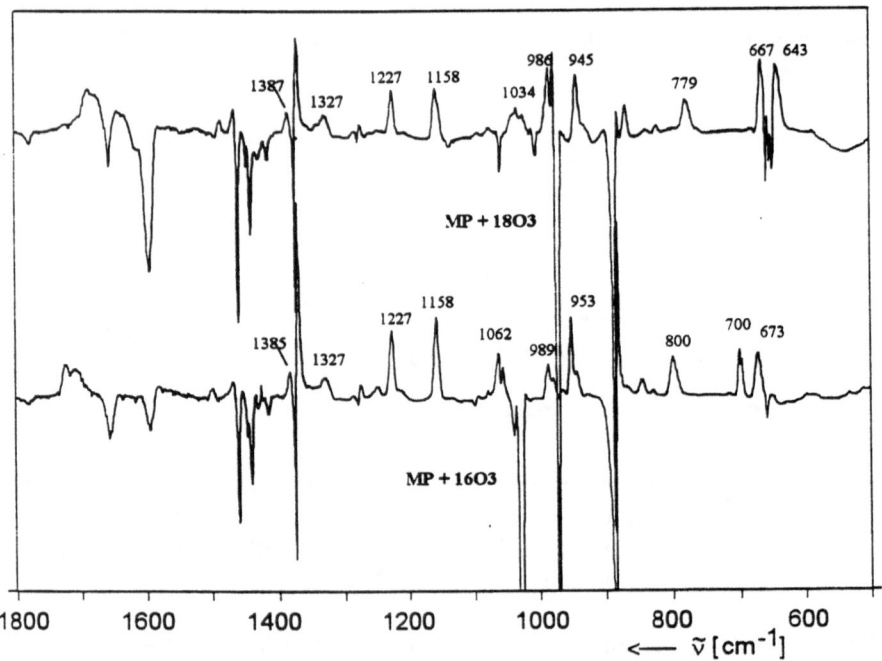

Fig. 2: IR spectrum of the thermal reaction (Xe matrix, 60 K) of 6 and $^{16}O_3$ or $^{18}O_3$, respectively. Top: absorptions growing in; bottom: absorptions disappearing, assigned to the primary ozonides.

The main products of the gas-phase ozonolysis of **6** are CO_2, formaldehyde, formic acid and acetone. Traces of acetic acid, hydroxy acetone, methanol, methylglyoxal and methane were also identified. Methyl acetate, the product of the rearrangement of acetone O oxide, is not formed. Under similar conditions tetramethylethylene produces methyl acetate, and thus we conclude that acetone O oxide is not a product of the ozonolysis of **6**. This is in contrast to the prediction of the preferred POZ cleavage and requests for an alternative mechanism to explain the formation of hydroxy acetone. The key step in this mechanism is the asynchronous ring opening of the POZ to yield a diradical. This diradical can be stabilized via cleavage to acetone O oxide and formaldehyde – the Criegee products – or by intramolecular abstraction of α- or ß-hydrogen atoms. This mechanism explains the formation of methylglyoxal, methanol, hydroxy acetone, and formaldehyde without recourse to acetone O oxide.

Dienes, such as butadiene or isoprene, react similarly with the formation of π-complexes as primary adducts. However, the subsequent chemistry is more complicated leading to methylvinylketone, methacrolein, formic acid, form-aldehyde, and peroxidic species. The mechanistic details of these reactions are still not completely understood and under current investigation in our laboratory.

The main results of our investigations are: (i) The first step in the alkene/ozone reaction is the formation of a π-complex with absorptions in the near UV or visible range. (ii) At 50–70 K this complex reacts to the primary ozonide (POZ) and, depending on substituents, traces of the secondary ozonide (SOZ). (iii) No carbonyl oxide was observed under any conditions used in our ozonolysis experiments. (iv) Some of the partially oxidized products formed are not in accordance with the *Criegee* mechanism and thus alternative mechanisms have to be considered.

Contribution to EUROTRAC and LACTOZ

The reaction of ozone and alkenes is sufficiently fast that it can compete with other removal processes and provide sinks for both ozone and alkenes in the troposphere. While kinetic data for a series of alkene/ozone reactions have been reported, not much is known about details of the reaction mechanisms, the role that carbonyl O oxides play, and the role that free radicals play in these processes. Our laboratory experiments provide the spectroscopic data (both infrared and UV/visible) that are important for the spectroscopic identification of Criegee intermediates in the troposphere. In addition, we were able to characterize secondary partially oxidized products (aldehydes, peroxides *etc.*) that are produced during the gas-phase ozonolysis. These products might lead to a net increase of ozone, if oxygen atoms are formed during their decomposition.

Acknowledgment

This research was supported by the *Bundesminister of Forschung and Technologie* (07EU761), the *Deutsche Forschungsgemeinschaft* and the *Fonds der Chemischen Industrie*.

References

1. B. J. Finlayson-Pitts, J. N. Pitts Jr., *Atmospheric Chemistry: Fundamentals and Experimental Techniques*, Wiley, New York 1986.
2. R. Criegee, *Angew. Chem.* **87** (1975) 765–771.
3. R. L. Kuczkowski, *Chem. Soc. Rev.* (1992) 79–83.
4. K. H. Becker, I. Barnes, A. Bierbach, F. Kirchner, L. Ruppert, E. Wiesen, K. Wirtz, F. Zabel, T. Zhu, *EUROTRAC Annual Report 1990*, Part 8, EUROTRAC ISS, Garmisch-Partenkirchen 1991, p. 95.
5. W. Sander, *Angew. Chem.* **102** (1990) 362–372; *Angew. Chem. Int. Ed. Engl.* **29** (1990) 344–354.
6. W. H. Bunnelle, *Chem. Rev.* **91** (1991) 335–362.
7. W. Sander, A. Patyk, G. Bucher, *J. Mol. Struct.* **222** (1990) 21–31.
8. W. Sander, *J. Org. Chem.* **54** (1989) 333–339.
9. W. Sander, *J. Org. Chem.* **53** (1988) 2091–2093.
10. W. Sander, *Spectrochim. Acta* **43**A (1987) 637–646.
11. G. Bucher, W. Sander, *Chem. Ber.* **125** (1992) 1851–1859.

12. S. Wierlacher, W. Sander, C. Marquardt, E. Kraka, D. Cremer, *Chem. Phys. Lett.* **222** (1994) 319–324.
13. R. D. Suenram, F. J. Lovas, *J. Amer. Chem. Soc.* **100** (1978) 5117–5122.
14. D. Cremer, T. Schmidt, W. Sander, P. Bischof, *J. Org. Chem.* **54** (1989) 2515–2522.
15. R. L. Kuczkowski, in: A. Padwa (ed), *Ozone and Carbonyl Oxides*, Vol. 2, Wiley, London 1984, p. 197.

3.19 Mechanistic and Kinetic Studies of Selected Oxidation Steps of Tropospheric Interest

R. N. Schindler

Institut für Physikalische Chemie der Universität, D-24098 Kiel, Germany

Summary

Reaction mechanisms and the kinetics of selected steps in the tropospheric oxidation of simple radical intermediates and of unsaturated organic compounds have been studied. As oxidants the nitrate radical NO_3, ozone O_3 and the peroxy radical CH_3O_2 were employed.

The primary interaction of NO_3 with radical species X at ambient temperatures yields rovibrationally excited products $XONO_2$ which in the gas phase immediately decompose to yield XO and NO_2. This way X = H, CH_3, HO, NO, HO_2, CH_3O_2, Cl and ClO are oxidised by NO_3 with high efficiency. This mechanism has been verified for X = H, HO and CH_3 by *ab initio* calculations on MP2-level of theory. Interaction of NO_3 with unsaturated organics starts through an electrophilic radical attack at the double bond forming a rovibrationally excited adduct radical with free rotation at the attacked bond. The adduct radical either spontaneously eliminates NO_2 to form a stable epoxide or becomes stabilised in air by formation of a nitroalkyl-peroxy radical.

O_3 alkene systems under atmospheric conditions were found to be a source for OH radicals. From experiments with an excess of CO added, source strengths of 0.50 ± 0.1, 0.37 and 0.24 ± 0.2 were obtained for TME, *trans*-2-butene and isoprene, respectively. Elucidation of the mechanism for OH generation is attempted by following intrinsic reaction pathways of the intermediates by model calculations.

A very sensitive and specific method of CH_3O_2 detection has been developed in the project. It is based on REMPI-MS techniques employing laser irradiation at two wavelengths. In EI-MS studies radical reaction of CH_3O_2 with other radicals were investigated including the species NO_3, halogen atoms, ClO and BrO.

Aims of the scientific research during the project

The activities of the Kiel group within LACTOZ concentrated on investigations of secondary atmospheric oxidation processes. The processes selected constitute elementary steps in reaction sequences leading to transformation and degradation of VOCs in the troposphere. The oxidants employed in our studies were the nitrate radical NO_3, ozone O_3, and in recent experiments the CH_3O_2 radical. It was the aim of the scientific research to elucidate reaction mechanisms operating in the transformation/degradation of unsaturated organic compounds as well as of short-lived species and to provide elementary kinetic data of the reactions studied for modelling purposes.

Principal scientific findings during the project

The principal findings shall be presented in three sections:

Investigations of NO₃ radical reactions

In kinetic investigations under flow conditions NO_3 radicals were generated preferentially by F atoms abstracting the hydrogen from HNO_3. Although this NO_3 source is very convenient and generally accepted by now, it must be handled with care, because of the fast secondary reactions between F and NO_3 in the system. The FO formed can lead to secondary side reactions which repeatedly offset the kinetic analysis. The kinetics of the NO_3 source has been dealt with in two contributions [1, 2]. Secondary oxidation reactions of NO_3 radicals with atomic and simple inorganic species, including the NO_3 self-reaction, have been reported [1–7]. The most tropospherically relevant of these studies was the investigation of reaction (1) which might serve as night-time OH source:

$$NO_3 + HO_2 \quad \rightarrow \quad OH + NO_2 + O_2 \qquad (1a)$$
$$\rightarrow \quad HNO_3 + O_2 \qquad (1b)$$

The branching ratio in (1) was determined by us to be $k(1a) / k(1b) \approx 4.4 \pm 0.8$ at low pressures, in reasonable agreement with two other low pressure studies. For a recent investigation of the reaction $NO_3 + CH_3O_2$ see below.

The majority of the kinetic investigations have dealt with the electrophilic attack of NO_3 on unsaturated organics [8–21]. A strong influence of methyl-substitutions at the double bond on reactivity became apparent in these studies.

Product analysis based on a new digital method of MS-data evaluation and mechanistic studies (see below) revealed that at low pressures epoxides are a predominant product of the reactions with NO_3 for all alkenes and dienes studied [10, 12, 14–18]. Besides the epoxides the corresponding NO_2 formed has also been identified by REMPI - MS [15, 19, 20]. It was found that the reaction sequence (2)–(3) is operative up to atmospheric pressure in all cases with large values of $k(2)$.

$$R_1R_2C{=}CR_3R_4 + NO_3 \quad \rightarrow \quad R_1R_2C(NO_3){-}CR_3R_4^* \qquad (2)$$
$$R_1R_2C(NO_3){-}CR_3R_4^* \quad \rightarrow \quad R_1R_2C(O)CR_3R_4 + NO_2 \qquad (3)$$

Statistical rate theory calculations were performed to quantitatively describe the observed pressure dependence [21]. Finally, epoxides formation with selected alkenes and isoprene was observed by FTIR spectroscopy. Under atmospheric conditions for tetramethylethylene an epoxide yield of ~20 % was observed [19].

This environmentally relevant finding is supported by recent results obtained by others. For unreactive alkenes only small epoxide yields were found.

Investigations of O_3 alkene reactions

The ozonolysis of isoprene and of selected butenes has been studied at 1 atm total pressure to quantify the strength of this night-time source in generating OH radicals and O as well as H atoms in the troposphere. In an earlier report from the Seinfeld group at Caltech, source strengths of 0.68 ± 0.15 and 0.45 ± 0.25 for OH and O atoms, respectively, were reported for each O_3 molecule consumed.

Experiments were carried out in the absence and in the presence of up to 20 % O_2. An excess of CO was used as scavenger. Model calculations indicated that the CO_2 formed in the reaction can be equated to the OH source strength. A value of 0.24 ± 0.02 was obtained in isoprene ozonolysis. In TME - ozonolysis this value was 0.50 ± 0.1 and in *trans*-2-butene 0.37 [22].] In the absence of O_2 the O atom source strength could be determined from oxirane formation to be ≤ 0.05 in all systems studied [23].

The OH radicals formed in ozonolysis are suggested to result in part from unimolecular decomposition of the intermediate Criegee radicals. *Ab initio* model calculations using G-2 theory have shown that for its decomposition thermodynamically accessible intrinsic reaction pathways can be plotted leading to the products HO and RCO. This reaction path is qualitatively supported by the detection of corresponding amounts of CO as product of the reaction. A different way to generate the Criegee radical CH_2O_2 is described in the next section (see below).

In the presence of O_2 there appear to exist also other routes of OH radical formation in ozonolysis. Here unimolecular decomposition of secondary alkoxy-peroxy radicals seem to be involved. For isoprene a sequence as given by (4)–(6) may be considered:

$$CH_2=C(CH_3)-CH=CH_2 + HO, O_2 \rightarrow CH_2=C(CH_3)-CH(OH)-CH_2O_2 \quad (4)$$
$$CH_2=C(CH_3)-CH(OH)-CH_2O_2 \rightarrow CH_2=C(CH_3)-CHO + CH_2OOH \quad (5)$$
$$CH_2OOH \rightarrow CH_2O + OH \quad (6)$$

The primary attack can also occur at the substituted double bend. In this case methylvinylketone (MVK) would result instead of methacrolein.

Investigations of CH_3O_2 reactions

Peroxy radicals are rather unreactive radicals and thus in the troposphere they are an abundant species. At ambient temperatures only their interaction with other open shell species needs to be considered. With NO they are easily converted to alkoxy radicals.

In PI-MS using ns lasers neither HO_2 nor CH_3O_2 yielded a characteristic parent ion signal. However, two-colour REMPI experiments make CH_3O_2 indirectly detectable through its photolysis product CH_3 generated at $\lambda = 248$ nm in the ion source [24, 25]. Photoionisation of CH_3 is carried out as a (2 + 1) process at $\lambda = 333$ nm. The spectrum shows rovibrational structures. This REMPI-detection is highly specific and also extremely sensitive. The detection limit is estimated to be $< 10^9$ cm^{-3}. Detection of HO_2 through its photolysis product H was also possible.

In recent experiments on the kinetics of the tropospherically relevant reaction (7) good agreement was found

$$CH_3O_2 + NO_3 \quad \rightarrow \quad CH_3O + NO_2 + O_2 \tag{7}$$

with data reported by other groups [26]. Together with the consumption of CH_3O_2 formation of CH_3O could be followed. Step (7) appears to be analogous to reaction (1) described above.

The reaction of CH_3O_2 with halogen atoms X (X = F, Cl, Br) constitutes an interesting source of Criegee radical formation in flow systems.

$$CH_3O_2 + X \quad \rightarrow \quad CH_2O_2 + HX \tag{8}$$

The exothermicity of this reaction varies from $\Delta H_R = -58.1$ kcal mol^{-1} for X = F to $\Delta H_R = -9.7$ kcal mol^{-1} for X = Br. The radical CH_2O_2 decomposes at low pressure spontaneously to yield predominantly CO_2 [27]. It appears that CH_2O_2 formed chemically activated – as in ozonolysis – follows a different decomposition pattern than CH_2O_2 formed kinetically hot in the abstraction process (8).

The peroxy radical CH_3O_2 is able to oxidise ClO to form ClOO. *Ab initio* calculations suggest that this reaction proceeds through direct dissociation of an intermediate CH_3OOCl collision complex [27]. The mechanism of CH_3OCl formation, which has been observed by several groups, is still not clear. A bimolecular substitution process can computationally be formulated employing the functional density approach.

Conclusions

Although not fully exploited because of the high diversity and complexity of laser experiments the introduction of multiphoton ionisation into mass spectrometry significantly expanded the applicability of this tool in the analysis of complex gas mixtures as the polluted atmosphere. Molecular specificity as obtained in high resolution vibrational spectroscopy combined with high sensitivity as known for ion counting methods makes this new technique unique in untangling reaction mechanisms [15, 19, 20, 24].

Acknowledgements

The work was supported through project 07 EU 709 4 of the BMBF, Bonn, and through project EV5V-CT93-0038 of the EC, Brussels.

References

1. M. M. Rahman, E. Becker, Th. Benter, R. N. Schindler, *Ber. Bunsenges. Phys. Chem.* **92** (1988) 91.
2. E. Becker, Th. Benter, R. Kampf, R. N. Schindler, U. Wille, *Ber. Bunsenges. Phys. Chem.* **95** (1991) 1168.
3. E. Becker, U. Wille, R. N. Schindler, in: K. H. Becker (ed.), *Atmospheric Oxidation Processes,* (1990) p. 209.
4. E. Becker, U. Wille, M. M. Rahman, R. N. Schindler, *Ber. Bunsenges. Phys. Chem.* **95** (1991) 1173.
5. E. Becker, M. M. Rahman, R. N. Schindler, *Ber. Bunsenges. Phys. Chem.* **96** (1992) 776.
6. P. Biggs, C. E. Canosa-Mas, P. S. Monks, R.P. Wayne, Th. Benter, R. N. Schindler, *Int. J. Chem. Kinet.* **25** (1993) 805.
7. A. S. Kukui, T. P. W. Jungkamp, R. N. Schindler, *Ber. Bunsenges. Phys. Chem.* **98** (1994) 1619.
8. Th. Benter, E. Becker, U. Wille, R. N. Schindler; in: R.A. Cox (ed), *Mechanisms of Gas Phase and Liquid Phase Chemical Transformations in Tropospheric Chemistry,* (1988), p. 155.
9. Th. Benter, R. N. Schindler; *Chem. Phys. Lett.* **145** (1988) 67.
10. I.T. Lancar, G. Poulet, G. LeBras, U. Wille, E. Becker, R. N. Schindler, in: K. H. Becker (ed), *Atmospheric Oxidation Processes,* (1990) p. 45.
11. Th. Benter, E. Becker, U. Wille, E. Canosa-Mas, S. J. Smith, S. J. Waygood, R. P. Wayne, *J. Chem. Soc. Faraday Trans.* **87** (1991) 2141.
12. U. Wille, E. Becker, R.N. Schindler, I.T. Lancar, G. Poulet, G. LeBras, *J. Atmos. Chem.* **13** (1991) 183.
13. Th. Benter, E. Becker, U. Wille, M. M. Rahman, R.N. Schindler, *Ber. Bunsenges. Phys. Chem.* **96** (1992) 769.
14. Th. Benter, M. Liesner, V. Sauerland, R.N. Schindler, in: R.A. Cox (ed), *Laboratory Studies on Atmospheric Chemistry, Air Pollution Research Report* **42** (1992) 207.
15. Th. Benter, M. Liesner, R.N. Schindler, in: J. Peeters (ed), *Chemical Mechanisms Describing Troposheric Processes, Air Pollution Research Report* **45** (1992) 119.
16. U. Wille, M.M. Rahman, R.N. Schindler, *Ber. Bunsenges. Phys. Chem.* **96** (1992) 833.
17. U. Wille, M.M. Rahman, R.N. Schindler, in: R. A. Cox (ed) *Laboratory Studies on Atmospheric Chemistry, Air Pollution Res. Report* **42** (1992) 219.
18. U. Wille, R.N. Schindler, *Ber. Bunsenges. Phys. Chem.* **97** (1993) 1447.
19. Th. Benter, M. Liesner, R.N. Schindler, H. Skov, J. Hjorth, G. Restelli, *J. Phys. Chem.* **98** (1994) 10492.
20. Th. Benter, M. Liesner, V. Sauerland, R.N. Schindler, *Fres. J. Anal. Chem.* **351** (1995) 489.
21. M. Olzmann, Th. Benter, M. Liesner, R.N. Schindler, *Atmos. Environ.* **28** (1994) 2677.

22.   R. Gutbrod, M. M. Rahman, R.N. Schindler, in: K.H. Becker (ed), *Tropospheric Oxidation Mechanisms, Air Pollution Research Report* **54**, EC, Luxembourg 1995, p.
23.   R. Gutbrod, M.M. Rahman, R.N. Schindler, *Atmos. Environ.* (1995) in press.
24.   Liesner, Th. Benter, R. N. Schindler, in: G. Restelli (ed) *Physico-Chemical Behaviour of Atmospheric Pollutants, Air Pollution Research Report* **50** (1993) 183.
25.   M. Liesner, *Ph.D. Thesis*, University of Kiel, 1995.
26.   A.S. Kukui, R.N. Schindler, to be published.
27.   T.P.W. Jungkamp, *Ph.D. Thesis*, University of Kiel, 1995.

3.20 Reactions of OH Radicals with Organic Nitrogen-Containing Compounds

Howard Sidebottom[1], Jack Tracy[2] and Ole John Nielsen[3]

[1]Department of Chemistry, University College Dublin, Dublin, Ireland
[2]Department of Chemistry, Dublin Institute of Technology, Dublin, Ireland
[3]Ford Forschungszentrum Aachen, Dennewartstraße 25, D-52068 Aachen, Germany

Summary

Rate constants for the reactions of OH radicals with n-nitroalkanes, n-alkyl nitrates and n-alkyl nitrites have been determined at 298 K and 1 atmosphere total pressure using both pulse radiolysis combined with kinetic spectroscopy and a conventional relative rate method. In order to provide more mechanistic information for these reactions, rate constants for the reaction of Cl atoms with these compounds were determined using the relative rate method. The data indicate that the reaction of OH radicals with these nitrogen-containing compounds involves both an abstraction and an addition channel. These results are discussed in terms of reactivity trends and compared with the data from the literature.

Aims of the research

The purpose of this research was to obtain kinetic data for the reaction of OH radicals with nitrogen-containing organic compounds under atmospheric conditions. The results give tropospheric lifetimes for these species and indicate their role in the long-range transport of odd nitrogen.

Principal scientific results

Introduction

Volatile organic nitrogen-containing compounds may act as more or less long-lived reservoirs for NO_x in the troposphere. Nitroalkanes are employed as propellants and industrial solvents leading to their potential release in the troposphere [1]. Alkyl nitrates can be formed in the oxidation of volatile organic compounds in the troposphere when peroxy radicals react with NO [2]. Small amounts of alkyl nitrites may be formed in the urban atmosphere by reaction of alkoxy with NO or via direct emission from the burning of alcohol-based fuels [3, 4]. The major atmospheric removal processes for these nitrogen-containing compounds are expected to be by photolysis or via reaction with OH radicals. Previously very little information concerning the kinetics and mechanisms for the reactions of these compounds with OH radicals has been available.

Experimental

Absolute experiments were performed in which a 30 ns pulse of 2 MeV electrons from an accelerator irradiated the gas mixtures to initiate the radical reactions. The decay of the radicals was recorded by use of transient UV spectroscopy, for OH at 309 nm. Hydroxyl radicals were produced by irradiation of 11 Torr H_2O mixed with 749 Torr Ar. Initially excited argon atoms, Ar*, are formed, and OH radicals are produced through the reaction:

Ar* + H_2O →   OH + H + Ar.

Formation of OH is rapid relative to the time scale of the OH decay. When adding up to 7.6 Torr of a substrate compound, RH, the OH decay obeys pseudo-first-order kinetics due to the reaction: RH + OH → products. A plot of the logarithm of the absorbency versus time is therefore linear. Plotting the slopes of these log plots, k', *versus* the RH concentrations gave linear plots with slopes equal to $k(OH + RH)$.

Relative rate experiments were carried out in a 50 litre FEP Teflon reaction chamber. The chamber was surrounded by blacklamps and sunlamps. Temperature was maintained at 298 K. Hydroxyl radicals were produced by photolysis of methyl nitrite in air with NO added or by the dark reaction of hydrazine with O_3. The generated OH radicals react with the substrate compound, RH, and the reference compound, REF:

$$OH + RH \quad \xrightarrow{k_1} \quad products \tag{1}$$

and

$$OH + REF \quad \xrightarrow{k_2} \quad products \tag{2}$$

Providing the substrate compound and the reference compound are consumed only by reaction with OH, k_1 can be determined from the expression:

$$\ln\left(\frac{[RH]_0}{[RH]_t}\right) = \frac{k_1}{k_2}\ln\left(\frac{[REF]_0}{[REF]_t}\right)$$

Chlorine atom rate constants can be derived analogously from the above equation. Chlorine atoms were generated directly from the photolysis of molecular chlorine. Gas chromatography (FID) was employed to carry out the quantitative analysis of the concentrations of substrate and reference compounds.

Results and discussion

The available kinetic data for the reactions of OH radicals and Cl atoms with *n*-nitroalkanes, *n*-alkyl nitrates and *n*-alkyl nitrites are given in Table 1 (section 3.14). Rate constant data derived in this work from pulse radiolysis and relative rate experiments for the reaction of OH radicals with these nitrogen-containing organics at atmospheric pressure and 298 K agree within experimental error.

Where comparisons are possible, agreement with other reported data for the reactions of both Cl atoms and OH radicals under these experimental conditions is, in general, also satisfactory.

It is of interest to compare the rate constants for the reaction of OH radicals and Cl atoms with n-nitroalkanes, n-alkyl nitrates and n-alkyl nitrites. It is apparent from the data that there is quite a dramatic difference in the deactivating effect of the $-NO_2$, $-ONO_2$ and $-ONO$ groups with respect to reaction with Cl atoms. Thus the reactivity of CH_3NO_2 is about 4 orders of magnitude less than that observed for the CH_3 group in C_2H_6 whilst in CH_3ONO the reduction in reactivity is only about a factor of 3. The reaction of Cl atoms with each of the three compounds presumably involves direct H atom abstraction, and the effect of the various functional groups on the observed reactivity is likely to be due to polar effects in the transition states for the reactions rather than changes in the overall enthalpy. Since Cl atoms are electrophilic the electron withdrawing abilities would appear to be in the order $-NO_2 > -ONO_2 > -ONO$. In all cases the deactivating effect of the functional group appears to extend to at least the β carbon atom. Obviously as the chain length increases the rate constant values begin to approach each other. For the reaction of OH radicals with n-nitroalkanes, n-alkyl nitrates and n-alkyl nitrites the reactivity trend is quite different from that found for the corresponding reactions with Cl atoms. In this case for all three functional groups the rate constants for reaction with the methyl compounds show relatively little variation, all falling within the range $(1.44–3.3) \times 10^{-13}$ cm^3 molecule^{-1} s^{-1} and also indicating that these compounds have about the same level of reactivity as C_2H_6 with respect to OH radical attack. The result is surprising, since if the reaction of OH radicals with these compounds involved mainly hydrogen atom abstraction, the presence of the strongly electron-withdrawing functional groups would be expected to considerably decrease the rate constants for abstraction by the electrophilic OH radical, as shown for the corresponding Cl atom reactions. These results indicate that there may be mechanistic differences between the reactions of OH radicals and Cl atoms with these compounds.

Reaction between the OH radical and CH_3NO_2, CH_3ONO_2 and CH_3ONO may involve both hydrogen atom abstraction and OH radical addition followed by the decomposition of the adduct, for example:

$$OH + CH_3ONO \longrightarrow CH_2ONO + H_2O \qquad (1a)$$

$$OH + CH_3ONO \rightleftharpoons CH_3\overset{\overset{\displaystyle OH}{|}}{O}NO \xrightarrow{\ M\ } CH_3O + NONO \qquad (1b)$$

It is suggested that the enhanced reactivity of these species with OH radicals relative to that shown for the reaction of Cl atoms is due to the importance of the addition channel at 1 atm pressure. Some support for this argument comes from the low values of the rate constants for reactions of OH radicals at low pressures. It is proposed that under low pressure conditions, reactions of OH radicals with CH_3NO_2, CH_3ONO_2 and CH_3ONO involve mainly abstraction, whereas at higher pressures the addition channel is the major reaction pathway. As the length of the side chain increases, the addition pathway becomes relatively less important as the deactivating effect of the functional groups on the abstraction process become less significant.

Conclusions

The atmospheric lifetimes of the n-nitroalkanes, n-alkyl-nitrates and n-alkyl nitrites due to loss by reaction with OH radicals can be estimated from the rate constant data determined in this work. Taking a tropospheric concentration of 1×10^6 molecules cm^{-3} for OH radicals in moderately polluted atmospheres, the rate constant data give lifetimes of approximately 1 month for CH_3NO_2, CH_3ONO_2 and CH_3ONO to around a few days for the n-C_5H_{11} derivatives. Estimates of the photodissociative lifetimes for alkyl nitrates of approximately 1 week have been reported [5], and hence these compounds could act as temporary reservoirs for nitrogen in the troposphere and may be important in long-range transport of odd nitrogen. Based on absorption cross-sectional data, photolysis lifetimes for nitroalkanes of several hours and for n-alkyl nitrites of a few minutes have been calculated [5]. Hence their atmospheric lifetimes will be dominated by photolysis and they will be rapidly removed during daylight hours.

Acknowledgements

We gratefully acknowledge financial support from the Commission of the European Communities.

References

1.   R. Liu, R.E. Huie, M.J. Kurylo, O.J. Nielsen, *Chem. Phys. Lett.* **167** (1990) 519.
2.   R. Atkinson, S.M. Aschmann, W.P.L. Carter, A.M. Winter, J.N. Pitts, Jr., *Int. J. Chem. Kinet.* **16** (1984) 1085.
3.   B.J. Finlayson-Pitts, J.N. Pitts, Jr., *Atmospheric Chemistry*, Wiley, New York 1986.
4.   P.L. Hanst, E.R. Stephens, *Spectroscopy* **4** (1989) 33.
5.   W.D. Taylor, T.D. Allston, M.J. Moscato, G.B. Fazekas, R. Kozlowski, G.A.Takacs, *Int. J. Chem. Kinet.* **12** (1980) 231.

3.21 Kinetics and Mechanism for the Reaction of Ozone with Cycloalkenes

Howard Sidebottom[1] and Jack Tracy[2]

[1]Department of Chemistry, University College Dublin, Dublin, Ireland
[2]Department of   Chemistry, Dublin Institute of Technology, Dublin, Ireland

Summary

Rate constants for the reaction of O_3 with a series of cycloalkenes have been investigated over the atmospherically important temperature range 240–324 K and 1 atmosphere total pressure. The rate data obtained in the present study are discussed in terms of substituent effects and ring strain for these cyclic systems. Hydroxyl radical formation yields were determined for the reactions of O_3 with cycloalkenes at 298 ± 2 K in an atmosphere of air. The results indicate that the OH yields are close to that obtained for the structurally similar molecule *cis*-2-butene providing support for the suggestion that OH radical yields in O_3-alkene systems depend largely on the degree of alkyl group substitution at the double bond site.

Aims of the research

The purpose of this research was to obtain kinetic data for the reaction of ozone with cycloalkenes and to determine the formation yields of hydroxyl radicals in these reactions under atmospheric conditions.

Principal scientific results

Introduction

In general the dominant loss process in the troposphere for organic compounds is through reaction with hydroxyl radicals. However, ozone also plays an important role in the chemistry of the Earth's atmosphere, its importance in the stratosphere and troposphere being particularly well documented [1–3]. Among the various classes of compounds present in the troposphere, unsaturated hydrocarbons are unusual in exhibiting significant reactivity with ozone as well as towards hydroxyl radical and nitrate radicals. Numerous potentially important roles of the ozone-alkene reaction have been recognised since these reactions can provide mutual sinks for both ozone and alkenes and concomitantly serve as sources of partially oxidised compounds and free radicals.

Rate constants for the reactions of O_3 with a large number of both anthropogenic and biogenic alkenes have been reported, mainly at room temperature. Kinetic studies on the gas-phase reactions of O_3 with cycloalkenes have been the subject of a number of studies although the effect of temperature on the rate constants has received little attention. Various kinetic and product studies have provided evidence that the gas-phase reactions of O_3 with alkenes lead to the formation of OH radicals [4–16], however, little quantitative data are available for cycloalkenes [15].

As part of an investigation into the reaction of ozone with alkenes, the temperature dependence of the rate constants for the reaction of O_3 with a series of cycloalkenes and the formation yields for OH radicals in these reactions have been obtained in the present work.

Experimental and results

All reactions were carried out in a 60 dm^3 Teflon reaction chamber housed in a commercial deep-freeze cabinet. Sub-ambient temperatures were obtained using a modified temperature control unit, while higher temperatures were obtained using hot air blowers. Reactants entered the chamber through a ¼" o.d. Teflon tube, placed along the centre of the bag, which was plugged at one end and perforated along its length. This allowed for rapid mixing of reactants ($\leq$ 1 min) as determined by gas chromatographic analysis of a test hydrocarbon. Ozone was prepared by passing zero grade oxygen (BOC) through an ozone generator (Monitor Labs) directly into the bag. Accurate concentrations of the hydrocarbons were added by placing known pressures into a calibrated volume and sweeping the reactants into the bag with zero grade N_2 (Air Products). Ozone concentrations were continuously monitored during the reaction by a Monitor Labs Model 8410 chemiluminescent ozone analyser. The signal was fed into a potentiometric chart recorder and to a microcomputer for data analysis. Quantitative analyses of hydrocarbons were carried out by gas chromatography (Shimadzu GC-8A) with flame ionisation detection. A 0.6 m × 6 mm o.d. Porapak Q column was used for analysis of ethene, *cis*-2-butene and 2-methyl-2-butene. Analyses of all other alkenes in the kinetic experiments were carried out on a 15 m × 0.54 mm i.d. SE-54 capillary column. In the OH radical formation yield experiments cyclohexanone was analysed on a 30 m × 0.54 mm i.d. carbowax column. The cycloalkenes and cyclohexane were quantitatively monitored using a 30 m × 0.54 mm i.d. SE-54 column.

Second-order rate constants were obtained by monitoring the increased rate of ozone decay in the presence of known excess concentrations of the alkene. In the presence of a cycloalkene, the processes for removing ozone are:

O_3 + wall	$\rightarrow$	loss of O_3	(1)
O_3 + cycloalkene	$\rightarrow$	products	(2)

hence

$-d[O_3] / dt$	=	$(k_1 + k_2 [\text{cycloalkene}]) [O_3]$	(i)
$-d(\ln[O_3]) / dt$	=	$k_2 [\text{cycloalkene}] + k_1$	(ii)

Table 1: Rate constants and Arrhenius parameters for the reaction of O_3 with cycloalkenes.

Alkene	$\dfrac{10^{15}\,A}{cm^3s^{-1}}$	$\dfrac{E}{cal\,mol^{-1}}$	$\dfrac{10^{18}k\ ^{a}}{cm^3s^{-1}}$	Reference
cis-2-butene	—	—	141	17
	—	—	161 ± 7	18
	3.11	1900 ± 107	126[b]	19
	—	—	138 ± 16	20
	$3.52^{+1.79}_{-1.19}$	1953 ± 229	130 ± 39	3
	—	—	129 ± 9	21
	3.4 ± 0.7	1945 ± 43	123 ± 18	22
	3.06 ± 0.15	1868 ± 83	131 ± 5	this work
2-methyl-2-butene	—	—	793	17
	—	—	493 ± 16	18
	6.34	1641 ± 155	397[d]	19
	$6.17^{+6.45}_{-3.16}$	1586 ± 389	423 ± 169[d]	3
	6.5 ± 1.3	1652 ± 86	425 ± 29	21
	5.21 ± 0.50	1459 ± 270	397 ± 12	22
			450 ± 15	this work
cyclopentene	—	—	813 ± 79	18
	50	2300	969	23
	—	—	275 ± 33	20
	—	—	497 ± 30	24
	—	—	624 ± 35	25
	—	—	655 ± 44	21
	1.6 ± 0.3	693 ± 87	491 ± 40	this work
1-methyl-1-cyclopentene	—	—	673 ± 99	this work
1-chloro-1-cyclopentene	0.443 ± 0.022	2005 ± 126	15 ± 0.5	this work
cyclohexene	—	—	160 ± 15	18
	12.6	2400	204	23
	—	—	104 ± 14	20
	—	—	151 ± 10	24
	—	—	78 ± 5	25
	—	—	75 ± 5	21
	2.6 ± 0.40	2113 ± 529	85 ± 8	this work
1-methyl-1-cyclohexene	5.25 ± 0.50	2066 ± 272	166 ± 12	this work
4-methyl-1-cyclohexene	2.16 ± 0.40	1892 ± 88	82 ± 3	this work
1-nitro-1-cyclohexene	—	—	1.2 ± 0.1	this work
cycloheptene	—	—	319 ± 36	20
	—	—	283 ± 15	25
	1.28 ± 0.40	981 ± 209	237 ± 21	this work
cis-cyclooctene	0.78 ± 0.03	432 ± 100	374 ± 11	this work
cis-cyclodecene	1.08 ± 0.57	2148 ± 338	29 ± 2	this work

[a] 298 ± 4 K;   [b] calculated from the Arrhenius expression.

Plots of the resulting pseudo-first-order rate constants against cycloalkene concentration were linear as expected. The Arrhenius parameters and measured room temperature rate constants obtained in this work are shown in Table 1.

In order to determine the OH formation yields from the reaction of O_3 with the cycloalkenes two different sets of experiments were carried out. Firstly, the loss of O_3 and the cycloalkene were monitored in experiments where $[O_3]_0$ and $[cycloalkene]_0$ were in the range 1 to 6 ppm. The results provided values for the stoichiometry of the reactions, $\Delta[cycloalkene] / \Delta[O_3]$. In a second series of experiments, sufficient cyclohexane was added to the reaction system in order to scavenge OH radicals formed in the ozone-cycloalkene reactions. In these experiments the loss of cycloalkene, O_3 and the yield of cyclohexanone, formed by the reaction of OH with cyclohexane were monitored. In order to determine the yield of cyclohexanone resulting from the reaction of OH with cyclohexane under the reaction conditions employed for the O_3/cycloalkene/cyclohexane experiments, a series of experiments was carried out in which the cyclohexane concentrations employed were such that the OH radicals react both with the cycloalkene and cyclohexane. Under these conditions the loss of cyclohexane due to reaction with OH and the formation of cyclohexanone could be determined. The results from these studies showed that $\Delta[cyclohexanone] / \Delta[cyclohexane] = 0.26 \pm 0.04$ for all the alkenes investigated.

Table 2: Hydroxyl radical formation yields from the gas-phase reactions of O_3 with ethene and a series of cycloalkenes at 298 ± 2 K in 1 atmosphere of air.

Alkene	$[alkene]_0$ ppm	$[O_3]_0$ ppm	$[c\text{-}C_6H_{12}]_0$ ppm	$\dfrac{\Delta[alkene]}{\Delta[O_3]}$	$\dfrac{\Delta[c\text{-}C_6H_{12}]}{\Delta[alkene]}$	OH radical yield
Ethene	0.7–2.5	1.6–6.1	0	1.0 ± 0.1	—	$< 0.1^a$
	0.7–2.5	1.6–2.1	500–1000	1.0 ± 0.1	< 0.01	$< 0.05^b$
Cyclopentene	1.0–6.0	2.0	0	1.46 ± 0.05	—	0.46 ± 0.05^a
	1.0–6.0	2.0	500–1000	1.0 ± 0.1	0.16 ± 0.02	0.62 ± 0.20^b
Cyclohexene	1.3–6.7	2.3	0	1.39 ± 0.02	—	0.39 ± 0.02^a
	1.3–6.7	2.3	500–1000	1.0 ± 0.1	0.10 ± 0.01	0.38 ± 0.11^b
Cycloheptene	1.0–6.4	3.1	0	1.44 ± 0.05	—	0.44 ± 0.05^a
	1.0–6.4	3.1	500–1000	1.0 ± 0.1	0.12 ± 0.01	0.46 ± 0.14^b
cis-cyclooctene	1.5–4.2	2.5	0	1.39 ± 0.05	—	0.39 ± 0.05^a
	1.5–4.2	2.5	500–1000	$\Delta 1.0 \pm 0.1$	0.09 ± 0.01	0.35 ± 0.10^b
cis-cyclodecene	1.1	1.3–2.3	0	1.40 ± 0.05	—	0.40 ± 0.05^a
	1.1	1.3–2.3	500–1000	1.0 ± 0.1	0.09 ± 0.01	0.35 ± 0.10^b

[a] based on reaction stoichiometry,
[b] based on $(c\text{-}C_6H_{10}O$ formed$) / (c\text{-}C_6H_{12}$ reacted$) = 0.26 \pm 0.04$, (determined experimentally (see text)).

This value can be used in conjunction with the data from experiments where the loss of cycloalkene and yields of cyclohexanone were determined in experiments where cyclohexane was present in high enough concentrations to scavenge all the OH radicals produced in the O_3-cycloalkene reactions. The OH formation yields were derived from both the reaction stoichiometries and from the amounts of cyclohexanone detected, and are shown in Table 2.

Discussion

Room temperature rate constants and Arrhenius parameters for the gas-phase reactions of ozone with cis-2-butene, 2-methyl-2-butene and a number of cycloalkenes are shown in Table 1 together with the literature values. The rate coefficients for cis-2-butene and 2-methyl-2-butene are in excellent agreement with the data evaluation of Atkinson and Carter [3]. The reported room temperature rate constants for the reaction of ozone with cyclopentene and cyclo-hexene show a considerable degree of scatter. The present results for cyclopentene provide support for the recent determinations by Bennett et al. [24], Nolting et al. [25], and Green and Atkinson [21], while the value for cycloheptene is slightly lower than the reported values [20] and [25]. No previous kinetic studies have been carried out on the reactions of O_3 with cis-cyclooctene and cis-cyclodecene.

Atkinson et al. [20] have previously suggested that variations in the room temperature rate constants for reaction with ozone with cycloalkenes, could at least in part, be due to differences in the ring-strain energies of the cycloalkenes. The initial step in the reaction of O_3 with an alkene is the formation of an ozonide in which the carbon-carbon double bond is replaced by a single bond. Presumably the stability of the transition state relative to the reactants depends to a certain extent on the difference in ring-strain in the two structures. The Arrhenius parameters determined in the present work provide support for this suggestion. The cyclo-hexene ring contains very little strain energy [26] and as expected, the Arrhenius parameters are very close to those observed for the structurally similar molecule cis-2-butene. In contrast cyclopentene is considerably more reactive and the data show that the increase in reactivity is largely due to the significant reduction in activation energy. This result indicates that the reduction in ring-strain in the transition state is a factor in determining the reaction rate. Presumably the increased reactivity of cycloheptene and cis-cyclooctene relative to that of cyclohexene is also the result of ring-strain, both of these species having about the same ring-strain as cyclopentene [26].

Rate constants for the reaction of O_3 with a number of substituted cycloalkenes were also determined in this work. From the results, the magnitude of the rate constant depends on the nature of the substituent and its position on the ring. As expected, substitution of a hydrogen atom attached at the 1 position in cyclo-pentene by an electron-withdrawing chlorine atom decreases the reactivity for reaction with the electrophilic ozone molecule. This would appear to be due to both a decrease in the pre-exponential factor and an increase in activation energy.

A similar large reduction in rate was observed in going from cyclohexene to 1-nitro-1-cyclohexene. Similarly, substituting a hydrogen atom by an electron-donating methyl group at the double bond site in cyclopentene and cyclohexene enhances the reactivity in comparison to the unsubstituted cycloalkene. The presence of substituents at sites remote from the double bond has little effect on the reactivity as can be seen by comparing the Arrhenius parameters for cyclo-hexene and 4-methyl-1-cyclohexene.

The reactions of O_3 with ethene and a series of cycloalkenes were investigated at 298 ± 2 K in air at atmospheric pressure in the presence of cyclohexane. The added concentrations of cyclohexane were sufficient to totally scavenge any OH radicals generated in the O_3-radical reactions. The OH radical formation yields were derived from both the reaction stoichiometries and from the amounts of cyclohexanone produced from the reaction of OH with cyclohexane, Table 2. The values of D[alkene] / D[O_3] are close to unity for experiments carried out in excess cyclohexane indicating that OH radicals are efficiently scavenged by cyclohexane. In the absence of cyclohexane, the ratio D[alkene] / D[O_3] is 1.42 ± 0.04 for all the cycloalkenes studied and is unity within experimental error for ethene. Based on the yields of cyclohexanone formed in experiments in which cyclohexane was present in high concentration, the OH radical formation yields for the cycloalkenes investigated are 0.43 ± 0.08. The yield of OH determined for the reaction of O_3 with ethene is < 0.1. Hence from the measured stoichiometries and the cyclohexanone yields the derived OH formation yields, are equal within ex-perimental error. The OH radical formation yields obtained in this work are lower than those previously obtained by Atkinson et al. [12,15] for ethene (0.12) and cyclohexene (0.68), who also used cyclohexane to scavenge OH radicals. Atkinson et al. [12,15] determined the OH yields from measurements of the sum of the cyclohexanone and cyclohexanol products formed, whereas, in this work a product of the O_3-alkene reaction was shown to interfere with the cyclohexanol measurements and hence OH yields were monitored using only the cyclohexanone product. The OH formation yields for reaction of O_3 with cycloalkenes are close to the value of around 0.4 found by Atkinson and Aschmann [15] and by Horie et al. [16] for the structurally similar molecule cis-2-butene providing support for the suggestion that OH radical formation yields in O_3-alkene systems depend largely on the degree of alkyl group substitution at the double bond site [15].

Conclusions

Kinetic studies on the gas-phase reactions of O_3 with cycloakenes as a function of temperature have provided a reliable kinetic database for these reactions which may be used as an important input into chemical models for the reactions of O_3 with alkenes under atmospheric conditions. The relatively high hydroxyl radical formation yields from the O_3-cyclohexene reactions indicate that reactions of O_3 with alkenes lead to the generation of OH radicals at night and hence allow the

OH-radical initiated oxidation of organic compounds to continue in the absence of sunlight.

Acknowledgements

We gratefully acknowledge financial support from the Commission of the European Communities.

References

1. R.P. Wayne, *Chemistry of Atmospheres*, Oxford University Press, Oxford 1985.
2. K.H. Becker, R.A. Cox, *Report* AP/54/84, CEC, Brussels 1986.
3. R. Atkinson, W.P.L. Carter, *Chem. Rev.* **84** (1984) 437.
4. Y.K. Wei, R.J. Cvetanovic, *Can. J. Chem.* **41** (1963) 913.
5. J.T. Herron, R.E. Huie, *J. Amer. Chem. Soc.* **99** (1977) 5430.
6. J.T. Herron, R.E. Huie, *Int. J. Chem. Kinet.* **10**, (1978)1019.
7. R.L. Martinez, J.T. Herron, R.E. Huie, *J. Amer. Chem. Soc.* **103** (1981) 3807.
8. R.L. Martinez, J. T. Herron, *J. Phys. Chem.* **91** (1987) 946.
9. H. Niki, P.D. Maker, C.M. Savage, L. P. Breitenbach, M. D. Hurley, *J. Phys. Chem.* **91** (1987) 941.
10. R.L. Martinez, J.T. Herron, *J. Phys. Chem.* **92** (1988) 4644.
11. R. Atkinson, D. Hasegawa, S.M. Aschmann, *Int. J. Chem. Kinet.* **22** (1990) 871.
12. R. Atkinson, S.M. Aschmann, J. Arey, B.J. Shorees, *J. Geophys. Res.* **97** (1992) 6065.
13. S. E. Paulson, R. C. Flagan, J. H. Seinfeld, *Int. J. Chem. Kinet.* **24** (1992) 103.
14. S. E. Paulson, J.H. Seinfeld, *Environ. Sci. Technol.* **26** (1992) 1165.
15. R. Atkinson, S. M. Aschmann, *Environ. Sci. Technol.* **27** (1993) 1357.
16. O. Horie, P. Neeb, G.K. Moortgat, *Int. J. Chem. Kinet.* **26** (1994) 1075.
17. R.A. Cox, S.A. Penkett, *J. Chem. Soc. Farad. Trans. 1.* **68** (1972) 1735.
18. S. M. Japar, C.H. Wu, H. Niki, *J. Phys. Chem.* **78** (1974) 2318.
19. R. E. Huie, J.T. Herron, *Int. J. Chem. Kinet. Symp.* **1** (1975) 165.
20. R. Atkinson, S.M. Aschmann, W.P.L. Carter, J.N. Pitts Jr., *Int. J. Chem. Kinet.* **15** (1983) 721.
21. C.R. Greene, R. Atkinson, *Int. J. Chem. Kinet.* **24** (1992) 803.
22. J.J. Treacy, M. El Hag, H.W. Sidebottom, *Ber. Bunsenges. Phys. Chem.* **96** (1992) 422.
23. S.A. Adeniji, J.A. Kerr, M.R. Williams, *Int. J. Chem. Kinet.* **13** (1981) 209.
24. P.J. Bennett, S.J. Harris, J.A. Kerr, *Int. J. Chem. Kinet.* **19** (1987) 609.
25. F. Nolting, W. Behnke, C.J. Zetzsch, *J. Atmos. Chem.* **6** (1988) 47.
26. S.W. Benson, *Thermochemical Kinetics*, Wiley, New York 1966.

3.22 Mechanisms for the Oxidation of C_5-C_7 Hydrocarbons

Peter Warneck, H.-J. Benkelberg, G. Heimann and R. Seuwen

Max-Planck-Institut für Chemie, Mainz, Germany

Summary

Product distributions resulting from the OH radical induced oxidation of the following hydrocarbons have been determined: 2-methyl-propane, 2,3-dimethyl-butane, 2-methyl-butane, n-pentane, cyclohexane, methylsubstituted 1-butenes, isoprene, toluene. Whenever possible, branching ratios for the self-reactions of alkylperoxy radicals and decomposition rate coefficients for alkoxyl radicals were derived.

Aims of research

Previous laboratory studies of the gas-phase oxidation of hydrocarbons, mainly in the C_1–C_4 size range, have established a general reaction mechanism involving alkylperoxy and alkoxyl radicals [1]. The present project was to determine for a number of hydrocarbons in the C_4–C_7 size range the distribution of products resulting from the OH radical-initiated oxidation, to establish the prevailing reaction mechanisms and to obtain branching ratios for reactions of alkylperoxy and alkoxyl radicals. Reactions were carried out in bulbs of two or ten litter capacity filled with air at atmospheric pressure. The hydrocarbon content was 100 – 1000 ppm. Hydroxyl radicals were generated by photodecomposition of hydrogen peroxide at 254 nm wavelength using a mercury lamp placed in a quartz finger in the centre of the bulb. Products were separated and quantified by means of gas chromatography. Although a substantial effort was made to obtain a complete product identification for each hydrocarbon, this goal has not been accomplished in all cases.

The reaction of OH radicals with alkanes results in the abstraction of a hydrogen atom followed by addition of oxygen whereby an alkylperoxy radical is formed; In the reaction with alkenes OH is added to the double bond followed by addition of oxygen, generally at the neighbouring carbon atom. The peroxy radicals then enter into a reaction sequence of the type

$$2 \, R_1CH(OO\cdot)R_2 \quad \rightarrow \quad 2 \, R_1CH(O\cdot)R_2 + O_2 \tag{1a}$$

$$2 \, R_1CH(OO\cdot)R_2 \quad \rightarrow \quad R_1CHOHR_2 + R_1COR_2 + O_2 \tag{1b}$$

$$R_1CH(O\cdot)R_2 + O_2 \quad \rightarrow \quad R_1COR_2 + HO_2 \tag{2a}$$

$$R_1CH(O\cdot)R_2 \, (+ O_2) \quad \rightarrow \quad R_1CHO + R_2OO \tag{2b}$$

$$R_1CH(O\cdot)R_2 \quad \rightarrow \quad R_1CHOHR_2 \tag{2c}$$

$$R_1CH(OO\cdot)R_2 + HO_2 \quad \rightarrow \quad R_1CH(OOH)R_2 + O_2 \tag{3}$$

followed by further reactions of the radicals produced in the decomposition and intramolecular hydrogen abstraction of the alkoxyl radical, reactions (2b) and (2c), respectively. Because these reactions produce new alkylperoxy radicals, they will enter into cross combination reactions with other peroxy radicals present in the reaction mixture. Higher hydrocarbons also allow hydrogen abstraction at more than one site of the carbon chain, so that several different primary alkylperoxy radical are produced and interact with each other. The complete reaction mechanism thus can be considerably more complex than the above simplified scheme indicates. Computer simulations were used as far as possible to assist in the evaluation of experimental data.

Principal scientific findings of the project

Detailed oxidation mechanisms for the individual hydrocarbons cannot be discussed for the lack of space. We state here mainly the product distributions observed (mole fraction unless otherwise indicated) and the principal conclusions derived for each substance studied.

2-Methyl-propane (isobutane)

Percent product distribution: acetone 24.5 ± 5.1, 2-methyl-2-propanol 18.8 ± 4.0, 2-methyl-2-hydroperoxypropane 36.7 ± 7.5, 2-methyl-propanal 14.0 ± 3.9, 2-methyl-propanol 4.4 ± 1.3, tertiary butylperoxide ≤ 1.7. The peroxy radicals involved are primary 2-methyl-1-propylperoxy, primary methylperoxy and tertiary 2-methyl-2-propylperoxy. The relatively large yield of tertiary butanol is due to the interaction between CH_3OO and tertiary butylperoxy radicals. Computer simulations based on the known rate coefficients for the self-reactions of these radicals [2] gave $k_{1pt} = 3 \times 10^{-14}$ cm^3 molecule^{-1} s^{-1} for the cross combination reaction. To simulate the observed ratio of primary alcohol and aldehyde requires a rate coefficient k_{1pt} μ 3×10^{-16} cm^3 molecule^{-1} s^{-1} for the interaction between 2-methyl-1-propylperoxy and tertiary 2-methyl-2-propyl-peroxy radicals. The oxidation mechanism is quantitatively well understood.

2,3-Dimethylbutane

Percent product distribution: acetone 46.8 ± 2.1, 2-propanol 7.5 ± 0.8, 2,3-dimethyl-2-hydroperoxybutane 29.3 ± 3.0, 2,3-dimethyl-2-butanol 2.2 ± 0.2, 2,3-dimethylbutanal 8.7 ± 0.8, 2,3-dimethyl-1-butanol 3.4 ± 0.5, unidentified 2.2 ± 0.2. The ratio of tertiary to primary hydrogen atom abstraction is 83:17. The reaction involves primary 2,3-dimethyl-butylperoxy, secondary 2-propylperoxy and tertiary 2,3-dimethyl-2-butylperoxy radicals. Large amounts of tertiary hydroperoxide are again formed due to the relatively rapid rate of the reaction between hydroperoxy and tertiary peroxy radicals compared to that of their self-reaction. Computer simulations of the product distribution allowed to estimate rate coefficients for the cross combination reactions of tertiary with primary and secondary alkylperoxy radicals, $k_{1pt} = 2 \times 10^{-16}$ and $k_{1st} = 1.7 \times 10^{-17}$ cm^3 molecule^{-1} s^{-1}, respectively. The branching ratios determined for the alkylperoxy radical self-reactions are shown in Table 1. These results have been published [3].

Table 1: Branching ratios $\hat{A} = k_{1a}/k_1$ for self-reactions of several alkylperoxy radicals.

Alkyl radical	Source	k_{1a}/k_1
2-propyl	2,3-dimethylbutane	0.39 ± 0.08
2,3-dimethyl-1-butyl	2,3-dimethylbutane	0.44 ± 0.07
2-pentyl	2-pentyliodide	0.47 ± 0.07
3-methyl-2-butyl	2-methylbutane	0.51 ± 0.08
3-pentyl	3-pentyliodide	0.38 ± 0.06
cyclo-hexyl	cyclohexane	0.42 ± 0.02
1-hydroxy-2-butyl	1-butene	0.75 ± 0.02
1-hydroxy-3-methyl-2-butyl	3-methyl-1-butene	0.83 ± 0.01
benzyl	toluene	0.76 ± 0.05

2-Methylbutane (isopentane)

Percent product distribution: acetone 27.6 ± 1.9, acetaldehyde 27.9 ± 1.9, ethanol 3.3 ± 0.3, 2-methyl-2-hydroperoxybutane 14.6 ± 1.1, 2-methyl-2-butanol 14.1 ± 1.7, 3-methyl-2-butanone 9.1 ± 0.8, 3-methyl-2-butanol 3.3 ± 0.3. Another product expected, 2-propanol, could not be quantified because its signal was overlapped by that of the parent hydro-carbon. Analysis of the product distribution suggests that the ratio of hydrogen abstraction probabilities at the primary, secondary and tertiary sites is 13:17:70. The production of ethanol indicates that ethylperoxy radicals are produced in addition to the initial primary 2-methyl-1-butylperoxy and 3-methyl-1-butylperoxy, secondary 3-methyl-2-butylperoxy and tertiary 2-methyl-2-butylperoxy radicals; and acetone can be produced from the decomposition of 2-methyl-2-butoxy as well as of 2-propylperoxy radicals generated by decomposition of the 3-methyl-2-butoxy radical. Computer simulations were performed to explore the effect of cross combination reactions between primary and tertiary peroxy radicals on the yield of 2-methyl-2-butanol. The same rate coefficient was applied for the reactions of all primary peroxy radicals. Good agreement with the experimental product distribution was obtained with $k_{1pt} = 3 \times 10^{-15}$ cm^3 molecule^{-1} s^{-1}. The branching ratio $\hat{A} = k_{1a}/k_1$ estimated for the self-reaction of 3-methyl-2-butylperoxy radicals is shown in Table 1.

n-Pentane

Percent product distribution: pentanal 3.4 ± 1.3, 1-pentanol 0.72 ± 0.06, 2-pentanone 21.6 ± 4.1, 2-pentanol 11.97 ± 0.39, 3-pentanone 20.7 ± 3.9, 3-pentanol 6.7 ± 0.39, butanal 0.2 ± 0.1, propanal 3.3 ± 0.8, acetaldehyde 5.2 ± 1.1, bifunctional products, not separated, 26.2 ± 4.8. Bifunctional products are expected to be formed as a result of intramolecular hydrogen abstraction, which the 1-pentoxyl and the 2-pentoxyl radicals are capable of undergoing. The probabilities for formation of 1-pentylperoxy, 2-pentyl-peroxy and 3-pentylperoxy radicals due to hydrogen abstraction by OH radicals are approximately 10:60:30. The corresponding C_5-alcohols are formed in the ratios 3.9%, 62.8% and 33.4%. The yield of 1-pentanol is significantly smaller than expected, particularly in relation to the yield of pentanal. The photolysis of 1-pentyliodide was used to prepare 1-pentylperoxy radicals directly. The product distribution in this case was pentanal 1.13 ± 0.1, 1-pentanol 1.0 ± 0.1, and at least two unidentified isomerization products 0.5 ± 0.05. These results show that in the self-reaction of 1-pentylperoxy radicals pentanal and 1-pentanol are formed in roughly equimolar amounts, and that the 1-pentoxyl radical formed in reaction (1a) largely isomerizes. In the oxidation of n-pentane, cross combination reactions with 2-pentylperoxy and 3-pentyl-peroxy radicals predominate. Computer simulations showed that a smaller branching ratio for 1-pentanol formation in the cross combination reactions, and a larger one for pentanal formation, can to some extent explain the lower yield of 1-pentanol and the higher yield of pentanal.

The product distribution obtained in the photolysis of 3-pentyliodide provided the branching ratio for the self-reaction of 3-pentylperoxy radicals and an estimate for the decomposition rate constant of the 3-pentoxyl radical (see Tables 1 and 2). Photolysis of 2-pentyliodide indicated considerable isomerization of the 2-pentoxyl radical. The 1,4-diols expected were not observed, however. Instead a complex spectrum of smaller fragments appeared, which proved difficult to analyze. The system obviously involves a greater number of different peroxy radicals that can interact with 2-pentylperoxy, which makes it difficult to isolate the self-reaction of this radical. The branching ratio given in Table 1 thus must be considered tentative. The ratios of pentanones/pentanols observed in the pentyliodide experiments differ considerably from those observed in the oxidation of n-pentane. Cross combination reactions of pentylperoxy radicals must largely be responsible for the differences. Further work is necessary to resolve these inconsistencies.

Table 2: Rate coefficients for the decomposition and isomerization of alkoxyl radicals. [a] The first pathway ($k_{2b} = k_{dec}$) yields aldehyde and an alkyl or hydroxyalkyl radical, the second pathway ($k_{2c} = k_{isom}$) yields a hydroxyalkyl radical.

alkoxyl radical	source	k_{dec} / s^{-1}	k_{isom} / s^{-1}
1-pentoxyl	1-pentyliodide	—	1.7×10^5
2-pentoxyl	2-pentyliodide	8.9×10^3	9.3×10^4 [b]
3-pentoxyl	3-pentyliodide	2.7×10^4	—
cyclo-hexoxyl	cyclohexane	4.2×10^3	—
1-hydroxy-2-butoxyl	1-butene	1.1×10^5	—
1-hydroxy-3-methyl-2-butoxyl	3-methyl-1-butene	6.8×10^4	—

[a] relative to reaction with oxygen, $k(O_2) = k_{2a} = 8 \times 10^{-15}$ cm^3 molecule^{-1} s^{-1}; [b] lower limit.

Cyclohexane

Percent product distribution: cyclo-hexanone 67.0 ± 1.2, cyclo-hexanol 29.2 ± 0.5, unidentified bifunctional compounds 3.8 ± 1.2. The cyclo-hexoxyl radical undergoes remarkably little ring cleavage. Values for the branching ratio of the cyclo-hexylperoxy radical self-reaction and for the decomposition rate coefficient of the cyclo-hexoxyl radical are shown in Tables 1 and 2 .

Methyl-substituted 1-butenes

The OH radical-induced oxidations of 1-butene, 3-methyl-1-butene and 2-methyl-1-butene were studied in order to quantify the 1,2-diols and 1-hydroxy-2-ketones expected to be formed as products.

1-Butene

Percent product distribution: propanal 50.6 ± 1.2, 2-hydroxy-butanal 7.0 ± 0.3, 1-hydroxy-butan-2-one 24.2 ± 0.7, 1,2 dihydroxybutane 18.2 ± 0.7. Computer assisted analysis of the product distribution showed that addition of the OH radical occurs to 26 % at the inner and to 74% at the outer position of the double bond. These reactions produced the corresponding primary and secondary hydroxy-alkylperoxy radicals. The branching ratio for the radical propagating channel of the self-reaction of the secondary peroxy radicals was determined to be $k_{1ssa}/k_{1ss} = 0.75 ± 0.02$; 28 % of the hydroxy-alkoxyl radical thus formed reacted with oxygen to produce hydroxyketone. If it is assumed that the rate coefficient for the reaction of the hydroxy-alkoxyl radical with oxygen is 8×10^{-15} cm^3 molecule^{-1} s^{-1}, the rate coefficient for the decomposition of this radical to produce propanal is 1×10^5 s^{-1}.

3-methyl-1-butene

Percent product distribution: 2-methyl-propanal 48.1 ± 1.1, 2-hydroxy-3-methyl-butanal 6.9 ± 0.34, 1-hydroxy-3-methyl-2-butanone 29.5 ± 0.61, 1,2-dihydroxy-3-methyl-butane 15.5 ± 1.1. Data evaluation in a manner similar to that for 1-butene showed 24 % of OH addition to occur at the inner and 76 % at the outer position of the double bond, $k_{1ssa}/k_{1ss} = 0.82 ± 0.02$, and 38% reaction of the hydroxy-alkoxyl radical with oxygen compared with decomposition. The rate coefficient for decomposition of 1-hydroxy-3-methyl-2-butoxyl was estimated to be 6.3×10^4 s^{-1}.

2-methyl-1-butene

Percent product distribution: butanone 80.8 ± 1.46, 2-hydroxy-2-methyl-butanal 4.6 ± 0.1, 1,2-dihydroxy-2-methyl-butane 14.3 ± 0.3. The addition of OH at the double bond of 2-methyl-1-butene occurs to 10.5 % at the inner and to 89.5 % at the outer position. This results mainly in the formation of a tertiary hydroxyalkyl peroxy radical, which carries no abstractable hydrogen at the

2-position. The formation of 1,2-diols thus results entirely from addition of OH at the inner position of the double bond.

Isoprene

Percent product distribution : Methylvinylketone (MVK) 49.0 ± 1.1, methacroleine (MAC) 44.3 ± 1.6, 3-methyl-furan (MFU) 6.7 ± 0.6 for isoprene mole fractions 400 ppm. The ratio of MVK to MAC under these conditions was 1.1 ± 0.6. In the presence of NO the ratio is close to 1.4 [4]. Computer simulations based on our results for methylsubstituted 1-butenes showed that the observed [MVK]/[MAC] ratios are obtained with essentially equal probabilities for OH attack at the two double bonds of isoprene. This result contradicts predictions based on OH reaction rate coefficients for various methylsubstituted butenes [5]. Similar to results obtained for the 1-butenes, diols and hydroxycarbonyl compounds were found in small yields among the products. These are not included in the above product distribution because they are difficult to quantify .

Toluene

Product distribution (percent reacted carbon): benzaldehyde 4.9 ± 1.1, benzylalcohol 2.3 ± 0.5, o-cresol 30.6 ± 6.5, p-cresol 11.1 ± 2.4, m-cresol ≤ 0.4, methyl-p-benzoquin-one 3.9 ± 0.8, unidentified 3.1 ± 0.6, formaldehyde 2.1 ± 0.7, acetaldehyde 1.1 ± 0.4, glyoxal 3.9 ± 1.3, methylglyoxal 4.9 ± 1.6, carbon monoxide 4.0 ± 1.0, carbon dioxide 8.9 ± 2.2. The total product yield was 81 ± 20 %; 56 % were ring-retaining products, 25 % resulted from ring cleavage. The distribution of ring cleavage products observed here under NO_x free conditions was similar to that reported previously in the presence of NO_x. The yield of cresols was higher. In particular, in comparison with previous results more p-cresol and less m-cresol was found. The self-reaction of hydroxy-methyl-cyclo-hexadienyl-peroxy (HMCHP) radicals is expected to produce dihydroxytoluenes; these compounds were not observed, however. A computer simulation indicated that internal rearrangement of the HMCHP radical must be rapid relative to reactions with other peroxy radicals. The branching ratio for the self-reaction of the benzylperoxy radical, $k_{1ppa}/k_{1pp} = 0.76 ± 0.05$ was determined by using chlorine atoms as reagent to produce the benzyl radical. In the oxidation of toluene induced by OH a smaller $\hat{A}$ value was obtained, which indicated that the branching ratio is influenced by reactions of benzylperoxy with other peroxy radicals.

Conclusions

Oxidation mechanisms for 2-methyl-propane, 2,3-dimethyl-butane, 2-methyl-butane, cyclohexane and methyl substituted 1-butenes have been fully delineated. The mechanism for n-pentane is understood although not entirely quantified. Progress has been made in deriving the mechanisms for the oxidation of isoprene and toluene, but in these cases a complete identification of all products has not been possible.

Acknowledgements

Financial support by the German Federal Ministry of Research and Technology is gratefully acknowledged. We thank Dr. Böge from the Institut für Troposphären-forschung, Leipzig, for the preparation of several compounds.

References

1.   R. Atkinson, *Atmos. Environ.* **24**A (1990) 1–41.
3.   G. Heimann, P. Warneck, *J. Phys. Chem.* **96** (1992) 8403–8409.
2.   P.D. Lightfoot, R.A. Cox, J.N. Crowley, M. Destriau, G.D. Hayman, M.E. Jenkin, G.K. Moortgat, F. Zabel, *Atmos. Environ.* **26**A (1992) 1805–1964.
4.   S.E. Paulson, R.C. Flagan, J. H. Seinfeld, *Int J. Chem Kinet.* **24** (1992) 79–101.
5.   J. Peeters, W. Boullart, J. Van Hoeymissen, in P.M. Borrell, P. Borrell, T. Cvitaš, W. Seiler (eds), *Proc. EUROTRAC Symp. '94*, EUROTRAC ISS, Garmisch-Partenkirchen 1994, pp. 110–114.

3.23 Studies of the Kinetics and Mechanisms of Interactions of Nitrate and Peroxy Radicals of Tropospheric Interest

Richard P. Wayne, Pete Biggs and Carlos E. Canosa-Mas

Physical Chemistry Laboratory, University of Oxford, South Parks Road, Oxford OX1 3QZ, UK

Summary

This report describes research conducted in Oxford during the years 1988–1994 aimed at improving understanding of the reactivity of the nitrate and peroxy radicals through laboratory investigations of kinetics and mechanisms. Methods utilised included conventional discharge-flow and laser-flash photolysis techniques, and a new stopped-flow system was devised to allow investigations of slow reactions of the NO_3 radical. Reactions of NO_3 with a variety of closed-shell molecules (alkanes, alkenes, aldehydes, *etc.*) were investigated, and the room-temperature rate coefficients and, where possible, Arrhenius parameters are reported here. Investigations of reactions with inorganic open-shell species included those with atomic I and the self-reaction of NO_3. In addition, a new reaction of NO_3 with molecular I_2 was discovered. Much effort was addressed towards the direct study of the reactions between organic radicals and NO_3, with the ultimate objective of understanding the products and kinetics of the interaction between CH_3O_2, $C_2H_5O_2$ (and other RO_2 species) and NO_3. A laser-induced fluorescence method was used to detect the RO species. The experiments make it clear that RO_2 does indeed react with NO_3, that one of the products is RO and that RO in turn reacts with NO_3 (in the laboratory system) to regenerate RO_2. Indirect studies, which involved examination of secondary reactions following H abstraction from organic substrates in the presence of O_2 supplemented the direct investigations. Reactions of acylperoxy radicals such as $CH_3CO.O_2$ were examined in this way (using aldehydes as starting materials); a more direct method of following these reactions was devised in which peroxyacetylnitrate (PAN) was pyrolysed in the presence of NO_3. The research on RO_2 radicals has been extended to measurements of the kinetics of the self-reactions of several substituted methylperoxy species. The experimental work on reaction kinetics has been supplemented by attempts to interpret the reactivity of NO_3 towards organic species using the framework of frontier orbital theory; calculations of orbital energies has permitted a variety of correlations to be evaluated and has provided further insight into the factors affecting reactivity.

Aims of the research

The primary aim of this project is the investigation of the kinetics and mechanisms of reactions of NO_3 with both radical and non-radical species. One objective is to gain an understanding of the reaction rates and product channels of the processes. The studies not only form a direct part of our interest in the physical chemistry of radical–radical interactions, as well as in the general aspects of atmospheric chemistry, but also help to identify patterns of reactivity that may have diagnostic and predictive value. In terms of the EUROTRAC project, the most important of the radical–radical reactions are likely to be those with organic peroxy radicals; included in this category are both alkylperoxy and acylperoxy species. Laboratory investigations need to encompass studies of R and RO radicals in addition to RO_2. Other problems in NO_3 chemistry that were aims of this research include possible interactions that are too slow to study by `conventional' time-resolved techniques; in this category fall the self-reaction of NO_3, which inevitably occurs in any experiment, reactions with some alkanes, and heterogeneous and thermal loss processes.

Principal scientific findings

Experimental methods

Most of the experiments reported here were conducted using conventional discharge-flow methods; linear flow velocities were typically of the order of 1 to 20 m s^{-1}. Reaction of F with HNO_3 was used to generate the NO_3 radicals. In the great majority of experiments, multiple-pass optical absorption at $\lambda = 662$ nm was used to monitor the radicals, although LIF was employed in a few instances. CH_3O and C_2H_5O radicals were monitored by LIF. One method for monitoring the RO_2 radicals involved conversion to the corresponding RO species, followed by LIF detection of the RO. Another technique that we employed was based on LIF or MS determination of the NO_2 generated on addition of excess NO to the gas flow.

In addition to the continuous-flow reactors, we developed a stopped-flow apparatus to study slow reactions of the NO_3 radical. In this apparatus, a series of solenoid valves is used to divert and isolate a flow of gas that contains the reaction mixture. These valves were designed and fabricated in this laboratory by the PI, and ensure that only glass is in contact with the flow. Concentrations of NO_3 are then followed as a function of time after the flow is cut off, the data being captured by computer. Figure 1 shows the apparatus in schematic form, while Fig. 2 illustrates decay curves for [NO_3] in the absence and presence of C_2H_4.

A modification to laser-flash photolysis was made in order to study the self-reaction of NO_3 radicals. Photolysis of F_2 in the presence of HNO_3 was used to generate NO_3. Laser pulses were repeated until a suitable amount of NO_3 had built up in the cell; irradiation was then terminated, and the decay of NO_3 followed in real time.

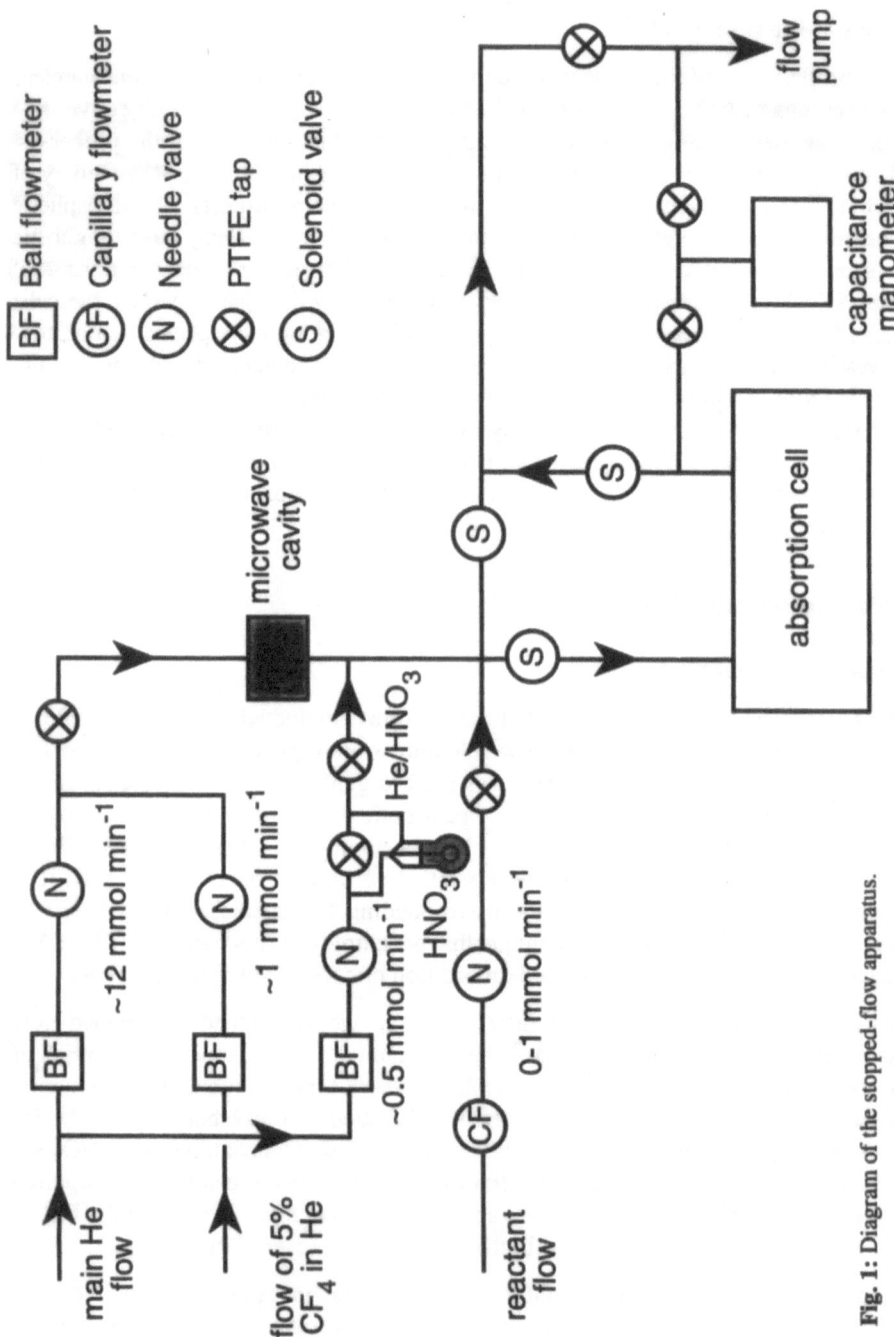

Fig. 1: Diagram of the stopped-flow apparatus.

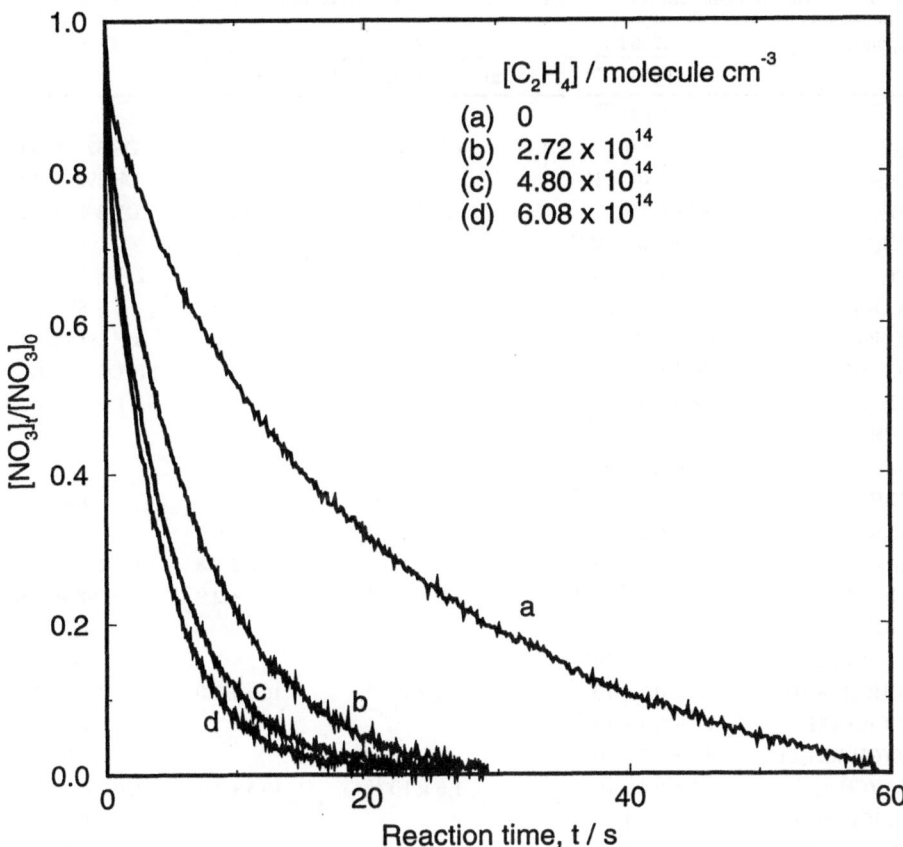

Fig. 2: Decay of NO_3 in the presence of C_2H_4.

Reactions of NO_3 with closed-shell organic species

Table 1 summarises the rate data obtained during the course of this project. Although there have been previous determinations of the rate of reaction of alkanes with NO_3 at room temperature, our work appears to be the first in which the temperature dependence of the reactions was studied. Our new measurements indicate that the reactions studied are not the major loss processes for alkanes in the atmosphere. They are, however, significant in the production of night-time nitric acid; the reaction with i-pentane makes this molecule the most important atmospheric alkane in terms of HNO_3 generation.

Since many urban areas contain relatively high concentrations of a wide range of alkanes and alkenes, it is desirable that structure–reactivity relationships be found that assist understanding of the urban night-time troposphere. These new data provide some of the necessary information for uncovering such relationships; a later section explores the work that we have done in this respect.

Table 1: Rate parameters for selected reactions of NO_3 investigated in this laboratory.

Compound	$k(298\ K)$ $\overline{cm^3\ molecule^{-1}\ s^{-1}}$	A $\overline{cm^3\ molecule^{-1}\ s^{-1}}$	E_a $\overline{kJ\ mol^{-1}}$	Technique
CH_4	$< 4 \times 10^{-19}$			SF
C_2H_6	8.3×10^{-19}	1.7×10^{-11}	41.8 ± 1.4	DF 373–553 K
$n\text{-}C_4H_{10}$	$0.45 \pm 0.06 \times 10^{-16}$	$2.5 \pm 0.6 \times 10^{-12}$	27.0 ± 0.7	DF 298–523 K
$i\text{-}C_4H_{10}$	$1.1 \pm 0.2 \times 10^{-16}$	$2.3 \pm 0.6 \times 10^{-12}$	24.6 ± 0.7	DF 298–523 K
CH_2Cl_2	$4.8 \pm 1.0 \times 10^{-18}$			SF
$CHCl_3$	$6.0 \pm 0.5 \times 10^{-17}$			SF
CH_3COCH_3	$8.5 \pm 2.5 \times 10^{-18}$			SF
CH_3CHO	$2.5 \pm 0.4 \times 10^{-15}$	$4.2 \pm 0.5 \times 10^{-12}$	14.9 ± 0.1	DF 298–473 K
CH_2ClCHO	$< 10^{-15}; > 6 \times 10^{-16}$		16 ± 3	SF, DF 293–473 K
$CHCl_2CHO$	$4 \pm 1 \times 10^{-17}$		23 ± 4	SF, DF 298–473 K
CCl_3CHO	$3 \pm 1 \times 10^{-18}$			SF
C_2H_4	$1.7 \pm 0.5 \times 10^{-16}$			SF
$1,1\text{-}C_2H_2Cl_2$	$1.2 \pm 0.3 \times 10^{-15}$	$4.7 \pm 2.4 \times 10^{-13}$	15.0 ± 2.0	DF
$1,2\text{-}C_2H_2Cl_2$	$1.2 \pm 0.5 \times 10^{-16}$	$4.5 \pm 0.5 \times 10^{-13}$	19.9 ± 0.3	DF
$1\text{-}C_4H_8$	$1.1 \pm 0.2 \times 10^{-14}$	2.5×10^{-13}	7.8 ± 0.8	DF 298–473 K
$ClCH_2CHCHCH_3$	$2.0 \pm 0.7 \times 10^{-14}$	6.0×10^{-13}	8.2 ± 2.9	DF 298–473 K
$ClCHCHCH_2CH_3$	$1.2 \pm 0.4 \times 10^{-14}$			DF
$CH_2CClCH_2CH_3$	$1.7 \pm 0.3 \times 10^{-14}$			DF
$CH_2CHCHClCH_3$	$3.0 \pm 0.7 \times 10^{-15}$	2.4×10^{-12}	16.6 ± 2.0	DF 298–473 K
$CH_3CClCHCH_3$	$1.1 \pm 0.4 \times 10^{-13}$			DF
$CH_2C(CH_3)CH_2Cl$	$9.0 \pm 2.3 \times 10^{-14}$			DF
$CHClC(CH_3)_2$	$2.5 \pm 0.4 \times 10^{-14}$	1.6×10^{-12}	10.5 ± 1.7	DF 298–473 K
$CH_2CHCHBrCH_3$	$4 \pm 1 \times 10^{-15}$			DF
$CH_2CHCH_2CH_2Br$	$5 \pm 1 \times 10^{-15}$			DF
$CH_3CHBrCHCH_3$	$1.3 \pm 0.1 \times 10^{-13}$			DF
$1\text{-}C_5H_{10}$	$(1.8 \pm 0.1) \times 10^{-14}$	9.1×10^{-13}	10.2 ± 1.1	DF 298–523 K
$1\text{-}C_6H_{12}$	$(1.5 \pm 0.1) \times 10^{-14}$	1.13×10^{-13}	10.9 ± 0.6	DF 298–523 K
I	$4.5 \pm 1.9 \times 10^{-10}$			DF
I_2	$1.5 \pm 0.5 \times 10^{-12}$			DF
O_3	$< 0.6 - 1 \times 10^{-19}$			SF
NO_3	$2.3 \pm 0.8 \times 10^{-16}$			LP
CH_3	$3.5 \pm 1.0 \times 10^{-11}$			DF
CH_3O	$2.3 \pm 0.7 \times 10^{-12}$			DF
CH_3O_2	$1.0 \pm 0.6 \times 10^{-12}$			DF
C_2H_5	$4.0 \pm 1.0 \times 10^{-11}$			DF
C_2H_5O	$3.5 \pm 1.0 \times 10^{-12}$			DF
$C_2H_5O_2$	$2.5 \pm 1.5 \times 10^{-12}$			DF
$CH_3CO.O_2$	$\sim 2 \times 10^{-11}$			DF

SF = stopped flow; DF = discharge flow; LP = laser photolysis

In analysing the results of the kinetic experiments, we paid considerable attention to the possible occurrence of secondary reactions. In the primary interaction of NO_3 with alkanes, RH, hydrogen abstraction yields the radical R. Thus, if oxygen is deliberately added to the reaction mixture, a source of RO_2 is available in the presence of the NO_3, so that information about interactions between the two radicals can be inferred from the detailed kinetic behaviour.

The data obtained in this way complements the information gained by the direct studies to be discussed later. In a converse manner, the data from the direct studies can be used to aid in the interpretation of secondary chemistry. Note that the experiments with CH_3CHO allow generation of $CH_3CO.O_2$ radicals.

Reaction of NO_3 with inorganic species

Data for four reactions are reported in Table 1. They are for the reactions with I, I_2, O_3, and the self-reaction with NO_3 itself. There is a possibility that iodine species could play a role in atmospheric chemistry close to the surface of oceans and in the emissions from nuclear power stations. The products of the reaction with I are thought to be $IO + NO_2$; the rate coefficient is surprisingly large, and the mechanism may be different from those for the reactions with the other halogens. Reaction of NO_3 with I_2 appears not to have been observed previously; the products seem to be $I + IONO_2$.

We have previously considered the possibility of an interaction between NO_3 and O_3. Detailed analysis of stopped-flow experiments yielded a value for the rate coefficient of between 0.6 and $1.0 \times 10^{-19} cm^3$ molecule^{-1} s^{-1}. However, this process might be heterogeneous, and we therefore suggest the upper limit given in the table. It seems unlikely that the reaction could be of atmospheric importance during the time during which substantial concentrations of NO_3 and O_3 coexist.

Studies of the self-reaction between NO_3 radicals have been conducted using both the stopped-flow and the `slow' laser-flash photolysis system. We prefer the results from the photochemical system, and these are reported in the table. The data are of importance in the analysis of laboratory data, especially for the slower reactions of NO_3.

Reactions of NO_3 with organic radicals

The prime requirement from the point of view of atmospheric chemistry is to discover the reaction channels and rates for the interaction of RO_2 with NO_3. However, for laboratory study, we also need equivalent information on the reaction of the related R and RO radicals. We discovered that R ($R = CH_3, C_2H_5$) reacts with NO_3 to yield RO; this process thus provides a convenient source for our studies of the RO radical. The peroxy radical, RO_2, can of course be generated by the addition of R to O_2; however, we also discovered that RO reacts with NO_3 to yield RO_2. Indeed, it turns out that RO_2 in turn reacts with NO_3 to regenerate RO, so that a cyclic process, represented for $R = CH_3$ by the equations

$$CH_3O + NO_3 \quad \rightarrow \quad CH_3O_2 + NO_2 \tag{1}$$
$$CH_3O_2 + NO_3 \quad \rightarrow \quad CH_3O + O_2 + NO_2 \tag{2}$$

operates. This finding is of critical importance to he interpretation of the observed concentration-time profiles, because there is a tendency for [RO] to approach a steady-state value at long contact times. Full numerical modelling of the reaction systems, using different sources of RO and RO_2 provides the rate coefficients reported in Table 1.

Because of the rapidity of the self-reaction of $RCO.O_2$ radicals and because RCO.O radicals are generally unstable, it is not possible to use a similar approach for the study of acyl peroxy radicals. Instead, we have used the thermal decomposition of PAN, $CH_3CO.O_2NO_2$, as an *in situ* source of $CH_3CO.O_2$. In our experiments, concentrations of PAN were measured by FTIR spectroscopy, and the decay of NO_3 in a (slow) flow system provides the kinetic information. It is clear that radical reactions, and not direct reaction with PAN, are responsible for the losses of NO_3. The current best fits to the experimental data were obtained with the rate coefficient given in Table 1 for the reaction

$$CH_3CO.O_2 + NO_3 \quad \rightarrow \quad CH_3CO.O + O_2 + NO_2 \tag{3}$$

and the value appears to be independent of temperature over the range $T = 300\text{--}423$ K.

The key conclusions from our studies of the reactions of RO_2 with NO_3 are:

(i) the interconversion of oxy (RO) and peroxy (RO_2) species through their reaction with NO_3 is critically necessary for the interpretation of the laboratory studies; and

(ii) the rate constants for the reaction of RO_2 with NO_3 are large enough to make this process a possible source of HO_2 and OH radicals in the troposphere at night. Since, in the presence of O_2, $CH_3CO.O$ radicals will be largely converted to CH_3O_2, this conclusion holds also for the acetyl peroxy radicals converted in reaction (3).

The self-reaction of RO_2 radicals

To complement the studies of the interaction between RO_2 and NO_3 radicals, we have carried out some studies on $RO_2 + RO_2$ systems, because this reaction, at least, must always compete with the reaction $RO_2 + NO_3$. We have probed the processes by following the second-order decays of RO_2 for several partially and fully halogenated methyl peroxy radicals. Allowance is made in the analysis for the reaction $RO_2 + HO_2$ in those cases where HO_2 is a secondary product. The preliminary rate coefficients obtained are displayed in Table 2.

Table 2: Rate coefficients for the self-reaction of selected RO_2 radicals.

RO_2	CH_2BrO_2	$CHBr_2O_2$	CH_2ClO_2	$CHCl_2O_2$	CF_3O_2
$10^{12}k(298 \text{ K}) / \text{cm}^3 \text{molecule}^{-1} \text{s}^{-1}$	2.0	2.5	1.0	4.0	2.1

Patterns of reactivity of the NO₃ radical

Work on correlations of rate coefficients with ionisation potentials for the reactions of NO₃ with a series of alkenes has been extended. Ionisation potentials are determined using a semi-empirical molecular orbital package (MOPAC 5.0). We are now able to estimate both room-temperature rate coefficients and Arrhenius parameters for the reaction of NO₃ with a wide range of alkenes, including those containing vinylic chlorine atoms (which we earlier found difficult to handle). In an effort to improve our understanding of the physical basis of our observations, *ab initio* calculations have been performed on a limited number of compounds. Such calculations should overcome some of the limitations of the semi-empirical methods when applied to elements such as chlorine and bromine.

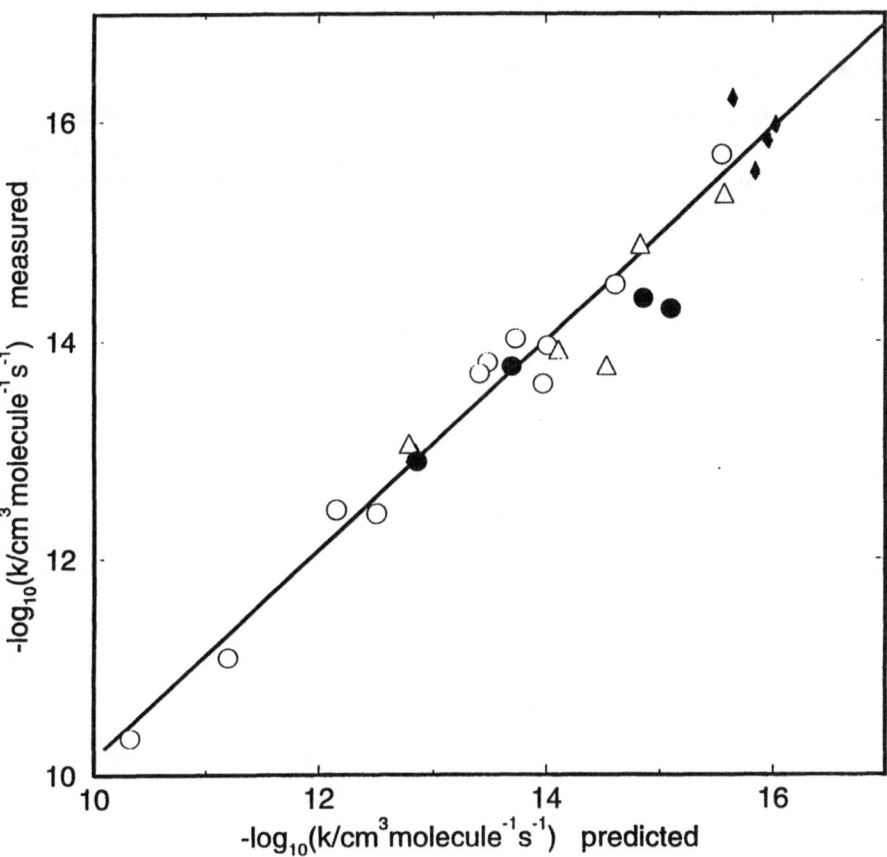

Fig. 3: Correlation for room-temperature rate coefficients in terms of *IP*.
○ alkenes and non-vinylic Cl-alkenes; ● Br-alkenes; △ Cl-alkenes with Cl at one end of double bond; ◆ Cl-alkenes with Cl at both ends of double bond.

Assessment of achievement of project

The work carried out within this project has provided answers to many of the questions posed at the outset. The data base for the kinetics of the reactions of NO_3 has been extended considerably, new reactions of NO_3 have been discovered, and correlations between reactivity and calculable molecular parameters have been found. The laboratory information required to evaluate the contribution of NO_3 chemistry to night-time chemical transformation in the troposphere, including both oxidation and formation of HNO_3, is now on a firmer footing in several respects. These objectives were included in those of LACTOZ, which itself conformed with the overall aims of EUROTRAC. EUROTRAC also seeks to foster international co-operation. We have actively sought collaboration with other groups within Europe, and have worked most successfully with teams from Bordeaux, Castilla La Mancha, Dublin, Gothenburg, Kiel, Mainz, Orléans, Risø and Wuppertal.

Acknowledgements to funding agencies

We gratefully acknowledge the support received from the Natural Environment Research Council, the Department of the Environment, the Commission of the European Communities, and the Institut Français du Pétrole, as well as the University of Oxford.

3.24 Time-Resolved Studies of the NO$_2$ and OH Formation in the Integrated Oxidation Chain of Hydrocarbons

R. Zellner, A. Hoffmann, W. Malms and V. Mörs

Institut für Physikalische Chemie, Universität Essen, D-45117 Essen, Germany

Summary

The oxidation of alkanes, alkenes and simple aromatics at 293 K under NO$_x$ rich tropospheric conditions has been studied using laser pulse initiation combined with cw laser long path absorption/LIF for the detection of OH and NO$_2$. In the case of aliphatic hydrocarbons the absolute yield and the kinetics of the formation of these products have been found to be sensitive indicators for the reaction behaviour of the oxy radicals RO. In combination with mechanistic simulations rate constants for individual reactions as well as branching ratios have been derived, which permit the evaluation of the compound specific NO/NO$_2$ conversion factors (NOCON - factors) for the first oxidation steps. In the case of benzene and toluene oxidation the results indicate that reaction of the primary formed X cyclohexa-dienyl radical (X = Cl, OH) with O$_2$ is the dominant pathway, although the rate coefficients were found to be lower than 2×10^{-16} cm^3/s.

Aims of the research

As has been shown in various experimental studies the excimer laser photolysis/ laser long path absorption/cw LIF technique developed in our laboratory within the LACTOZ program provides reliable information on branching ratios and rate coefficients in hydrocarbon oxidation under simulated NO$_x$ rich tropospheric conditions. As an systematic extension of our previous work on the alkane series C$_1$ – C$_6$ in the present period it was aimed to study:

- the oxidation of long chain hydrocarbons (C$_7$, C$_8$) and selected isomers (*iso*-butane) and cyclic alkanes (*c*-hexane) to refine our understanding of oxy radical reaction pathways and possibly identify mechanistic characteristics. Moreover, branching ratios and integrated rate coefficients for oxy radical decomposition/isomerisation reactions were to be determined to extract substance specific NO/NO$_2$ conversion factors from this information.
- the oxidation of simple alkenes (ethene, propene and isoprene), in order to derive characteristic NOCON factors for this class of substances.
- the oxidation mechanism of simple aromatics including the addition/abstraction ratio in the initial oxidation step, as well as rate coefficients and branching ratios in the consecutive reactions of O$_2$ with the X cyclohexadienyl-adduct (X = Cl, OH)

Principal experimental results

The oxidation of the hydrocarbons has been studied by observing the temporal evolution of OH and NO_2 following the pulse initiation of the oxidation chain using Cl atoms generated via excimer laser photolysis of Cl_2 at $l = 351$ nm. A detailed description of the experimental technique can be found elsewhere [1] and is not repeated here.

The temporal behaviour of the formation of NO_2 and OH reflects the time constant of the rate determining steps of their formation. Together with a complete chemical model and suitable adjustments of modelled and observed profiles individual rate constants were extracted. This data corresponds to an overall NO_2 (and therefore O_3) yield per RH molecule entering the oxidation process.

The significance of NO_2 and OH profile measurement in the oxidation of selected alkanes and alkenes may be seen from consideration of the generalised reaction scheme as presented in Fig. 1.

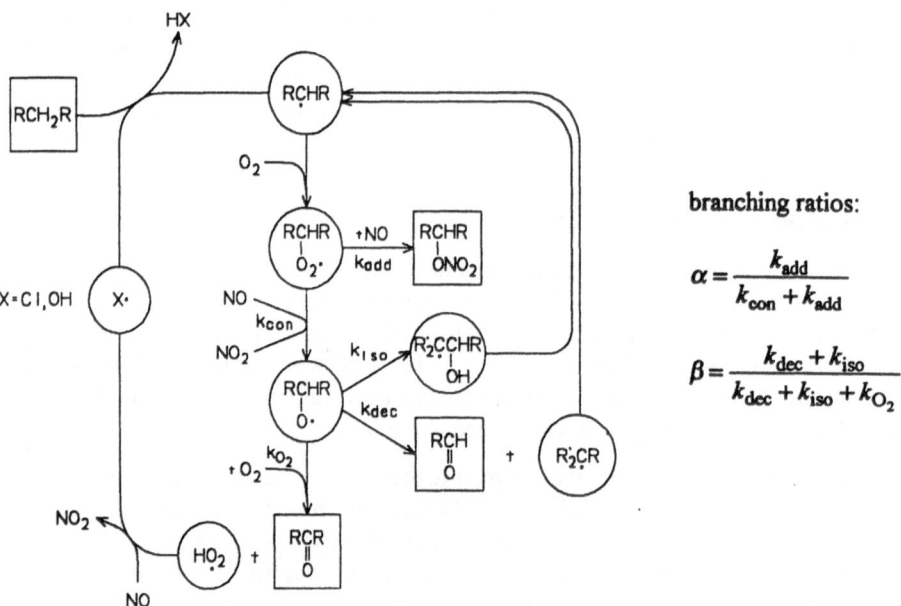

Fig. 1: Generalised oxidation scheme for aliphatic hydrocarbons.

Decomposition or isomerization of the oxy radicals leads to the formation of another alkyl radical, followed by additional NO to NO_2 conversion. The generation of OH radicals is coupled with the formation of HO_2 radicals in the course of the reaction of the oxy radicals with O_2 and, therefore, reflects the significance of this pathway. The experiments have been performed at $p = 50$ mbar and $T = 293$ K under different conditions characterised by a variation of NO $\{(0.5-2.0) \times 10^{14} \text{ cm}^{-3}\}$ and $O_2 \{(3.0-12.0) \times 10^{17} \text{ cm}^{-3}\}$ concentration.

Alkanes

In Table 1 the experimental results for the alkanes investigated are summarised together with the branching ratio a and b (as defined in Fig. 1) and the ratios DNO_2 / DOH and $DNO_2 / \Delta RH$ (equivalent to the NOCON factor) extrapolated to lower tropospheric conditions. The NOCON factors have been determined neglecting the pressure dependence of the decomposition/isomerisation rate constants. C_1 to C_6-alkanes have been investigated within LACTOZ I.

Table 1: Summary of results obtained for the oxidation of alkanes

C_nH_{n+2}	$\dfrac{k_{dec} + k_{iso}}{s^{-1}}$	$a*$	b	DNO_2 / DOH $p(O_2) = 0.2$ bar	DNO_2 / DRH $p(O_2) = 0.2$ bar
CH_4	—	—	—	2.0	2.0
C_2H_6	$\leq 1\times10^2$	0.0	≤ 0.01	2.0	2.0
C_3H_8	$\leq 4\times10^2$	0.02	≤ 0.01	2.01	1.96
n-C_4H_{10}	3.8×10^3	0.04	0.10	2.10	2.01
i-C_4H_{10}	1.1×10^3	0.09	0.03	2.03	1.85
n-C_5H_{12}	7.1×10^3	0.06	0.17	2.17	2.04
n-C_6H_{14}	3.3×10^4	0.12	0.49	2.49	2.19
c-C_6H_{12}	3.0×10^3	0.07	0.07	2.07	1.92
n-C_7H_{16}	4.5×10^4	0.24	0.54	2.54	1.93
n-C_8H_{18}	4.7×10^4	0.37	0.56	2.56	1.61

* weighted average value for all different oxy radicals

The branching ratio β generally increases with the chain length of the aliphatic hydrocarbon. The rate constants for the reactions of the oxy radicals with O_2 appear to be independent of the nature of the oxy radical (excl. CH_3O) with $k(O_2) = (8 \pm 2) \times 10^{-15} \text{ cm}^3/\text{s}$, whereas the overall rates of decomposition and isomerisation depend both on the chain length and the structure of the radical.

Alkenes

As can be seen from Fig. 1 the oxidation of alkenes generally corresponds to the alkane oxidation. However, it has to be considered that in this case the chlorine atom, which is used as a primary oxidant, predominantly adds to the double bond. Therefore, chlorine containing intermediates are formed. However, as a consequence of the oxidation mechanism only b-Cl oxy radicals are formed which are not expected to show significant deviations from the reactivity of their non-substituted analogues. In Table 2 the experimental results for alkenes are summarised when $k(O_2)$ is fixed to $8 \times 10^{-15} \text{ cm}^3/\text{s}$.

Table 2: Summary of results obtained for the oxidation of alkenes.

C_nH_m	$\dfrac{k_{dec} + k_{iso}}{s^{-1}}$	$a*$	b	DNO$_2$ / DOH $p(O_2) = 0.2$ bar	DNO$_2$ / DRH $p(O_2) = 0.2$ bar
C_2H_4	$\leq 2 \times 10^3$	0.0	≤ 0.05	2.0	2.0
C_3H_6	$\leq 2 \times 10^3$	0.024	≤ 0.05	2.0	1.96
Isoprene	1.2×10^4	0.061	0.34	2.34	1.91

* weighted average value for all different oxy radicals

Based on the values for the parameters a and b for any aliphatic hydrocarbon, its oxidation to the primary carbonyl compound may be summarised by the overall mechanistic equation

$$\text{OH} + \text{RH} + [a+(1-a)(2+b)]\,\text{NO}$$
$$\rightarrow a\,\text{RONO}_2 + (1-a)\,[(1+b)\,\text{Ald.} + (2+b)\,\text{NO}_2] + \text{H}_2\text{O}$$

which may be a useful approach to a condensed oxidation mechanism in tropospheric modelling. In general, the results obtained

- are in agreement with literature data [3–5] concerning the fact, that the significance of decomposition and/or isomerisation reactions of the oxy radicals increases with the number of C atoms;
- show, that the relation DNO$_2$ / DRH, which is related to the ozone formation potential lies below 2.2 for all hydrocarbons investigated when the results are extrapolated to lower tropospheric conditions.

Aromatics

The oxidation of benzene and toluene has been studied on the basis of a simplified reaction scheme presented in Fig. 2. The specific results obtained are collected in Table 3.

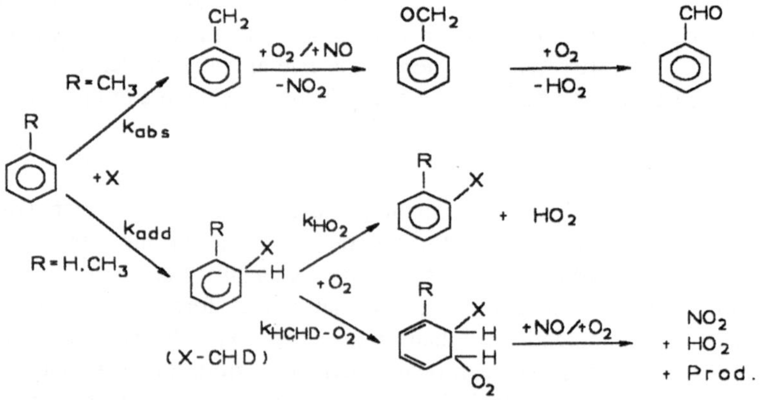

Fig. 2: Generalised oxidation scheme for simple aromatic hydrocarbons.

Table 3: Summary of results obtained for the oxidation of aromatic hydrocarbons.

aromatic	initiating radical	initiation reactions k_{add}/k_{abs}	X-CHD reactions $\frac{\sum k(O_2)}{10^{-16} cm^3 s^{-1}}$	$\frac{k(HO_2)}{k(XCHD\text{-}O_2)}$
benzene	Cl	—	1.2 ± 0.6	0.50 ± 0.25
toluene	OH	≥ 6	0.4–2.0	insensitive

The results for the rate coefficients and the branching ratio for the initial addition/abstraction reactions of toluene with OH are in agreement with previous measurements [7]. For benzene and toluene the values for the rate coefficients of the reactions of X–CHD with O_2 tend to be lower than previously reported by Zetzsch *et al.* [7]. However, it has to be recognised that the data of the Cl atom initiated oxidation of benzene is based on the investigation of chlorine and hydroxyl containing hexadienyl radicals. A separation of the O_2 reactions of the different radicals was not possible.

In the OH initiated oxidation of toluene a rough determination of the rate coefficient of $k(O_2)$ was possible. The biexponential decay profiles of OH and the extend of NO_2 formation appeared to be sensitive to the overall X CHD loss. However, the relatively fast recycling of OH due to formation of HO_2 during both degradation pathways made it impossible to deduce a branching ratio for these reactions.

Conclusions

The present experiments were aimed at a quantification of NO/NO_2 conversion factors in the first oxidation step of a series of alkanes, alkenes and aromatics. Moreover, the experimental approach, namely time-resolved measurement of OH and NO_2 evolution together with numerical simulation permitted the extraction of individual rate coefficients for the decomposition/isomerisation of oxy radicals. It has been shown that these rate coefficients increase with the size of the molecule up to a limiting value of ~ 10^5 s^{-1} at 298 K. The NOCON factors derived may be used to extract condensed mechanistic equations for the primary oxidation steps of hydrocarbons. The values derived in this work are integrated quantities comprising weighted averages of all fractional yields of detailed mechanisms as initiated by the attack of the primary oxidant to the various positions of the parent compound.

Together with the rate coefficients of primary oxidant attack (which correspond to the lifetime of the parent compound) NOCON factors may be used in atmospheric chemical - dynamic modelling to derive photochemical ozone creation potentials (POCP - values) for the individual hydrocarbons. POCP- values are expected to be the primary assessment factors in photochemical smog abatement strategies.

Acknowledgement

Support of this work by the "Bundesministerium für Forschung und Technologie" within the EUROTRAC subproject LACTOZ is gratefully acknowledged.

References

1.   A. Hoffmann, V. Mörs, R. Zellner, *Ber. Bunsenges. Phys. Chem.* **96** (1992) 437.
2.   K. H. Becker *et al.*, in: *EUROTRAC Ann. Rep.* 1990, Part 8, EUROTRAC ISS, Garmisch-Partenkirchen 1991.
3.   R. Atkinson, *Atmos. Environ.* **24**A (1990) 1.
4.   W. P. L. Carter, A. C. Lloyd, J. L. Sprung, J. N. Pitts, *Int. J. Chem. Kinet.* **9** (1979) 45.
5.   S. Dobe, T. Berces, F. Marta, *Int. J. Chem. Kinet.* **18** (1986) 329.
6.   R. Atkinson, *J. Phys. Chem. Ref. Data* **13** (1984) 1.
7.   R. Knispel, R. Koch, M. Siese, C. Zetzsch, *Ber. Bunsenges. Phys. Chem.* **94** (1990) 1375.

3.25 Adduct Formation of OH with Aromatics and Unsaturated Hydrocarbons and Consecutive Reactions with O_2 and NO_x to Regenerate OH

Cornelius Zetzsch, R. Koch, B. Bohn, R. Knispel, M. Siese and F. Witte

Fraunhofer-Institut für Toxikologie und Aerosolforschung, Nikolai-Fuchs-Straße 1, D-30625 Hannover, Germany

Summary

Using time-resolved resonance fluorescence detection of OH, rate constants for the removal of aromatic'–OH adducts from the thermal equilibrium between OH and adduct were measured [1–4] for the aromatics benzene, toluene, m- and p-xylene, phenol, m-cresol, aniline and naphthalene. In the case of O_2 as scavenger, the rate constants (cm^3/s) fall into two ranges depending on whether the aromatic is hydroxyl-substituted (phenol: 3×10^{-14}, m-cresol: 8×10^{-14}) or not (2×10^{-16} to 2×10^{-15}). In additional smog chamber studies [5] on benzene, toluene, and p-xylene at low O_2 concentrations the FP/RF results were confirmed. Although the reactions with NO_2 as scavenger are fast (2.5 to 5×10^{-11}), they are unimportant at atmospheric conditions (based on our results, however, high-NO_x smog-chamber studies had to be reinterpreted [6]). NO turned out to be unreactive against aromatic'–OH adducts, and upper limits (of a few 10^{-14}) were obtained.

Cycling experiments (FP/RF measurements in the systems OH/unsaturated HC/O_2/NO) with some of the aromatics [3, 7] (at temperatures, where the back-decomposition of the adduct is slow) and also with isoprene, showed that the O_2 reactions of the adducts lead to the formation HO_2 (detected by conversion to OH: $HO_2 + NO \rightarrow OH + NO_2$) with high yield and without detectable delay (< 50 ms). Cycling of OH with acetylene [8–10] is rapid in the absence of NO.

Aims of the research

For most unsaturated hydrocarbons, addition of OH is the first and rate-limiting step of the photochemical reaction chain. In the case of aromatics, which are emitted from automobiles, forest fires and fuel wood burning [11], the addition reaction is reversible at atmospheric temperatures. The effective rate constant for removal of the aromatic depends on consecutive reactions of the adduct. Prior to LACTOZ, consecutive reactions with O_2 had not been detected for benzene'–OH ([12], $k < 2 \times 10^{-16}$ cm^3/s) and toluene'–OH ([13] $k < 10^{-15}$ cm^3/s), so that Atkinson *et al.* concluded from their product studies [14, 15] that the corresponding NO_2 reactions [12, 16] dominate under their experimental conditions and possibly even in the troposphere. The project was undertaken to decide what is the second reaction step under tropospheric conditions and to gain further insight into the mechanism. From the products of aromatic oxidation and their properties one may eventually explain the high ozone formation potential of aromatics found in smog chambers and extrapolate it to tropospheric conditions.

Results

The method of flash photolysis/resonance fluorescence (FP/RF) and the details of our setup have been described elsewhere [17]. Reactions of the adduct aromatic–OH are determined from biexponential time profiles of OH, monitored by resonance fluorescence on a 1-to-600 ms timescale. The slope of the first exponential is dominated by the back and forth reaction between OH and the adduct, whereas the title reactions add to the final decay rate, common to OH and the adduct. The reaction mechanism is given in Fig. 1.

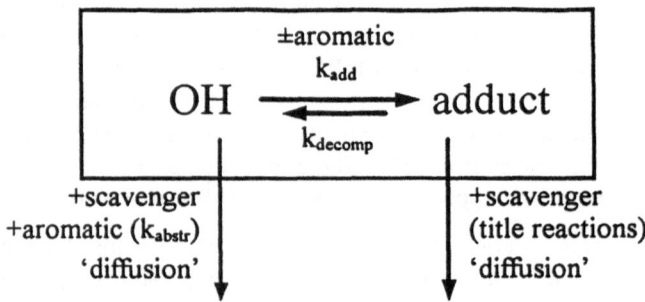

Fig. 1: The title reactions, adduct + scavenger, are studied by evaluating biexponential time profiles of OH. The structure of the adduct resembles cyclohexadienyl.

A set of decays recorded at constant temperature and pressure and various concentrations of hydrocarbon, R, and scavenger, S, is evaluated as a whole by fitting to them simulated decays calculated according to the 2nd order differential equation:

$$d[OH]/dt \quad = \quad -a[OH] + b[adduct]$$
$$d[adduct]/dt \quad = \quad c[OH] - d[adduct]$$

with concentration-dependent coefficients parameterised as follows

$$a \quad = \quad a_0 + (k_{abstr} + k_{add})\,[R] \; + a_S\,[S]$$
$$b \cdot c \quad = \quad k_{decomp} \cdot k_{add}\,[R]$$
$$d \quad = \quad k_{adduct\text{-}diff} + k_{decomp} + d_R\,[R] + d_S\,[S]$$

(inspection of the OH-part of the solution [1] shows that two of the four coefficients a to d occur always as product bc, *i.e.*, only three parameters are obtainable from a single biexponential decay of OH – plus its absolute intensity, which has no significance in pseudo-first-order kinetics). The fit parameters a_S and d_S are the rate constants for reactions of OH and adduct with the scavenger. The parameter d_R, denoting a reaction of the adduct with the aromatic reactant was never found to be of significance. The three remaining fit parameters, a_R, bc_R and d_0, are combinations of four rate constants: abstraction, adduct formation and decomposition and a loss of the adduct not leading back to OH, slightly faster than adduct diffusion alone (4 to 10 s^{-1} instead of < 1 s^{-1}). Especially at the lower temperatures, where k_{decomp} is slow, this degeneracy leads to increased uncertainties for the rate constants involved, but it does not influence the evaluation of the title reactions.

Pure aromatics

Rate constants for adduct formation and decomposition and for the abstraction pathway were obtained from experiments in the absence of NO_x and O_2. Values for the sum of both pathways ($k_{abstr} + k_{add}$, unaffected by the degeneracy) are given in Table 1. They are in good agreement with literature data [18]. The rate constants k_{decomp} are very similar for the studied aromatics, in the range 1 to 10 s^{-1} at 298 K, with activation energies of 8000 to 9000 K, except for naphthalene–OH ($k_{decomp} = 14$ s^{-1} at 400 K). Earlier estimates of k_{decomp} [18] based on the observation of non-exponential decays of OH at faster timescales, *i.e.*, higher temperatures, are in reasonable agreement with our results. Our branching ratios for the abstraction pathway (see also Table 1) tend to be somewhat lower than is extrapolated from high-temperature data [18] and, if combined with them, extrapolate to room temperature values in agreement with branching ratios derived from product studies [14, 15, 19, 20].

Table 1: Rate constants at room temperature and formal activation energies for the total OH reactivity of aromatics, $k_{OH} = k_{abstr} + k_{add}$, obtained from measurements in 130 mbar Ar (slightly below the high-pressure limit) and within the indicated temperature range. Rate constants for the abstraction pathway are given at temperatures were the observed losses are not dominated by the unresolved loss of adduct not leading back to OH.

	$\dfrac{k_{OH}(298\ K)}{10^{-12}\,cm^3\,s^{-1}}$	$\dfrac{E/R}{K}$	$\dfrac{T\text{-range}}{K}$	$\dfrac{k_{abstr}}{10^{-13}\,cm^3\,s^{-1}}$	$\dfrac{T}{K}$
benzene	1.05	−150	298–353	< 0.3	340
naphthalene	27.1	−630	295–352	< 2	420
toluene	6.38	−900	299–339	5.7 ± 1	340
p-xylene	15.8	−590	274–334	24 ± 3	390
m- xylene	21.1	−470	303–357	20 ± 2	380
phenol	24.7	−530	245–353	55 ± 7	340
m-cresol	56	−500	264–353	51 ± 5	380
aniline	100	−680	277–344	−	

We also studied the reaction of OH with isoprene [7] (250 to 440 K) and with acetylene [9] (294 to 353 K). For isoprene we obtained the expression $k_{OH} = 0.97 \times 10^{-10}$ $(T/298 \text{ K})^{-1.36}$ cm^3/s. The decays were monoexponential up to 440 K, i.e., no indication for decomposition of the adduct. For acetylene we obtained the Arrhenius expression $k_{OH} = 1.2 \times 10^{-12}$ $e^{-204 \text{ K}/T}$ cm^3/s and confirmed the pressure dependence in good agreement with literature data

Adduct + NO$_2$

In the case of NO$_2$ as scavenger the differential equation is supplemented by an source term of OH due to the conversion of H atoms (from the H$_2$O photolysis) by H + NO$_2$ → OH + NO leading to triexponential decays. Fig. 2 shows the measurements on p-xylene. While the determination of three exponentials from individual decays is too unstable, the use of the analytical solution [M. Siese in ref. 1] in a global fit leads to the rate constants given in Table 16, Chapter 2 of the general introduction. The reactions are fast – several 10^{-11} cm^3/s – and show a weak dependence on type of aromatic and temperature. Literature values on benzene [12, 21, 22] are considerably scattered, whereas the single literature value on toluene [23] is in reasonable agreement with our results.

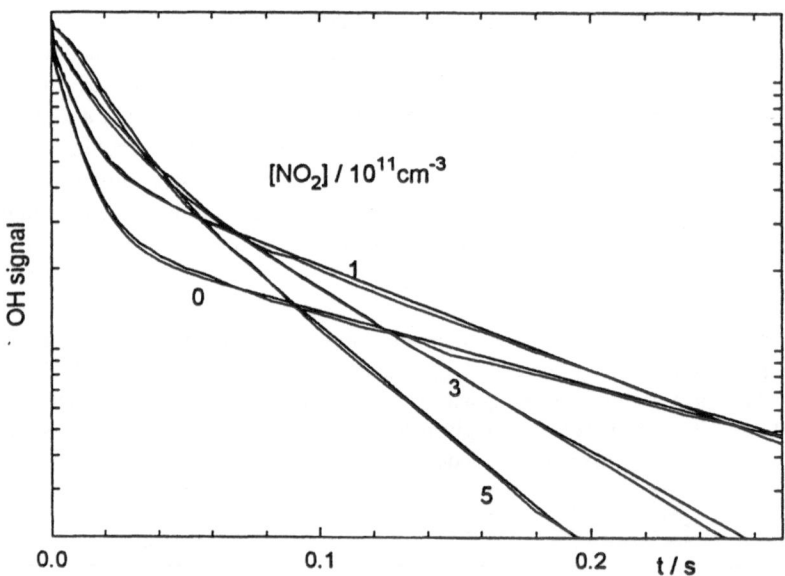

Fig. 2: Decays of OH in the presence of p-xylene (5.1 × 10^{12} cm^{-3}) and increasing amounts of NO$_2$ at 317 K and 130 mbar of Ar. The continuous lines indicate the result of a triexponential global fit according to the mechanism of Fig. 1 plus an OH source due to conversion of H atoms by NO$_2$. A minor discrepancy at $t < 50$ ms has been reduced by allowing [H]$_0 \neq$ [OH]$_0$: 20 % extra H atoms are assumed to stem from the photolysis of p-xylene. This degree of freedom has virtually no influence on the value of the target reaction $-k(p\text{-xylene} \cdot -OH + NO_2) = (3.5 \pm 0.5) \times 10^{-11}$ cm^3/s.

Adduct + **NO**

This reaction was found to be slow compared to NO reactions with other organic radicals: Upper limits between 0.3 and 1×10^{-13} cm^3/s were obtained by careful consideration of interfering reactions: NO$_2$ impurities in the NO are undesirable for two reasons. (i) The fast reaction adduct + NO$_2$ mimics a NO reactivity – probably the cause for a higher rate constant in the literature [12], k(benzene –OH + NO) = 1×10^{-12} cm^3/s. We successfully purified the NO with FeSO$_4$. (ii) Although the concentration of NO$_2$ was measured (in the absence of aromatic) to be at most 10^{-4} relative to NO, *i.e.*, well below [OH]$_0$, the effect of the OH source term (H + NO$_2$) may still be important in the presence of aromatic, because of the lower level of OH during the late part of the biexponential decays. Model calculations with [NO$_2$] = 10^{-4} [NO] showed that the final slope of the decays may even *decrease* with NO. Finally, the unavoidable reaction OH + NO + M $\rightarrow$ HONO + M ($\sim 10^{-12}$ cm^3/s at our pressure and temperature) is formally independent (see above a_S and d_S), but increases the uncertainty.

Adduct + **O$_2$**

In order to measure these low rate constants (with [O$_2$] limited to below 10^{17} cm^{-3} because of quenching of the OH fluorescence) slow decays in the absence of O$_2$ are required, *i.e.*, a clean set-up, pure inert gas, a low level of radicals to avoid radical-radical reactions (here: S/N > 100 at 10^{10} cm^{-3}) and a non-reactive source of OH (here: flash photolysis of H$_2$O). In the absence of reactants the decay rate of OH is as low as 3 to 4 s^{-1}, with a stability of ±0.2 s^{-1}.

Up to some 10^{17} cm^{-3} of O$_2$ has been applied to considerably steepen the final exponential curve. The loss of fluorescence intensity due to quenching was compensated by averaging several hundred shots. An initial concentration of O atoms (about ten times the OH concentration at 10^{17} cm^{-3} O$_2$) is estimated by comparing the absorption cross sections of H$_2$O and O$_2$. Because of the low flash energy, 'O + radical' reactions are still unimportant, but a possible OH formation by 'O + aromatic' would lead to triexponential decays with the rate of the third exponential, up to 100 s^{-1}, dominated by the competing three-body reaction with O$_2$ (forming the less reactive ozone). Since the observed decays do not deviate significantly from the model, upper limits are estimated for that OH forming reaction: 0.2, 1, 8, and 10×10^{-14} cm^3/s for benzene, toluene, *p*-xylene, and aniline, respectively, and about collision frequency for phenol and *m*-cresol (the trend being due to the concentrations of aromatic and O$_2$ applied when measuring the title reactions).

Fig. 3 shows decays in the presence of 3×10^{13} cm^{-3} of toluene and increasing amounts of O$_2$ together with the results of individual fits (broken lines) and of the global fit (full lines) according to the mechanism of Fig. 1.

The obtained rate constants for the O$_2$ reaction of the adducts divide into two ranges depending on whether the aromatic is hydroxyl-substituted or not

(Table 16, Chapter 2). At tropospheric conditions, the O_2 reaction dominates – even in the case of benzene.

The competition with the thermal back-decomposition of the adduct, k_{decomp}, was studied by lowering the O_2 concentration in chamber experiments on benzene, toluene, and p-xylene (with photolysis of H_2O_2 as NO_x-free source of OH) [5]. At low levels of O_2, the removal of the aromatics (observed by cryo-focusing/ GC/FID) corresponds to the abstraction channel, and increases to the sum of abstraction and addition in the limit of high O_2. In Table 2, ratios k_{decomp} / $k(O_2)$ obtained from such measurements are compared to those calculated from our absolute rate constants. That an earlier chamber study on toluene and p-xylene [24] shows much less O_2 dependence, has been attributed [19] to rather high concentrations of hydrocarbons, H_2O_2, and thus radicals in their experiments leading to further loss processes of the adduct.

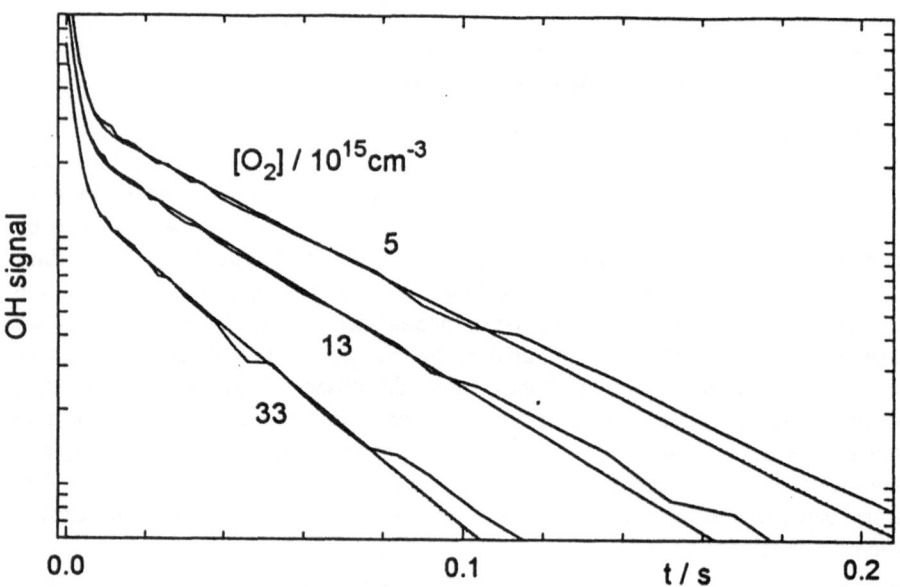

Fig. 3: Decays of OH in the presence of toluene (85×10^{12} cm^{-3}) and increasing amounts of O_2 at 339 K and 134 mbar of Ar. The biexponential individual fits (dotted lines) and the global fit according to the mechanism in Fig. 1 (continuous lines) yield the identical result, $k(\text{toluene} -OH + O_2) = (5.6 \pm 0.6) \times 10^{-16}$ cm^3/s.

Table 2: Ratios k_{decomp} / ($k(O_2)\cdot10^{16}$ cm^{-3}) obtained from smog chamber data [5] compared to those calculated with our absolute rate constants (300 K).

	FP/RF	smog chamber
benzene	2.3	1.8
toluene	1.1	1.2
p-xylene	0.52	0.36

Cycling experiments product of the reaction of O_2 with the adduct. The resulting cycle of odd-hydrogen species allows us

- to measure $k_{add + O_2}$ at temperatures too low for the direct regeneration of OH by thermal decomposition of the adduct,
- to determine the yield of HO_2 (a possible loss reaction would be ROO + NO $\rightarrow$ RONO$_2$ instead of the chain propagation step $\rightarrow$ RO + NO$_2$, see the case of isoprene below),
- to detect delays in the formation of HO_2.

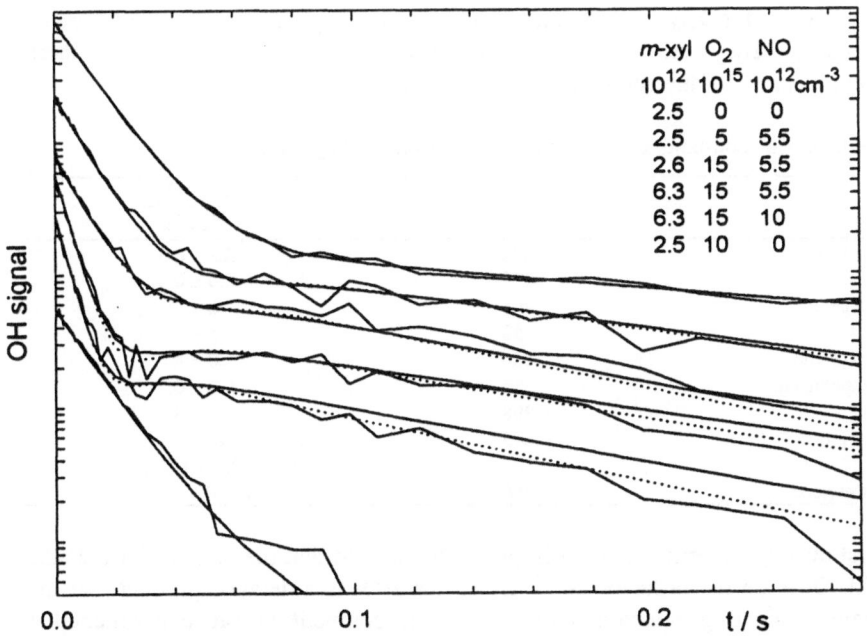

Fig. 4: Cycling of OH in the presence of varying concentrations of *m*-xylene, O_2, and NO at 286 K and 255 mbar of He. The curves are arbitrarily shifted for clarity. The weak second exponential in the absence of O_2 (topmost curve) is due to the thermal decomposition of the adduct – which is scavenged by the addition of O_2 (lowest curve). With the addition of NO the signal comes up again (to about 4 times the level without O_2; compare curves 1 and 3 at $t = 0.1$ s respective to their initial intensities). The full and dotted lines are calculated by two alternative global models (see text).

Fig. 4 shows recent measurements with m-xylene. As with benzene and toluene, the measurements are fairly well reproduced by a model including only the prompt formation of HO_2 with unit yield. Based on the depth and width of the dip between the initial and final exponential decay, we can state that, if HO_2 is not formed *via* the prompt path, any further reaction steps should cause delays not larger than 20 ms (based on our experimental conditions), *i.e.*,

- any unimolecular step should be faster than 50 s^{-1}
- any reaction step with NO should be faster than $k(HO_2 + NO)$
- any reaction step with O_2 should be faster than $k(add + O_2)$.

In the case of m-xylene, however, the dependence of the final signal level on NO concentration could be fitted only by either allowing the rate constant of $HO_2 + NO \rightarrow OH + NO_2$ to settle at half the literature value (full lines in Fig. 4) or by inserting a further reaction with NO into the cycle: $RO_2 + NO \rightarrow RO + NO_2$ (followed by the fast step $RO + O_2 \rightarrow HO_2$). The latter alternative allows us to slightly improve the modelling of the final decay rates by introducing the loss reaction $RO_2 + NO \rightarrow RONO_2$ (dotted lines in Fig. 4).

The results for $k_{add + O_2}$ obtained from cycling experiments are summarised in Table 3. They are in good agreement with those obtained by thermal decomposition of the adduct (at slightly higher temperatures).

Table 3: Rate constants $k(adduct+O_2)$ derived from cycling experiments.

	T/K	$\dfrac{k(add + O_2)}{10^{-16} cm^3 s^{-1}}$
benzene	290	2.3 ± 0.3
	273	2.0 ± 0.3
toluene	295	6 ± 1
	258	5 ± 1
m-xylene	286	20 ± 4
naphthalene	336	0.8 ± 0.3
	298	8 ± 3
isoprene	345	0.6×10^4
	300	2.0×10^4
acetylene	296	$(4.6 \pm 0.4) \times 10^4$

First cycling experiments with isoprene [8] showed that the reaction of the adduct with O_2 is on the order of 10^{-12} cm^3/s, that HO_2 is a delayed product and that a chain-terminating reaction with NO is in competition (at experimental NO concentrations) with a chain-propagating reaction with O_2.

In case of acetylene, OH is regenerated directly in a reaction of the adduct with O_2 [25]. We reinvestigated this reaction using FP/RF [8,9] and obtained a slightly negative temperature coefficient (294–353 K) and no pressure dependence (16–260 mbar Argon). The OH regeneration yield was found to be 86 % at room temperature and 260 mbar. However, at oxygen mixing ratios exceeding 1 % (studied by laser photolysis/cw uv laser long-path absorption [10]) this yield decreases markedly – from 82 % at 6 mbar O_2 (in 1 bar N_2) to 70 % in synthetic air to 57 % in pure oxygen. The measurements are in accord with a model where the OH yield is lower in the reaction of O_2 with not fully thermalised adducts.

Achievements

In the atmosphere the reactions of the adducts aromatic–OH with O_2 dominate over their NO_2 reactions (and the reactions with NO are totally unimportant). However, this competition is at work in smog chamber experiments in the presence of NO_x. Product studies from benzene, toluene [14] and the xylene isomers [15,6] show a dependence on NO_2 of the nitroaromatic yields in rough agreement with our absolute rate data.

Acknowledgements

Support from the BMBF (grant 07 EU 705 A/O) and the CEC (grants STEP 0007-C and EV5V-CT93-0309) is gratefully acknowledged.

References

1. C. Zetzsch, R. Koch, M. Siese, F. Witte, P. Devolder, in: G. Restelli, G. Angeletti (eds), *Physico-Chemical Behaviour of Atmospheric Pollutants, Air Pollution Research Report* **23**, CEC DG XII, EUR 12542, Brussels 1990, p. 320.
2. R. Knispel, R. Koch, M. Siese, C. Zetzsch, *Ber. Bunsenges. Phys. Chem.* **94** (1990) 1375.
3. R. Koch, J. Nowack, M. Siese, C. Zetzsch, in: R.A. Cox (ed), *Laboratory studies on atmospheric chemistry, Air Pollution Research Report* **42**, EUR 14756, CEC DG XII, Brussels 1991, p. 47.
4. C. Zetzsch, R. Knispel, R. Koch, M. Siese, in: J. Peeters (ed), Chemical mechanisms describing tropospheric processes, Air Pollution Research Report **45**, EUR 14884, CEC DG XII, Brussels 1992, p. 107.
5. M. Elend, C. Zetzsch, in: R.A. Cox (ed), *Laboratory Studies on Atmospheric Chemistry, Air Pollution Research Report* **42**, EUR 14756, CEC DG XII, Brussels 1991, p. 123.
6. R. Atkinson, S. M. Aschmann, *Int. J. Chem. Kinet.* **26** (1994) 929.
7. M. Siese, R. Koch, C. Fittschen, C. Zetzsch, in: P.M. Borrell, P. Borrell, T. Cvitaš, W. Seiler (eds), *Proc. EUROTRAC Symp. '94*, SPB Acad. Publ. bv, The Hague 1994, p. 115.
8. R. Koch, M. Siese, C. Fittschen, C. Zetzsch, presented at the LACTOZ/HALIPP meeting, Leipzig, 1994 (in press).
9. M. Siese, C. Zetzsch, *Z. Phys. Chem.* **188** (1995) 75.
10. B. Bohn, M. Siese, C. Zetzsch, presented at the *Faraday Discussion* **100**, Norwich, 1995.
11. S.D. Piccot, J.J. Watson, J.W. Jones, *J. Geophys. Res.* **97** (1992) 9897.
12. R. Zellner, B. Fritz, M. Preidel, *Chem. Phys. Lett.* **121** (1985) 412.
13. R.A. Perry, R. Atkinson, J.N. Pitts Jr., *J. Phys. Chem.* **81** (1977) 296.

14. R. Atkinson, S.M. Aschmann, J. Arey, W.P.L. Carter, *Int. J. Chem. Kinet.* **21** (1989) 801.

15. R. Atkinson, S.M. Aschmann, J. Arey, *Int. J. Chem. Kinet.* **23** (1991) 77.

16. N. Bourmada, P. Devolder, L.R. Sochet, *Chem. Phys. Lett.* **149** (1988) 339.

17. A. Wahner, C. Zetzsch, *J. Phys. Chem.* **87** (1983) 4945, and ref. 1–4.

18. R. Atkinson, *J. Phys. Chem. Ref. Data, Monograph* **1** (1989).

19. K.H. Becker, I. Barnes, A. Bierbach, M. Martin-Reviejo, E. Wiesen, in: J. Peeters (ed), *Chemical Mechanisms Describing Tropospheric Processes, Air Pollution Research Report* **45**, EUR 14884, CEC DG XII, Brussels 1992, p. 87.

20. R. Seuwen, P. Warneck, in: J. Peeters (ed), *Chemical Mechanisms Describing Tropospheric Processes, Air Pollution Research Report* **45**, EUR 14884, CEC DG XII, Brussels 1992, p. 95.

21. A. Goumri, J.P. Sawerysyn, J.F. Pauwels, P. Devolder, in: G. Restelli, G. Angeletti (eds), *Physico-Chemical Behaviour of Atmospheric Pollutants, Air Pollution Research Report* **23**, CEC DG XII, EUR 12542, Brussels 1990.

22. E. Bjergbakke, P. Pagsberg, presented at the LACTOZ/HALIPP meeting, Leipzig, 1994 (in press).

23. A. Goumri, J.F. Pauwels, P. Devolder, *Canad. J. Chem.* **69** (1991) 1057

24. I. Barnes, K.H. Becker, A. Bierbach, E. Wiesen, p. 183 in: see ref. 3.

25. V. Schmidt, G. Y. Zhu, K. H. Becker and E. H. Fink, *Ber. Bunsenges. Phys. Chem.* **89** (1985) 321.

Chapter 4

LACTOZ Publications 1988–1995

1988

Benter, Th., E. Becker, U. Wille, R. N. Schindler;
Absolute rate constants for the reaction of NO_3 with alkenes,
in: R. A. Cox (ed), *Mechanisms of Gas Phase and Liquid Phase Chemical Transformations in Tropospheric Chemistry, Air Pollution Research Report* **17**, EUR 12035, CEC, Brussels 1988, pp. 155–159.

Benter, Th., R. N. Schindler;
Absolute rate coefficients for the reaction of NO_3 radicals with simple dienes,
Chem. Phys. Lett. **145** (1988) 67–70.

Boodaghians, R.B., C.E. Canosa-Mas, P.J. Carpenter, R.P. Wayne;
The reactions of NO_3 with OH and H,
J. Chem. Soc., Faraday Trans. 2, **84** (1988) 931

Canosa-Mas, C.E., R.P. Wayne;
Some reactions of NO_3 with atoms and radicals,
in: O.J. Nielsen, R.A. Cox (eds), *Tropospheric NO_x Chemistry – Gas Phase and Multiphase Aspects, Air Pollution Research Report* **9**, EUR 11440, CEC, Brussels 1988, p. 84.

Canosa-Mas, C.E., S. J. Smith, S. Toby, R. P. Wayne;
Reactivity of the nitrate radical towards alkynes and some other molecules,
J. Chem. Soc., Faraday Trans. 2, **84** (1988) 247.

Canosa-Mas, C.E., S. J. Smith, S. Toby, R. P. Wayne;
Temperature dependences of the reactions of the nitrate radical with some alkynes and with ethylene,
J. Chem. Soc., Faraday Trans. 2, **84** (1988) 263.

Canosa-Mas, C.E., S.J. Smith, S.J. Waygood, R.P. Wayne;
Stoicheiometries and absolute concentrations in radical experiments,
in: R. A. Cox (ed), *Mechanisms of Gas Phase and Liquid Phase Chemical Transformations in Tropospheric Chemistry, Air Pollution Research Report* **17**, EUR 12035, CEC, Brussels 1988, p. 17.

Daumont, D., A. Barbe, J. Brion, J. Malicet;
New absolute absorption cross-sections of O_3 in the 195–350 nm spectral range,
in: R.D. Bojkov, P. Fabian (eds), *Ozone in the Atmosphere, Proc. Quadrennial Ozone Symp.*, Deepak Publ., Hampton 1990, pp. 710–711.

Devolder, P., N. Bourmada, J. P. Sawerysyn;
Rate constants of the reactions of OH with benzene and toluene in the fall-off range,
in: R. A. Cox (ed), *Mechanisms of Gas Phase and Liquid Phase Chemical Transformations in Tropospheric Chemistry, Air Pollution Research Report* **17**, EUR 12035, CEC, Brussels 1988, p. 117.

Hall, I.W., R.P. Wayne, R.A. Cox, M.E. Jenkin, G.D. Hayman;
Kinetics of the reaction of NO_3 with HO_2,
J. Phys. Chem. **92** (1988) 5049.

Jenkin, M.E., R.A. Cox, G.D. Hayman, L.J. White;
Kinetic study of the reactions of CH_3O_2 + CH_3O_2 and CH_3O_2 + HO_2 using molecular modulation spectroscopy,
J. Chem. Soc., Faraday Trans. 2 **84** (1988) 913.

Mellouki, A., G. Le Bras, G. Poulet;
Kinetics of the reaction of NO_3 with OH and HO_2,
J. Phys. Chem. **148** (1988) 231.

Nielsen, O.J., H.W. Sidebottom, D.J. O'Farrell, M. Donlon,, J. Treacy;
Absolute and relative rate constants for the gas-phase reaction of OH radicals with CH_3NO_2, CD_3NO_2 and $CH_3CH_2CH_3$ at 295 K and 1 atm,
Chem. Phys. Lett. **146** (1988) 197–203.

Rahman, M. M., E. Becker, Th. Benter, R. N. Schindler;
A gas-phase investigation of the system $F+HNO_3$ and the determination of absolute rate constants for the reaction of the NO_3 radical,
Ber. Bunsenges. Phys. Chem. **92** (1988) 91–100.

Tsalkani, N., A. Mellouki, G. Poulet, G. Toupance, G. Le Bras;
Rate constant measurement for the reaction of OH and Cl with peroxyacetyl nitrate at 298 K,
J. Atmos. Chem. **7** (1988) 409.

Witte, F., C. Zetzsch;
Adduct formation and its unimolecular decay in the reaction of OH with toluene,
in: R.A. Cox (ed), *Mechanisms of gas phase and liquid phase chemical transformations in tropospheric chemistry, Air Pollution Research Report* **17**, EUR 12035, CEC, Brussels 1988, p. 115.

1989

Bächmann, K., J. Hauptmann;
Determination of organic peroxides,
in: H.W. Georgii (ed), *Mechanisms and Effects of Pollutant Transfer into Forests*, Kluwer Academic Publ., Dordrecht 1989.

Becker, E., U. Wille, R.N. Schindler;
An Investigation of the system Cl + NO$_3$ by mass spectrometry,
in: K.H. Becker (ed), *Atmospheric Oxidation Processes,* pp. 209–211.

Becker, K.H., K. Wirtz;
Gas phase reactions of alkyl nitrates with hydroxyl radicals under tropospheric conditions in comparison with photolysis,
J. Atmos. Chem. **9** (1989) 419

Becker, K.H., K.J. Brockmann, J. Bechara;
Tunable diode-laser measurements of CH$_3$OOH absorption cross-sections near 1320 cm^{-1},
Geophys. Res. Lett. **16** (1989) 1367.

Burrows, J.P., G.K. Moortgat, G.S. Tyndall, R.A. Cox, M.E. Jenkin, G.D. Hayman, B. Veyret;
Kinetics and mechanism of the photooxidation of formaldehyde. 2. Molecular modulation studies,
J. Phys. Chem. **93** (1989) 2375–2382.

Canosa-Mas, C.E., P.J. Carpenter, R.P. Wayne;
The reaction of NO$_3$ with atomic oxygen,
J. Chem. Soc., Faraday. Trans. 2 **85** (1989) 697.

Canosa-Mas, C.E., S.J. Smith, S. Toby, R.P. Wayne;
   Laboratory studies of the reactions of the nitrate radical with chloroform, methanol,
   hydrogen chloride and hydrogen bromide,
   J. Chem. Soc., Faraday. Trans. 2 **85** (1989) 709.

Heiss, A., K. Sahetchian;
   Reactions of *n*-butoxy and *s*-butoxy radicals in presence of oxygen,
   in: *Proc. Joint Meeting of the British and French Sections of the Combustion Institute*,
   Rouen 1989, p. 93.

Hjorth, J., F. Cappellani, C.J. Nielsen, G. Restelli
   Determination of the $NO_3 + NO_2 \rightarrow NO + O_2 + NO_2$ rate constant by infrared diode
   laser and Fourier transform spectroscopy,
   J. Phys. Chem. **93** (1989) 5458.

Horie, O., G.K. Moortgat;
   A new transitory product in the ozonolysis of *trans*-2-butene at atmospheric pressure,
   Chem. Phys. Lett. **156** (1989) 39–46 and **158** (1989) 178.

Lançar, I.T., G. Poulet, G. Le Bras, U. Wille, E. Becker, R.N. Schindler;
   On the kinetics and the mechanism of the reaction between NO_3 and isoprene,
   in: K. H. Becker (ed), *Atmospheric Oxydation Processes*, p. 45–49.

Malicet, J., J. Brion, D. Daumont;
   Temperature dependence of the absorption cross-section of ozone at 254 nm,
   Chem. Phys. Lett. **158** (1989) 293.

Mellouki, A., G. Poulet, G. Le Bras, R. Singer, J. Burrows, G.K. Moortgat;
   Discharge flow kinetic study of the reactions of NO_3 with Br, BrO, HBr and HCl,
   J. Phys. Chem. **93** (1989) 8017.

Moortgat, G.K., B. Veyret, R. Lesclaux;
   Absorption spectrum and kinetics of reactions of the acetylperoxy radical,
   J. Phys. Chem. **93** (1989) 2362–2368.

Moortgat, G.K., B. Veyret, R. Lesclaux;
   Kinetics of the reaction of HO_2 with $CH_3C(O)O_2$ in the temperature range
   253–368 K,
   Chem. Phys. Lett. **160** (1989) 443–447.

Moortgat, G.K., R.A. Cox, G. Schuster, J.P. Burrows, G.S. Tyndall;
   Peroxy radical reactions in the photooxidation of CH_3CHO,
   J. Chem. Soc., Faraday Trans. 2, **85** (1989) 809–829.

Nielsen, O.J., H.W. Sidebottom, D.J. O'Farrell, M. Donlon, J. Treacy;
Rate constants for the gas-phase reactions of OH radicals and Cl atoms with
$CH_3CH_2NO_2$, $CH_3CH_2CH_2NO_2$, $CH_3CH_2CH_2CH_2NO_2$ and $CH_3CH_2CH_2CH_2CH_2NO_2$
Chem. Phys. Lett. **156** (1989) 312–318.

Nielsen, O.J., H.W. Sidebottom, L. Nelson, J.J. Treacy, D.J. O'Farrel;
An absolute and relative study of the reaction of OH radicals with dimethyl sulfide,
Int. J. Chem. Kinet. **21** (1989) 1101.

Nielsen, O.J., M. Donlon, D. J. O'Farell, J. Treacy, H. Sidebottom;
Kinetics and mechanism for the reaction of hydroxyl radicals with nitrogen containing
compounds,
in: *Tenth Int. Symp. on Gas Kinetics – Abstracts of Papers*, Swansea 1989, p. A39.

Rattigan, O., H.W. Sidebottom, J. Treacy, O.J. Nielsen;
The reaction of hydroxyl radicals and chlorine atoms with sulphur dioxide,
Proc. Roy.Irish Acad. **89**B (1989) 353–361.

Veyret, B., R. Lesclaux, M.T. Rayez, J.C. Rayez, R.A. Cox, G.K. Moortgat;
Kinetics and mechanism of the photooxidation of formaldehyde. 1. Flash photolysis
study,
J. Phys. Chem. **93** (1989) 2368–2374.

Wassell, P.T., J. Ballard, W.B. Johnston, R.P. Wayne;
Laboratory spectroscopic studies of atmospherically important radicals using Fourier
transform spectroscopy,
J. Atmos. Sci. **8** (1989) 63.

Zabel, F., A. Reimer, K. H. Becker, E. H. Fink;
Thermal decomposition of alkyl peroxynitrates,
J. Phys. Chem. **93** (1989) 5500.

1990

Bächmann, K., J. Hauptmann;
Determination of organic peroxides,
in: G. Restelli, G. Angeletti (eds), *Fifth European Symp. on Physico-Chemical
Behaviour of Atmospheric Pollutants*, Kluwer Academic Publ., Dordrecht 1990.

Bächmann, K., W. Schmied, M. Przewosnik;
Determination of traces of aldehydes and ketones in the troposphere via solid phase
derivatisation with DNSH,
Fresenius Z. Anal. Chem. **33**S (1990) 464

Bagley, J.A., P. Biggs, C.E. Canosa-Mas, M.R. Little, A. D. Parr, S. J. Smith,
S.J. Waygood, R.P. Wayne;
Temperature dependence of reactions of the nitrate radical with alkanes,
in: G. Restelli, G. Angeletti (eds), *Fifth European Symp. on Physico-Chemical
Behaviour of Atmospheric Pollutants*, Kluwer Academic Publ., Dordrecht 1990,
pp. 328–333.

Bagley, J.A., P. Biggs, C.E. Canosa-Mas, M.R. Little, A. D. Parr,S. J. Smith,
S.J. Waygood, R.P. Wayne;
Temperature dependence of reactions of the nitrate radical with alkanes,
J. Chem. Soc., Faraday Trans. **86** (1990) 2109.

Barnes, I., V. Bastian, K.H. Becker, T. Zhu;
Kinetics and products of the reactions of NO_3 with monoalkenes, dialkenes and
monoterpenes,
J. Phys. Chem. **94** (1990) 2414.

Becker, K.H., K.J. Brockmann, J. Bechara;
Production of hydrogen peroxide in forest air by reaction of ozone with terpenes,
Nature **346** (1990) 256

Biggs, P., A.C. Brown, C.E. Canosa-Mas, P.J. Carpenter, R.P. Wayne;
The reactions of NO_3 with atoms and radicals,
in: G. Restelli, G. Angeletti (eds) *Fifth European Symp. on Physico-Chemical
Behaviour of Atmospheric Pollutants*, Kluwer Academic Publ., Dordrecht 1990,
pp 204–209.

Biggs, P., C.E. Canosa-Mas, M. Joseph, P. Monks, R. P. Wayne;
The self-reaction of the NO_3 radical,
in: G. Restelli, G. Angeletti (eds), *Fifth European Symp. on Physico-Chemical
Behaviour of Atmospheric Pollutants*, Kluwer Academic Publ., Dordrecht 1990,
pp 408–413.

Cariati, F., B. Rindone, J. Hjorth, G. Restelli;
Products and Mechanism of the Gas Phase Reaction between the Nitrate Radical and
Arenes,
in: G. Restelli, G. Angeletti (eds), *Fifth European Symp. on Physico-Chemical
Behaviour of Atmospheric Pollutants*, Kluwer Academic Publ., Dordrecht 1990,
pp 400–407.

Cox, R.A., J. Munk, O.J. Nielsen, P. Pagsberg;
Ultraviolet absorption spectra and kinetics of acetonyl and acetonyl peroxy radicals,
Chem. Phys. Lett. **173** (1990) 206.

Crowley, J.N., J.P. Burrows, G.K. Moortgat;
Room temperature rate coefficient for the reaction between CH_3O_2 and NO_3,
in: G. Restelli, G. Angeletti (eds), *Fifth European Symp. on Physico-Chemical Behaviour of Atmospheric Pollutants*, Kluwer Academic Publ., Dordrecht 1990, pp. 371–376.

Crowley, J.N., J.P. Burrows, G.K. Moortgat, G. Poulet, G. Le Bras;
Room temperature rate coefficient for the reaction of CH_3O_2 with NO_3,
Int. J. Chem. Kinet. **22** (1990) 673–681.

Elfers, G., F. Zabel, K.H. Becker;
Determination of the rate constant ratio for the reactions of the ethylperoxy radical with NO and NO_2,
Chem. Phys. Lett. **168** (1990) 14.

Goumri, A., J.F. Pauwels, J.P. Sawerysyn, P. Devolder;
Reaction rates at (297 ± 3) K with O_2, NO and NO_2 of the *p*-fluorobenzyl and the *m*-fluorobenzyl radicals by discharge flow / laser induced fluorescence,
Chem. Phys. Lett. **171** (1990) 303.

Goumri, A., J.P. Sawerysyn, J.F. Pauwels, P. Devolder;
Tropospheric oxidation of toluene: reactions of some intermediate radicals,
in: G. Restelli, G. Angeletti (eds), *Fifth European Symp. on Physico-Chemical Behaviour of Atmospheric Pollutants*, Kluwer Academic Publ., Dordrecht 1990, pp 315–319.

Hartmann, D., J. Karthauser, J. P. Sawerysyn, R. Zellner;
Kinetics and HO_2 product yield of the reaction $C_2H_5O + O_2$ between 295 and 411 K,
Ber. Bunsenges. Phys. Chem. **94** (1990) 639.

Hartmann, D., J. Karthauser, J.P. Sawerysyn, R. Zellner;
Kinetics and HO_2 product yield of the reaction $C_2H_5O + O_2$ between 295 and 411 K,
Ber. Bunsenges. Phys. Chem. **94** (1990) 639.

Hartmann, D., J. Karthauser, R. Zellner;
Laser photofragment emission: A novel technique for the study of gas phase reactions of the CH_3O_2 radical,
in: G. Restelli, G. Angeletti (eds), *Proc. 5th European Symp. on Physico-chemical Behaviour of Atmospheric Pollutants*, Kluwer Academic Publ., Dordrecht 1990, pp. 352–358.

Hjorth, J., C. Lohse, C.J. Nielsen, H. Skov, G. Restelli;
Products and mechanisms of the reaction of NO_3 with a series of alkenes,
J. Phys. Chem. **94** (1990) 7494.

Horie, O., G.K. Moortgat;
Ozonolysis of alkenes under atmospheric conditions,
in: R.D. Bojkov, P. Fabian (eds), *Ozone in the Atmosphere, Proc. Quadrennial Ozone Symp.*, Deepak Publ., Hampton 1990, pp. 698–701.

Horie, O., J.N. Crowley, G.K. Moortgat;
Methylperoxy self-reaction: Products and branching ratio between 223 and 333 K,
J. Phys. Chem. **94** (1990) 8198–8203.

Horie, O., J.N. Crowley, G.K. Moortgat;
Methylperoxy self-reaction: products and branching ratio between 223 and 333 K,
in: G. Restelli, G. Angeletti (eds), *Fifth European Symp. on Physico-Chemical Behaviour of Atmospheric Pollutants*, Kluwer Academic Publ., Dordrecht 1990, pp. 341–346.

Jenkin, M.E., R.A. Cox;
Laboratory studies of $HOCH_2CH_2O_2$ radicals produced by the photolysis of 2-iodoethanol in the presence of O_2,
in: G. Restelli, G. Angeletti (eds), *Fifth European Symp. on Physico-Chemical Behaviour of Atmospheric Pollutants*, Kluwer Academic Publ., Dordrecht 1990, pp. 383–393.

Kirchner, F., F. Zabel, K.H. Becker;
Determination of the rate constant ratio for the reactions of the acetylperoxy radical with NO and NO_2,
Ber. Bunsenges. Phys. Chem. **94** (1990) 1379.

Knispel, R., R. Koch, M. Siese, C. Zetzsch;
Adduct formation of OH radicals with benzene, toluene, phenol and consecutive reactions of the adducts with NO_x and O_2,
Ber. Bunsenges. Phys. Chem. **94** (1990) 1375.

Lightfoot, P.D., B. Veyret, R. Lesclaux;
Flash photolysis study of the $CH_3O_2 + HO_2$ reaction between 248 and 573 K,
J. Phys. Chem. **94** (1990) 708.

Lightfoot, P.D., P. Roussel, B. Veyret, R. Lesclaux;
Flash photolysis study of the spectra and self-reactions of neopentylperoxy and *t*-butylperoxy radicals,
J. Chem. Soc. Faraday Trans. **86** (1990) 2927.

Lightfoot, P.D., R. Lesclaux, B. Veyret;
Flash photolysis study of the $CH_3O_2 + CH_3O_2$ reaction: rate constants and branching ratio from 248 to 573 K,
J. Phys. Chem. **94** (1990) 700.

Liu, R., R.E. Huie, M. J. Kurylo, O.J. Nielsen; .
The gas phase reactions of hydroxyl radicals with a series of nitroalkanes over the
temperature range 240–400 K,
Chem. Phys. Lett. **167** (1990) 519.

Madronich, S., R.B. Chatfield, J.G. Calvert, G.K. Moortgat, B. Veyret, R. Lesclaux;
A photochemical origin of acetic acid in the troposphere,
Geophys. Res. Lett. **17** (1990) 2361–2364.

Nelson, L., I. Shanahan, H.W. Sidebottom, J. Treacy, O.J. Nielsen;
Kinetics and mechanism for the oxidation of 1,1,1-trichloroethane,
Int. J. Chem. Kinet. **22** (1990) 577.

Nielsen, O.J., H.W. Sidebottom, L. Nelson, O. Rattigan, J.J. Treacy, D.J. O'Farrell;
Rate constants for the reaction of OH radicals and Cl atoms with diethyl sulfide,
di-n-propyl sulfide and di-n-butyl sulfide,
Int. J. Chem. Kinet. **22** (1990) 603.

Nielsen, O.J., L. Nelson, O. Rattigan, H. Sidebottom, J. Treacy:
Absolute and relative rate constants for the reaction of hydroxyl radicals and chlorine
atoms with a series of aliphatic alcohols and ethers,
Int. J. Chem. Kinet. **22** (1990) 1111.

Nielsen, O.J., M. Donlon, H. Sidebottom, J. Treacy;
Reactions of OH radicals with alkyl nitrates,
in: G. Restelli, G. Angeletti (eds), *Fifth European Symp. on Physico-Chemical
Behaviour of Atmospheric Pollutants*, Kluwer Academic Publ., Dordrecht 1990
pp. 309–314.

Nielsen, O.J., O. Jorgensen, M. Donlon, H.W. Sidebottom, D.J. O'Farrell, J. Treacy;
Rate constants for the gas phase reactions of OH radicals with nitroethene,
3-nitropropene and 1-nitrocyclo-hexene at 298 K and 1 atm,
Chem. Phys. Lett. **168** (1990) 319.

Platt, U., G. Le Bras, G. Poulet, J.P. Burrows, G.K. Moortgat;
Peroxy radicals from night-time reaction of NO_3 with organic compounds,
Nature **348** (1990) 147–149.

Raber, W., K. Reinholdt, G. K. Moortgat;
Photooxidation study of Methylethylketone and Methylvinylketone,
in: G. Restelli, G. Angeletti (eds), *Fifth European Symp. on Physico-Chemical
Behaviour of Atmospheric Pollutants*, Kluwer Academic Publ., Dordrecht 1990,
pp. 364–370.

Simon, F.G., W. Schneider, G.K. Moortgat;
UV-absorption spectrum of the methylperoxy radical and the kinetics of its
disproportionation reaction at 300 K,
Int. J. Chem. Kinet. **22** (1990) 791–813.

Wängberg, I., E. Ljungström, J. Hjorth, G. Ottobrini;
FTIR studies of reactions between the nitrate radical and chlorinated butenes,
J. Phys. Chem. **94** (1990) 8036.

Wängberg, I., S. Langrova, E. Ljungström;
Heterogeneous transformation of nitrogen compounds: PAN decomposition,
in: G. Restelli, G. Angeletti (eds), *Fifth European Symp. on Physico-Chemical
Behaviour of Atmospheric Pollutants*, Kluwer Academic Publ., Dordrecht 1990, p. 334.

Zetzsch, C., R. Koch, M. Siese, F. Witte, P. Devolder;
Adduct formation of OH with benzene and toluene and reaction of the adducts with NO
and NO_2,
in: G. Restelli, G. Angeletti (eds), *Fifth European Symp. on Physico-Chemical
Behaviour of Atmospheric Pollutants*, Kluwer Academic Publ., Dordrecht 1990,
pp. 320–327.

Zhang, Z.Y., J. Peeters;
The reaction of OH with C_2H_2 and the subsequent reaction of hydrovinyl with O_2,
in: G. Restelli, G. Angeletti (eds), *Fifth European Symp. on Physico-Chemical
Behaviour of Atmospheric Pollutants*, Kluwer Academic Publ., Dordrecht 1990,
pp. 377–382.

1991

Bächmann, K., J. Hauptmann;
Chemiluminescence – fluorescence – polarography. Three ways for the determination of
organic hydroperoxides,
in: P.M. Borrell, P. Borrell, T. Cvitaš, W. Seiler (eds), *Proc. EUROTRAC Symp. '90*,
SPB Academic Publ., The Hague 1991, pp. 1064–1068, pp. 373–375.

Bächmann, K., J. Hauptmann;
Investigation on the photolytical production of peroxy compounds
in: K.H. Becker (ed), *Atmospheric Oxidation Processes, Air Pollution Research Report*
33, CEC, Brussels 1991, pp. 72–75.

Barnes, I., K.H. Becker, A. Bierbach, E. Wiesen;
Mechanisms of the OH radical initiated oxidation of aromatics in the absence of NO_x,
in: R.A. Cox (ed), *Laboratory Studies on Atmospheric Chemistry, Air Pollution
Research Report* **42**, CEC, Brussels 1991, pp. 183–189.

Barnes, I., K.H. Becker, L. Ruppert;
FTIR-product study of the self reaction of 2-hydroxyethylperoxy radicals
($HOCH_2CH_2O_2$),
in: R.A. Cox (ed), *Laboratory Studies on Atmospheric Chemistry, Air Pollution Research Report* **42**, CEC, Brussels 1991, pp. 89–93.

Barnes, I., K.H. Becker, T. Zhu;
Atmospheric fate of organic difunctional nitrates: Reactions with OH radicals and photolysis,
in: K.H. Becker (ed), *Atmospheric Oxidation Processes, Air Pollution Research Report* **33**, CEC, Brussels 1991, pp. 63–71.

Barnes, I., V. Bastian, K.H. Becker, A. Bierbach, E. Wiesen;
Mechanisms of the OH radical initiated oxidation of xylenes and toluene in the absence of NO_x,
in: K.H. Becker (ed), *Atmospheric Oxidation Processes, Air Pollution Research Report* **33**, CEC, Brussels 1991, pp. 96–102.

Bechara, J., K. H. Becker, K. J. Brockmann, W. Thomas;
H_2O_2 formation by the ozonolysis of biogenic alkenes,
in: K.H. Becker (ed), *Atmospheric Oxidation Processes, Air Pollution Research Report* **33**, CEC, Brussels 1991, pp. 33–40.

Bechara, J., K.H. Becker, K.J. Brockmann;
La formation d'eau oxygénée dans l'ozonolyse des alcènes
Pollution Atmospherique No Spécial **33e** (1991) 21.

Becker, E., Th. Benter, R. Kampf, R.N. Schindler, U. Wille;
A redetermination of the rate constant of the reaction $F + HNO_3 \rightarrow HF + NO_3$,
Ber. Bunsenges. Phys. Chem. **95** (1991) 1168–1173.

Becker, E., U. Wille, M.M. Rahman, R.N. Schindler;
An investigation of the reaction of NO_3 radicals with Cl and ClO,
Ber. Bunsenges. Phys. Chem. **95** (1991) 1173–1179.

Becker, K.H.;
Introduction to LACTOZ: Laboratory Studies of chemistry related to tropospheric ozone,
in: P.M. Borrell, P. Borrell, T. Cvitaš, W. Seiler (eds), *Proc. EUROTRAC Symp. '90*, SPB Academic Publ., The Hague 1991, pp. 369–372.

Becker, K.H., H. Geiger, P. Wiesen;
Kinetic study of the OH radical chain in the reaction system
$OH + C_2H_4 + NO + air$
Chem. Phys. Lett. **184** (1991) 256

Becker, K.H., I. Barnes, V. Bastian, K. Wirtz , T. Zhu;
Nitrate formation in the atmospheric oxidation of VOC,
in: P.M. Borrell, P. Borrell, T. Cvitaš, W. Seiler (eds), *Proc. EUROTRAC
Symp. '90*, SPB Academic Publ., The Hague 1991, pp. 377–378.

Becker, K.H., K. J. Brockmann, J. Bechara;
Terpenes and forest decline
Nature **352** (1991) 672.

Benter, Th., E. Becker, U. Wille, R.N. Schindler, C.E. Canosa-Mas, S.J. Smith,
S.J. Waygood, R.P. Wayne;
Nitrate radical reactions: Interactions with alkynes,
J. Chem. Soc., Faraday Trans. **87** (1991) 2141–2145.

Biggs, P., A.A. Boyd, C.E. Canosa-Mas, D.M. Joseph, R.P. Wayne;
A discharge stopped-flow technique for time-resolved studies of slow radical reactions
in the gas phase
Meas. Sci. Technol. **2** (1991) 675.

Biggs, P., C.E. Canosa-Mas, R.P. Wayne;
Studies of the kinetics and mechanisms of interactions of the nitrate radical of
tropospheric interest,
in: P.M. Borrell, P. Borrell, T. Cvitaš, W. Seiler (eds), *Proc. EUROTRAC
Symp. '90*, SPB Academic Publ., The Hague 1991, pp. 379–382.

Biggs, P., C.E. Canosa-Mas, R.P. Wayne, I. Barnes, H. Sidebottom;
General trends in the reactivity of the NO_3 radical in the gas phase,
in: K.H. Becker (ed), *Atmospheric Oxidation Processes, Air Pollution Research Report*
33, CEC, Brussels 1991, pp. 212–217.

Bingemann, D., D. Hartmann, A. Hoffmann, J. Karthäuser , R. Zellner;
Recent advances in laboratory detection methods of RO_2 and RO radicals,
in: P.M. Borrell, P. Borrell, T. Cvitaš, W. Seiler (eds), *Proc. EUROTRAC
Symp. '90*, SPB Academic Publ., The Hague 1991, pp. 383–386.

Boyd, A.A., C.E. Canosa-Mas, A.D. King, R.P. Wayne, M.R. Wilson;
Use of a stopped-flow technique to measure the rate constants at room temperature for
reactions between the nitrate radical and various organic species,
J. Chem. Soc., Faraday Trans. **87** (1991) 2913.

Bridier, I., F. Caralp, H. Loirat, R. Lesclaux, B. Veyret, K.H. Becker, A. Reimer, F. Zabel;
Kinetic and theoretical studies of the reactions $CH_3C(O)O_2 + NO_2 + M$
$\leftrightarrow CH_3C(O)O_2NO$ between 248 and 393 K and between 30 and 760 Torr,
J. Phys. Chem. **95** (1991) 3594.

Canosa-Mas, C.E., S.J. Smith, S.J. Waygood, R.P. Wayne;
Study of the temperature dependence of the reaction of the nitrate radical with propene
J. Chem. Soc., Faraday Trans. **87** (1991) 3473.

Crowley, J.N., D. Bauer, G.K. Moortgat;
  Self-reaction kinetics and UV spectra of ethyl- and bromoethyl-peroxy radicals,
  in: R.A. Cox (ed), *Proc. Laboratory Studies of Atmospheric Chemistry:
  CEC/EUROTRAC Discussion Meeting,* York 1991, pp. 101–104.

Crowley, J.N., F. Simon, O. Horie, W. Schneider , G.K. Moortgat;
  Spectrum, kinetics and mechanism of CH_3O_2 radical reactions,
  in: P.M. Borrell, P. Borrell, T. Cvitaš, W. Seiler (eds), *Proc. EUROTRAC
  Symp. '90,* SPB Academic Publ., The Hague 1991, pp. 387–389.

Crowley, J.N., F.G. Simon, J.P. Burrows, G.K. Moortgat, M.E. Jenkin, R.A. Cox;
  The HO_2 radical UV spectrum measured by molecular modulation, UV-diode array
  spectroscopy
  J. Photochem. Photobiol. A: Chem. **60** (1991) 1–10.

Elend, M., C. Zetzsch;
  The influence of oxygen on the apparent rate constant for the reaction of OH with
  aromatics in smog chamber experiments in nitrogen at one atmosphere,
  in: R.A. Cox (ed), *Laboratory Studies on Atmospheric Chemistry, Air Pollution
  Research Report* **42**, CEC, Brussels 1991, pp. 123–126.

Goumri, A., J.F. Pauwels, J.P. Sawerysyn , P. Devolder;
  Tropospheric oxidation of aromatics: reaction rates of some intermediate radicals in the
  oxidation of toluene initiated by OH radical,
  in: P.M. Borrell, P. Borrell, T. Cvitaš, W. Seiler (eds), *Proc. EUROTRAC
  Symp. '90,* SPB Academic Publ., The Hague 1991, pp. 391–392.

Goumri, A., J.F. Pauwels, P. Devolder;
  Rate of the OH + C_6H_6 + He reaction in the fall-off range by discharge flow and OH
  resonance fluorescence,
  Can. J. Chem. **69** (1991) 1057.

Heimann, G., P. Kutsenogiy , P. Warneck;
  Products and pathways in the oxidation of saturated and unsaturated hydrocarbons with
  OH-radicals,
  in: P.M. Borrell, P. Borrell, T. Cvitaš, W. Seiler (eds), *Proc. EUROTRAC
  Symp. '90,* SPB Academic Publ., The Hague 1991, p. 393.

Heimann, G., P. Warneck;
  Product studies in the reaction between OH radicals and *n*-pentane
  in: K.H. Becker (ed), *Atmospheric Oxidation Processes, Air Pollution Research Report*
  33, CEC, Brussels 1991, pp. 86–90

Heiss, A., J. Tardieu de Maleissye, V. Viossat, K. Sahetchian, I.G. Pitt;
  Reactions of primary and secondary butoxy radicals in oxygen at atmospheric pressure,
  Int. J. Chem. Kinet. **23** (1991) 607.

Hjorth, J., N.R. Jensen, O. Lohse, J. Notholt, H. Skov, G. Restelli;
A spectroscopic study of the equilibrium $N_2O_5 + M \leftrightarrow NO_3 + NO_2 + M$ and the kinetics
of the $N_2O_5/NO_3/NO_2/O_3$ air system,
Int. J. Chem. Kinet. **24** (1991) 51.

Hjorth, J., N.R. Jensen, O. Lohse, J. Notholt, H. Skov, G. Restelli;
Gas phase reactions of interest in nighttime tropospheric chemistry,
in: P.M. Borrell, P. Borrell, T. Cvitaš, W. Seiler (eds), *Proc. EUROTRAC
Symp. '90*, SPB Academic Publ., The Hague 1991, pp. 395–396.

Horie, O., G.K. Moortgat;
Decomposition pathways of the excited Criegee intermediates in the ozonolysis of
simple alkanes,
Atmos. Environ. **25**A (1991) 1881–1896.

Horie, O., G.K. Moortgat;
The analysis of reaction products in the oxidation reactions of hydrocarbons by means
of matrix-isolation FTIR spectroscopy,
Fresenius J. Anal. Chem. **340** (1991) 641–645.

Horie, O., J.N. Crowley, D. Bauer, G.K. Moortgat;
The kinetics and mechanism of the reaction system $CH_3COO_2-CH_3O_2-HO_2$ between
233 and 333 K. Absorption spectra of HO_2 and $C_2H_5O_2$,
in: K.H. Becker (ed), *Atmospheric Oxidation Processes, Air Pollution Research Report*
33, CEC, Brussels 1991, pp. 109–114.

Jenkin, M.E., R.A. Cox, G.D. Hayman;
Laboratory studies of peroxy radicals produced in the tropospheric oxidation of volatile
organic compounds,
in: P.M. Borrell, P. Borrell, T. Cvitaš, W. Seiler (eds), *Proc. EUROTRAC
Symp. '90*, SPB Academic Publ., The Hague 1991, pp. 397–401.

Jenkin, M.E., R.A. Cox, M. Emrich, F.G. Simon, G.K. Moortgat;
Mechanism of the Cl atom initiated oxditation of acetone in air,
in: K.H. Becker (ed), *Atmospheric Oxidation Processes, Air Pollution Research Report*
33, CEC, Brussels 1991, pp. 81–90.

Kirchner, F., F. Zabel, K.H. Becker;
Determination of the rate constant ratio for the reactions of the benzoylperoxy radical
with NO and NO_2,
in: R.A. Cox (ed), *Proc. Laboratory Studies of Atmospheric Chemistry:
CEC/EUROTRAC Discussion Meeting*, York 1991, pp. 61–64.

Kirchner, F., G. Elfers, F. Zabel, K.H. Becker;
Reactions of ethylperoxy and acetylperoxy radicals under NO_x rich conditions,
in: P.M. Borrell, P. Borrell, T. Cvitaš, W. Seiler (eds), *Proc. EUROTRAC
Symp. '90*, SPB Academic Publ., The Hague 1991, pp. 403–405.

Koch, R., J. Nowack, M. Siese, C. Zetzsch;
Analysis of biexponential decays of OH in the presence of aromatics and O_2: New results on phenol,
in: R.A. Cox (ed), *Laboratory Studies on Atmospheric Chemistry, Air Pollution Research Report* **42**, CEC, Brussels 1991, pp. 47–50.

Lançar, I.T., A. Mellouki, G. Poulet;
Kinetics of the reactions of hydrogen iodide with hydroxyl and nitrate radicals,
Chem. Phys. Lett. **177** (1991) 554.

Lançar, I.T., V. Daele, G. Le Bras, G. Poulet;
Etude de la réactivité des radicaux NO_3 avec le diméthyl-2,3 butène-2,
le butadiène-1,3 et le diméthyl-2,3 butadiène-1,3
J. Chim. Phys. **88** (1991) 1777.

Lightfoot, P.D., A.A. Jemie-Alade;
The temperature dependence of the HO_2 and CH_3O_2 UV absorption spectra,
J. Photochem. Photobiol. A: Chem. **59** (1991) 1.

Lightfoot, P.D., P. Roussel, R. Lesclaux;
A flash photolysis study of the $CH_3O_2 + CH_3O_2$ and $CH_3O_2 + HO_2$ reactions
600–719 K: The unimolecular decomposition of methylhydroperoxide
J. Chem. Soc., Farad. Trans. **87** (1991) 3213.

Meller, R., W. Raber, J.N. Crowley, M.E. Jenkin, G.K. Moortgat;
The UV-visible absorption spectrum of methylglyoxal by conventional spectroscopy and modulated photolysis,
J. Photochem. Photobiol. A: Chem. **60** (1991) 163–171.

Moortgat, G.K., O. Horie, C.B. Zahn;
Ozonolysis of VOC: recent studies by matrix isolation FTIR spectroscopy,
in: *Pollution Atmosphérique, Quatrièmes entretiens du Centre Jacques Cartier sur Réactivité Chimique de L'Atmosphère et Mesure des Polluants Atmosphèriques*,
Grenoble 1990, *Special Issue* **33** (1991) 29–44.

Murrells, T.P., M.E. Jenkin, S.J. Shalliker, G.D. Hayman;
Laser flash photolysis study of the UV spectrum and kinetics of reactions of $HOCH_2CH_2O_2$ radicals
J. Chem. Soc., Faraday Trans. **87** (1991) 2351.

Nielsen, O.J.;
Rate constants for the gas-phase reactions of OH radicals with CH_3CHF_2 (HFC-152a) and $CHCl_2CF_3$ (HCFC-123) over the temperature range 295–388 K,
Chem. Phys. Lett. **187** (1991) 286.

Nielsen, O.J., H.W. Sidebottom, M. Donlon, J. Treacy;
Rate constants for the gas-phase reactions of OH radicals and Cl atoms with
n-alkyl nitrites at atmospheric pressure,
Int. J. Chem. Kinet. **23** (1991) 1095.

Nielsen, O.J., H.W. Sidebottom, M. Donlon, J. Treacy;
An absolute and relative rate study of the gas-phase reactions of OH radicals and
Cl atoms with *n*-alkyl nitrates,
Chem. Phys. Lett. **178** (1991) 163.

Nielsen, O.J., J. Munk, G. Locke, T.J. Wallington;
UV absorption spectra and kinetics of the self reaction of CH_2Br and CH_2BrO_2 radicals
in the gas phase at 298K,
J. Phys. Chem. **95** (1991) 8714.

Nielsen, O.J., M. Donlon, H.W. Sidebottom, J. Treacy;
Reactions of OH radicals with alkyl nitrates and nitroalkanes,
in: P.M. Borrell, P. Borrell, T. Cvitaš, W. Seiler (eds), *Proc. EUROTRAC
Symp. '90*, SPB Academic Publ., The Hague 1991, pp. 407–411.

Peeters, J., C. Vinckier, I. Langhans, J. Vertommen;
Reactions of peroxy radicals in the tropospheric hydrocarbon/NO chain,
in: P.M. Borrell, P. Borrell, T. Cvitaš, W. Seiler (eds), *Proc. EUROTRAC
Symp. '90*, SPB Academic Publ., The Hague 1991, pp. 413–416.

Peeters, J., J. Vertommen;
Rate constants of the reactions of CF_3O_2, i-$C_3H_7O_2$, t-$C_4H_9O_2$ and
n-$C_4H_9O_2$ with NO,
in: R.A. Cox (ed), *Laboratory Studies on Atmospheric Chemistry, Air Pollution
Research Report* **42**, CEC, Brussels 1991, pp. 105–111.

Poulet, G., I.T. Lançar, A. Mellouki, G. Le Bras;
Laboratory studies of NO_3 reactions with dienes and with nighttime
Peroxy Radicals,
in: P.M. Borrell, P. Borrell, T. Cvitaš, W. Seiler (eds), *Proc. EUROTRAC
Symp. '90*, SPB Academic Publ., The Hague 1991, pp. 417–418.

Raber, W., K. Reinholdt , G.K. Moortgat;
Photooxidation study of methylethylketone and methylvinylketone,
in: P.M. Borrell, P. Borrell, T. Cvitaš, W. Seiler (eds), *Proc. EUROTRAC
Symp. '90*, SPB Academic Publ., The Hague 1991, pp. 419–421.

Raber, W., R. Meller, G.K. Moortgat;
Photooxidation study of some carbonyl compounds of atmospheric interest,
in: K.H. Becker (ed), *Atmospheric Oxidation Processes, Air Pollution Research Report*
33, CEC, Brussels 1991, pp. 91–95.

Rindone, B., F. Cariati, G. Restelli, J. Hjorth;
The gas-phase reaction between the nitrate radical and arenes,
Fresenius J. Anal. Chem. **339** (1991) 673

Rowley, D.M., P.D. Lightfoot, R. Lesclaux, T.J. Wallington;
The UV absorption spectrum and self-reaction of cyclohexylperoxy radicals,
J. Chem. Soc., Farad. Trans. **87** (1991) 3221.

Sahetchian, K., A. Heiss , J. Tardieu de Maleissye;
Laboratory study of reactions of primary and secondary butoxy radicals in oxygen at
atmospheric pressure,
in: P.M. Borrell, P. Borrell, T. Cvitaš, W. Seiler (eds), *Proc. EUROTRAC
Symp. '90*, SPB Academic Publ., The Hague 1991, pp. 423–425.

Vertommen, J., I. Langhans, J. Peeters;
Rate constant of the reactions of *t*-butylperoxy and iso-propylperoxy radicals with nitric
oxide,
in: K.H. Becker (ed), *Atmospheric Oxidation Processes, Air Pollution Research Report*
33, CEC, Brussels 1991, pp. 115–120.

Veyret, B., P.D. Lightfoot, R. Lesclaux;
Kinetics and mechanism of organic peroxy radical reactions important in tropospheric
chemistry,
in: P.M. Borrell, P. Borrell, T. Cvitaš, W. Seiler (eds), *Proc. EUROTRAC
Symp. '90*, SPB Academic Publ., The Hague 1991, pp. 427–428.

Wayne, R.P., I. Barnes, P. Biggs, J.P. Burrows, C.E. Canosa-Mas, J. Hjorth,
G. Le Bras, G.K. Moortgat, D. Perner, G. Poulet, G. Restelli, H. Sidebottom;
The nitrate radical: physics, chemistry and the atmosphere,
Atmos. Environ. **25**A (1991) 1–203; *Air Pollution Research Report* **31**, CEC, Brussels
1991.

Wille, U., E. Becker, R. N. Schindler, I.T. Lançar, G. Poulet, G. Le Bras;
A discharge flow mass-spectrometric study of the reaction between
the NO_3 radical and isoprene,
J. Atmos. Chem. **13** (1991) 183.

Wille, U., E. Becker, R.N. Schindler, I.T. Lancar, G. Poulet, G. LeBras;
A discharge flow mass-spectrometric study of the reaction between the
NO_3 radical and isoprene,
J. Atmos. Chem. **13** (1991) 183–193.

Zetzsch, C., R. Knispel, R. Koch, J. Nowack, M. Siese;
Direct study of tropospheric consecutive reactions of OH-attack on aromatics:
Competition between NO_2 and O_2,
in: K.H. Becker (ed), *Atmospheric Oxidation Processes, Air Pollution Research Report*
33, CEC, Brussels 1991, pp. 103–109.

Zetzsch, C., R. Knispel, R. Koch, M. Siese;
Adduct formation of OH radicals and consecutive reactions of the adducts with NO_x and O_x,
in: P.M. Borrell, P. Borrell, T. Cvitaš, W. Seiler (eds), *Proc. EUROTRAC Symp. '90*,
SPB Academic Publ., The Hague 1991, pp. 429–433.

Zhu, T., I. Barnes, K.H. Becker;
Relative rate study of the gas-phase reaction of hydroxyl radicals with difunctional
organic nitrates at 298 K and atmospheric pressure,   ·
J. Atmos. Chem. **13** (1991) 301.

1992

Aird, R.W.S., C.E. Canosa-Mas, D.J. Cook, E. Ljungström, G. Marston,
R.P. Wayne;
Kinetics of the reactions of the nitrate radical with a series of halogenobutenes,
J. Chem. Soc., Faraday Trans. **88** (1992) 1093.

Arnold, B.R., J.C. Scaiano, G.F. Bucher, W.W. Sander;
Laser flash photolysis studies on 4-oxocyclohexa-2,5-dienylidene,
J. Organ. Chem. **57** (1992) 6470.

Bächmann, K., J. Hauptmann, J. Polzer, P. Schütz;
Determination of organic alkyl- and hydroxy hydroperoxides
Fresenius J. Anal. Chem. **342** (1992) 809–812.

Becker, K.H., I. Barnes, L. Ruppert;
A product study of the OH-radical initiated oxidation of isoprene in the
absence of NO_x,
in: J. Peeters (ed), *Chemical Mechanisms Describing Tropospheric Processes, Air
Pollution Research Report* **45**, CEC, Brussels 1992, pp. 193–198.

Bauer, D., J.N. Crowley, G.K. Moortgat;
The UV absorption spectrum of the ethylperoxy radical and its self-reaction. Kinetics
between 218 and 333 K,
J. Photochem. Photobiol. A: Chem. **65** (1992) 329–344.

Becker, E., M.M. Rahman, R.N. Schindler;
Determination of rate constants for the gas phase reaction of NO_3 with H, OH and HO_2
radicals at 298 K,
Ber. Bunsenges. Phys. Chem., **96** (1992) 776–783.

Becker, K. H., I. Barnes, A. Bierbach, M. Martin-Reviejo, E. Wiesen;
OH-radical initiated degradation of aromatic hydrocarbons in the absence of NO_x,
in: J. Peeters (ed), *Chemical Mechanisms Describing Tropospheric Processes, Air Pollution Research Report* **45**, CEC, Brussels 1992, pp. 87–94.

Benter, Th., Becker, E., U. Wille, M.M. Rahman, R.N. Schindler;
The Determination of Rate Constants for the Reaction of Some Alkenes with the NO_3 Radical,
Ber. Bunsenges. Phys. Chem. **96** (1992) 769–775.

Benter, Th., M. Liesner, R.N. Schindler;
A REMPI-study of NO_3 and epoxide formation in the reaction of styrene with NO_3 radicals at low pressure,
in: J. Peeters (ed), *Chemical Mechanisms Describing Tropospheric Processes, Air Pollution Research Report* **45**, CEC, Brussels 1992, pp. 119–124.

Benter, Th., M. Liesner, V. Sauerland, R.N. Schindler;
A REMPI-study of the reaction of isoprene with NO_3,
in: R.A. Cox (ed), *Air Pollution Research Report* **42**, E. Guyot SA, Brussels 1992.

Berges, M.G.M., P. Warneck;
Product quantum yields for the 350 nm photo-decomposition of pyruvic acid in air,
Ber. Bunsenges. Phys.Chem. **96** (1992) 413–416.

Bierbach, A., I. Barnes, K.H. Becker;
Rate coefficients for the gas-phase reactions of hydroxyl radicals with furan, 2-methylfuran, 2-ethylfuran and 2,5-dimethylfuran at 300 ± 2 K,
Atmos. Environ. **26**A (1992) 813.

Bierbach, A., I. Barnes, K.H. Becker;
Atmospheric chemistry of unsaturated dicarbonyls,
in: J. Peeters (ed), *Chemical Mechanisms Describing Tropospheric Processes, Air Pollution Research Report* **45**, CEC, Brussels 1992, pp. 237–242.

Biggs, P., C.E. Canosa-Mas, R.P. Wayne,
Reactivity of the NO_3 radical towards halogenated species,
in: R.A. Cox (ed), *Air Pollution Research Report* **42**, E. Guyot SA, Brussels 1992, pp. 83–87.

Bridier, I., R. Lesclaux, B. Veyret;
Flash photolysis kinetic study of the equilibrium $CH_3O_2 + NO_2 = CH_3O_2NO_2$
Chem. Phys. Lett. **191** (1992) 259.

Bucher, G., W. Sander;
Carbonyl *O*-oxides and dioxiranes: The influence of substituents on spectroscopic properties,
Chem. Ber. **125** (1992) 1851.

Canosa-Mas, C.E., P.S. Monks, R.P. Wayne;
    The temperature dependence of the reaction of the nitrate radical with but-1-ene
    J. Chem. Soc., Faraday Trans. **88** (1992) 11.

Chambers, R.M., A.C. Heard, R.P. Wayne;
    Inorganic gas-phase reactions of the nitrate radical: $I_2 + NO_3$ and $I + NO_3$,
    J. Phys. Chem. **96** (1992) 3321.

Crowley, J.N., G. K. Moortgat;
    2-Bromoethylperoxy and 2-bromo-1-methylpropylperoxy radicals:
    ultraviolet absorption spectra and self-reaction rate constants at 298 K,
    J. Chem. Soc., Faraday Trans. **88** (1992) 2437–2444.

Daumont, D., J. Brion, J. Charbonnier, J. Malicet;
    Ozone U.V. spectroscopy. I. Absorption cross-sections at room temperature,
    J. Atmos. Chem. **15** (1992) 145.

Ellermann, T., O.J. Nielsen, H. Skov;
    Absolute rate constants for the reaction of NO_3 radicals with a series of dienes
    at 295 K,
    Chem. Phys. Lett. **200** (1992) 224.

Gourmi, A., L. Elmaimouni, J.P. Sawerysyn, P. Devolder;
    Reactions rates at (297 ± 3) K of four benzyl-type radicals with O_2, NO and NO_2 by
    discharge flow/laser induced fluorescence,
    J. Phys. Chem. **96** (1992) 5395.

Helleis, F., J.N. Crowley, G.K. Moortgat;
    The reaction between CD_3O_2 and NO_3: A kinetic study at 298 K,
    in: J. Peeters (ed), *Chemical Mechanisms Describing Tropospheric Processes, Air
    Pollution Research Report* **45**, CEC, Brussels 1992, pp. 205–210.

Hjorth, J.;
    Night-time tropospheric chemistry: Kinetics and mechanisms of reactions of the NO_3
    radical with volatile organic compounds,
    Trends Phys. Chem. **27** (1992) 107.

Hoffmann, A., V. Mors, R. Zellner;
    A novel laser-based technique for the time resolved study of integrated hydrocarbon
    oxidation mechanisms,
    Ber. Bunsenges. Phys. Chem. **96** (1992) 437.

Horie, O., G.K. Moortgat;
    Reactions of $CH_3C(O)O_2$ radicals with CH_3O_2 and HO_2 between 263 and 333 K,
    J. Chem. Soc., Faraday Trans. **88** (1992) 3305–3312.

Horie, O., G.K. Moortgat;
Photolysis of ketene-oxygen mixtures between 253 K and 233 K in relation to the
formation of Criegee intermediates,
Ber. Bunsenges. Phys. Chem. **96** (1992) 404–408.

Kirchner, F., F. Zabel, K.H. Becker;
Kinetic behaviour of benzylperoxy radicals in the presence of NO and NO_2,
Chem. Phys. Lett. **191** (1992) 169.

Kirchner, F., F. Zabel, K.H. Becker;
Thermal stability of substituted methyl peroxynitrates,
in: J. Peeters (ed), *Chemical Mechanisms Describing Tropospheric Processes, Air
Pollution Research Report* **45**, CEC, Brussels 1992, pp. 215–220.

Lançar, I.T., G. Le Bras, G. Poulet;
Redetermination de la constante de vitesse de la reaction CH_4 + OH et son implication
atmosphérique,
C. R. Acad. Sci. Paris **315** (ser. 11) (1992) 1487.

Langer, S., I. Wangberg, E. Ljungström;
Heterogeneous transformation of peroxyacetylnitrate
Atmos. Environ. **26**A (1992) 3089.

Lightfoot, P.D., R.A. Cox, J.N. Crowley, M. Destriau, G.D. Hayman, M.E. Jenkin,
G.K. Moortgat, F. Zabel;
Organic peroxy radicals: kinetics, spectroscopy and tropospheric chemistry,
Atmos. Environ. **26**A (1992) 1805–1961.

Maric, D., J.P. Burrows, G.K. Moortgat;
A study of the formation of N_2O in the reaction of NO_3 (A^2E') with N_2,
J. Atmos. Chem. **15** (1992) 157–168.

Moortgat, G.K., J.N. Crowley, D. Bauer;
Molecular modulation spectroscopy: recent studies involving peroxy radical reactions,
in: *Proc. Int. Symp. on Environmental Sensing*, Berlin 1992.

Notholt, J., J. Hjorth, F. Raes;
Formation of HNO_2 on aerosol surfaces during foggy periods in the presence of NO and
NO_2,
Atmos. Environ. **26**A (1992) 211.

Peeters, J., V. Pultau;
Rate constant of CF_3CHFO_2 + NO,
in: J. Peeters (ed), *Chemical Mechanisms Describing Tropospheric Processes, Air
Pollution Research Report* **45**, CEC, Brussels 1992, pp. 225–230.

Peeters, J., J. Vertommen, I. Langhans;
Rate constants of the reactions of CF_3O_2, i-$C_3H_7O_2$, and t-$C_4H_9O_2$ with NO,
Ber. Bunsenges. Phys. Chem. **96** (1992) 431.

Raber, W., G.K. Moortgat;
A study of the photooxidation of methylglyoxal in air,
in: J. Peeters (ed), *Chemical Mechanisms Describing Tropospheric Processes, Air Pollution Research Report* **45**, CEC, Brussels 1992, pp. 149–154.

Rahman, M.M., E. Becker, U. Wille, R.N. Schindler;
Determination of rate constants for the reaction of the CH_2SH radical with O_2, O_3 and NO_2 at 298 K,
Ber. Bunsenges. Phys. Chem. **96** (1992) 783–787.

Rowley, D.M., P.D. Lightfoot, R. Lesclaux, T.J. Wallington;
Ultraviolet absorption spectrum and self-reaction of cyclopentylperoxy radicals,
J. Chem. Soc., Farad. Trans. **88** (1992) 1369.

Rowley, D.M., R. Lesclaux, P.D. Lightfoot, B. Nozieres, T.J. Wallington, M.D. Hurley;
Kinetic and mechanistic studies of the reactions of cyclopentylperoxy and cyclohexylperoxy radicals with HO_2,
J. Phys. Chem. **96** (1992) 4889.

Rowley, D.M., R. Lesclaux, P.D. Lightfoot, K. Hughes, M.D. Hurley, S. Rudy, T.J. Wallington;
A kinetic and mechanistic study of the reaction of neopentylperoxy radicals with HO_2,
J. Phys. Chem. **96** (1992) 7043.

Skov, H., J. Hjorth, C. Lohse, N.R. Jensen, G. Restelli;
Products and mechanism of the reactions of the nitrate radical (NO_3) with isoprene, 1,3-butadiene and 2,3-dimethyl-1,3-butadiene in air,
Atmos. Environ. **26**A (1992) 2771.

Treacy, J., M. El Hag, D. O'Farrell, H. Sidebottom;
Reactions of ozone with unsaturated organic compounds,
Ber. Bunsenges. Phys. Chem. **96** (1992) 422.

Wallington, T.J., J.M. Andino, A.R. Potts, O.J. Nielsen;
Pulse radiolysis and Fourier transform infrared study of neopentyl peroxy radicals in the gas phase at 297 K,
Int. J. Chem. Kinet. **24** (1992) 649.

Wallington, T.J., M.M. Maricq, T. Ellermann, O.J. Nielsen
Novel method for the measurement of gas phase peroxy radical absorption spectra,
J. Phys. Chem. **96** (1992) 982.

Wängberg, I., E. Ljungström, B.E.R. Olsson, J. Davidsson;
The temperature dependence of the reaction $NO_3 + NO_2 \rightarrow NO + NO_2 + O_2$ in the range from 296 to 332 K,
J. Phys. Chem. **96** (1992) 7640.

Wille, U., M.M. Rahman, R. N. Schindler;
Oxirane formation in the reaction of NO_3 radicals with alkenes,
Ber. Bunsenges. Phys. Chem., **96** (1992) 833–835.

Wille, U., M.M. Rahman, R.N. Schindler;
Reactivity of NO_3 radicals towards selected satured and unsatured epoxides,
in: R.A. Cox (ed), *Air Pollution Research Report* **42**, E. Guyot SA, Brussels 1992.

Zetzsch, C., R. Knispel, R. Koch, M. Siese;
Absolute rate constants of secondary steps in the hydroxyl-initiated degradation of aromatics: Fate of the adducts in the presence of O_2 and NO_x,
in: J. Peeters (ed), *Chemical Mechanisms Describing Tropospheric Processes, Air Pollution Research Report* **45**, CEC, Brussels 1992, pp. 107–111.

1993

Bächmann, K., I. Haag, A. Röder, U. Sprenger, K.H. Steeg;
Single rain drop analysis,
in: P.M. Borrell, P. Borrell, T. Cvitaš, W. Seiler (eds), *Proc. EUROTRAC Symp. '92*, SPB Academic Publ., The Hague 1993, p. 624.

Bächmann, K., J. Hauptmann, J. Polzer;
Oxidation of C_4 and C_5 hydrocarbons – analysis of the reaction products by GC/MS and HPLC,
in: P.M. Borrell, P. Borrell, T. Cvitaš, W. Seiler (eds), *Proc. EUROTRAC Symp. '92*, SPB Academic Publ., The Hague 1993, pp. 404–407.

Barnes, I., K.H. Becker, L. Ruppert;
FTIR product study of the self reaction of b-hydroxyethylperoxy radicals,
Chem. Phys. Lett. **203** (1993) 295.

Barnes, I., K.H. Becker, T. Zhu;
Near UV absorption spectra and photolysis products of difunctional organic nitrates,
J. Atmos. Chem. **17** (1993) 353.

Becker, K.H., F. Kirchner, F. Zabel;
Thermal stability of peroxynitrates,
in: H. Niki, K.H. Becker (eds), *Proc. NATO Advance Research Workshop, Tropospheric Chemistry of Ozone in the Polar Regions, NATO ASI subseries: Global Environmental Change*, Springer-Verlag, Heidelberg 1993, p. 351.

Becker, K.H., I. Barnes, F. Kirchner, L. Ruppert, F. Zabel;
Atmospheric chemistry of peroxy radicals. Part 1. Thermal decomposition of peroxynitrates. Part 2. Self-reaction of alkylperoxy radicals – a product study,
in: P.M. Borrell, P. Borrell, T. Cvitaš, W. Seiler (eds), *Proc. EUROTRAC Symp. '92*, SPB Academic Publ., The Hague 1993, pp. 394–398.

Becker, K.H., J. Bechara, K.J. Brockmann;
Studies on the formation of H_2O_2 in the ozonolysis of alkenes,
Atmos. Environ. **27**A (1993) 57.

Bednarek, G., P. Biggs, B. Cabanas, C. E. Canosa-Mas, G. Marston, R. P. Wayne;
A study of the kinetics and mechanisms of some reactions of organic compounds and radicals with NO_3,
in: P.M. Borrell, P. Borrell, T. Cvitaš, W. Seiler (eds), *Proc. EUROTRAC Symp. '92*, SPB Academic Publ., The Hague 1993, pp. 411–415.

Biggs, P., C.E. Canosa-Mas P.S. Monks, R.P. Wayne, Th. Benter, R.N. Schindler;
The Kinetics of the Nitrate Radical Self Reaction,
Int. J. Chem. Kinet. **25** (1993) 805–817.

Biggs, P., C.E. Canosa-Mas, J.-M. Fracheboud, D.E. Shallcross, R.P. Wayne;
Reactions of methyl, methoxy and methylperoxy radicals,
in: J. Peeters (ed), *Chemical Mechanisms Describing Tropospheric Processes, Air Pollution Research Report* **45**, E. Guyot SA, Brussels 1993, pp. 131–135.

Biggs, P., C.E. Canosa-Mas, J.M. Fracheboud, A.D. Parr, D.E. Shallcross, R.P. Wayne, F. Caralp;
Investigation of the pressure dependence between 1 and 10 Torr of the reactions of NO_2 with CH_3 and CH_3O,
J. Chem. Soc., Faraday Trans. **89** (1993) 4163.

Biggs, P., C.E. Canosa-Mas, P.S Monks, R.P. Wayne, Th. Benter, R.N. Schindler;
The kinetics of the nitrate radical self-reaction,
Int. J. Chem. Kinetics **25** (1993) 805.

Bridier, I., B. Veyret, R. Lesclaux, M.E. Jenkin;
Flash photolysis study of the UV spectrum and kinetics of reactions of the acetonyl peroxy radical,
J. Chem. Soc., Faraday Trans. **89** (1993) 2993.

Brion, B., B. Coquart, D. Daumont, A. Jenouvrier, J. Malicet, M.F. Mérienne;
High resolution laboratory absorption cross sections of O_3 and NO_2,
in: P.M. Borrell, P. Borrell, T. Cvitaš, W. Seiler (eds), *Proc. EUROTRAC Symp.* '92, SPB Academic Publ., The Hague 1993, pp. 423–426.

Brion, J., A. Chakir, D. Daumont, J. Malicet, C. Parisse;
High-resolution laboratory absorption cross section of ozone. Temperature effect,
Chem. Phys. Lett. **213** (1993) 610.

Chiodini, G., B. Rindone, F. Cariati, S. Polessello, G. Restelli, J. Hjorth;
Comparison between the gas-phase and the solution reaction of the nitrate radical and methylarenes,
Environ. Sci. Technol. **27** (1993) 1659.

Crowley, J.N., D. Bauer, G.K. Moortgat;
UV spectra and self-reaction kinetics of ethyl- (218–333 K), 2-bromo-ethyl (298 K) and 3-bromo-*sec*-butyl (298 K) peroxy radicals,
in: P.M. Borrell, P. Borrell, T. Cvitaš, W. Seiler (eds), *Proc. EUROTRAC Symp.* '92, SPB Academic Publ., The Hague 1993, p. 417.

Crowley, J.N., J.P. Burrows, G.K. Moortgat, G. Poulet, G. Le Bras;
Optical detection of NO_3 and NO_2 in "Pure" HNO_3 vapor, the liquid phase decomposition of HNO_3,
Int. J. Chem. Kinet. **26** (1993) 795–803.

Curley, M., J. Treacy, H. Sidebottom;
Reaction of ozone with cycloalkenes,
in: P.M. Borrell, P. Borrell, T. Cvitaš, W. Seiler (eds), *Proc. EUROTRAC Symp.* '92, SPB Academic Publ., The Hague 1993, pp. 381–384.

Daële, V., D. Johnstone, G. Laverdet, G. Le Bras, G. Poulet;
Laboratory studies of NO_3 reactions of atmospheric importance,
in: P.M. Borrell, P. Borrell, T. Cvitaš, W. Seiler (eds), *Proc. EUROTRAC Symp.* '92, SPB Academic Publ., The Hague 1993, pp. 408–410.

Dingenen, R. Van, N.R. Jensen, J. Hjorth, F. Raes;
Laboratory study of aerosol formation from the night-time oxidation of DMS,
in: P.M. Borrell, P. Borrell, T. Cvitaš, W. Seiler (eds), *Proc. EUROTRAC Symp.* '92, SPB Academic Publ., The Hague 1993, p. 416.

Eberhard, J., C. Müller, D.W. Stocker, J.A. Kerr;
The photo-oxidation of diethyl ether in smog chamber experiments simulating tropospheric conditions: product studies and proposed mechanism,
Int. J. Chem. Kinet. **25** (1993) 639.

Eberhard, J., M. Semadeni, D.W. Stocker, J.A. Kerr;
Photo-oxidation of VOCs under simulated tropospheric conditions,
in: P.M. Borrell, P. Borrell, T. Cvitaš, W. Seiler (eds), *Proc. EUROTRAC Symp. '92*, SPB Academic Publ., The Hague 1993, pp. 367–371.

Elmaimouni, L., R. Minetti, J.P. Sawerysyn, P. Devolder;
Kinetics and thermochemistry of the reaction of benzyl radical with O_2: Investigations by discharge flow/laser induced fluorescence between 393 and 433 K,
Int. J. Chem. Kinet. **25** (1993) 99.

Fenter, F.F., V. Catoire, R. Lesclaux, P.D. Lightfoot;
The ethylperoxy radical: Its ultraviolet spectrum, self-reaction and reaction with HO_2, each studied as a function of temperature,
J. Phys. Chem. **97** (1993) 3530–3538.

Goumri, A., L. Elmamouni, J.-P. Sawerysyn, P. Devolder;
Tropospheric oxidation of toluene derivatives: Branching ratios between abstraction and addition and reaction rates of some benzyl type radicals,
in: P.M. Borrell, P. Borrell, T. Cvitaš, W. Seiler (eds), *Proc. EUROTRAC Symp. '92*, SPB Academic Publ., The Hague 1993, pp. 363–366.

Hjorth, J., F. Cappellani, G. Restelli;
A TDL and FTIR study of the reaction $HO_2 + NO_3 \rightarrow HO + NO_2 + O_2$,
in: P.M. Borrell, P. Borrell, T. Cvitaš, W. Seiler (eds), *Proc. EUROTRAC Symp. '92*, SPB Academic Publ., The Hague 1993, p. 393.

Jenkin, M.E., G.D. Hayman, T.J. Wallington, M.D. Hurley, J.C. Ball, O.J. Nielsen, T. Ellerman;
Kinetic and mechanistic study of the self reaction of $CH_3OCH_2O_2$ radicals at room temperature,
J. Phys. Chem. **97** (1993) 11712.

Jenkin, M.E., R.A. Cox, M. Emrich, G.K. Moortgat;
Mechanisms of the Cl atom initiated oxidation of acetone and hydroxyacetone in air,
J. Chem. Soc., Faraday Trans. **89** (1993) 2983–2991.

Jenkin, M.E., T.P. Murrells, S.J. Shalliker, G.D. Hayman;
Kinetics and product study of the self reactions of allyl and allyl peroxy radicals at 296 K,
J.Chem. Soc., Faraday Trans. **89** (1993) 433.

Langer, S., E. Ljungström, I. Wängberg;
Rates of reaction between the nitrate radical and some aliphatic esters,
J. Chem. Soc., Faraday Trans. **89** (1993) 425.

Le Bras, G., C. Golz, U. Platt;
Production of peroxy radicals in the DMS oxidation during night-time,
in: G. Restelli, G. Angelletti (eds), *Dimethylsulphide: Ocean Atmosphere and Climate,*
Kluwer Academic Publishers, Dordrecht 1993, pp. 251.

Lesclaux, R.;
Oxidation of volatile organic compounds under low NO_x conditions: Reactions of
peroxy radicals,
in: P.M. Borrell, P. Borrell, T. Cvitaš, W. Seiler (eds), *Proc. EUROTRAC*
Symp. '92, SPB Academic Publ., The Hague 1993, p. 346.

Liesner, M., Th. Benter, R.N. Schindler;
A TOF-mass spectrometric study of flash photolyzed $CH_4/O_2/Cl_2$ mixtures applying
electron impact and multiphoton ionization,
in: G. Restelli (ed.) *Physico-Chemical behaviour of Atmospheric Pollutants,*
Air Pollution Report **50** (1993) 183–189.

Ljungström, E., I. Wängberg, S. Langer;
Absolute rate coefficients for the reaction nitrate radicals and some cyclic alkenes,
J. Chem. Soc., Faraday Trans. **89** (1993) 2977.

Marston, G., P.S. Monks, C.E. Canosa-Mas, R.P. Wayne;
Correlations between rate parameters and calculated molecular properties in the reaction
of the nitrate radical with alkenes,
J. Chem. Soc., Faraday Trans., **89** (1993) 899.

Nielsen O.J., J. Sehested;
Upper limits for the rate constants of the reactions of CF_3O_2 and CF_3O radicals with
ozone at 295 K,
Chem. Phys. Lett. **213** (1993) 433.

Nielsen, O.J., T. Ellermann, M. Donlon, H.W. Sidebottom, J. Treacy;
Reactions of OH radicals with organic nitrogen containing Compounds,
in: P.M. Borrell, P. Borrell, T. Cvitaš, W. Seiler (eds), *Proc. EUROTRAC*
Symp. '92, SPB Academic Publ., The Hague 1993, pp. 377–380.

Orlando, J.J., G.S Tyndall, G.K. Moortgat, J.G. Calvert;
Quantum yields for NO_3 photolysis between 570 and 635 nm,
J. Phys. Chem. **97** (1993) 10996–11000.

Peeters, J., J. Vertommen, I. Langhans;
Rate constants of the reactions of CF_3O_2, i-$C_3H_7O_2$, t-$C_4H_9O_2$ and
s-$C_4H_9O_2$ with NO,
in: P.M. Borrell, P. Borrell, T. Cvitaš, W. Seiler (eds), *Proc. EUROTRAC*
Symp. '92, SPB Academic Publ., The Hague 1993, pp. 399–403.

Polzer, J., K. Bächmann;
Sensitive determination of alkylhydroperoxides by HRGC/MS and HRGC/FID,
J. Chromatogr. **653**A (1993) 283–291.

Sahetchian, K., A. Heiss, R. Rigny, J. Tardieu de Maleissye;
Laboratory study of reactions of alkoxy radicals in oxygen,
in: P.M. Borrell, P. Borrell, T. Cvitaš, W. Seiler (eds), *Proc. EUROTRAC
Symp. '92*, SPB Academic Publ., The Hague 1993, pp. 389–392.

Sander, W., G. Bucher, P. Komnick, J. Morowictz, P. Bubenitschek, P.G. Jones,
A. Chrapkowski;
Structure and spectroscopic properties of *p*-benzoquinone diazides,
Chem. Ber. **126** (1993) 2101.

Sander, W., M. Träubel, G. Bucher, A. Kirschfeld, S. Wierlacher;
Mechanisms of the oxidative cleavage of unsaturated organic trace contaminants in the
atmosphere,
in: P.M. Borrell, P. Borrell, T. Cvitaš, W. Seiler (eds), *Proc. EUROTRAC
Symp. '92*, SPB Academic Publ., The Hague 1993, pp. 385–388.

Sehested, J., O.J. Nielsen;
Absolute rate constants for the reaction of CF_3O_2 and CF_3O radical with NO
at 295 K,
Chem. Phys. Lett. **206** (1993) 369.

Sehested, J., O.J. Nielsen, T.J. Wallington;
Absolute rate constants for the reaction of NO with a series of peroxy radicals in the gas
phase at 298 K,
Chem. Phys. Lett. **213** (1993) 457.

Sehested, J., T. Ellerman, O.J. Nielsen, T.J. Wallington, M.D. Hurley;
UV absorption spectrum, and kinetics and mechanism of the self reaction of $CF_3CF_2O_2$
radicals in the gas phase at 295 K,
Int. J. Chem. Kinet. **25** (1993) 701.

Semadeni, M., D.W. Stocker, J.A. Kerr;
Further studies of the temperature dependence of the rate coefficients for the reactions
of OH with a series of ethers under simulated atmospheric conditions,
J. Atmos. Chem. **16** (1993) 79.

Seuwen, R., P. Warneck;
Carbon balancing and branching ratios for the gas phase oxidation of toluene by OH
radicals in the absence of NO_x
in: J. Peeters (ed), *Chemical Mechanisms Describing Tropospheric Processes*,
Air Pollution Research Report **45**, Guyot, Brussels 1993, pp.95–100.

Sidebottom, H. W.;
Atmospheric NO_y – implications of kinetic and mechanistic studies,
in: P.M. Borrell, P. Borrell, T. Cvitaš, W. Seiler (eds), *Proc. EUROTRAC Symp. '92*, SPB Academic Publ., The Hague 1993, pp. 335–338.

Wallington, T.J., M.D. Hurley, J.C. Ball, M.E. Jenkin;
FTIR product study of the reaction of $CH_3OCH_2O_2$ + HO_2,
Chem. Phys. Lett. **41** (1993) 211.

Wallington, T.J., T. Ellermann, O.J. Nielsen;
Atmospheric chemistry of dimethyl sulfide: UV spectra and self reaction kinetics of CH_3SCH_2 and $CH_3SCH_2O_2$ radicals, and kinetics of the reactions
CH_3SCH_2 + O_2 → $CH_3SCH_2O_2$, and $CH_3SCH_2O_2$ + NO → CH_3SCH_2O + NO_2
J. Phys. Chem. **97** (1993) 8442.

Wängberg, I.;
Mechanisms and products of the reactions of NO_3 with cycloalkenes,
J. Atmos. Chem. **17** (1993) 229.

Wayne, R.P;
Nitrogen and nitrogen compounds in the atmosphere,
in: T.P. Burt, A.L. Heathwaite, S.T. Trudgill (eds), *Nitrate: Processes, Patterns and Management*, J. Wiley & Sons, Chichester 1993, pp. 23–28.

Weißenmayer, M., J. Burrows, R. Gall, D. Hastie, A. Ladstätter-Weißenmayer, M. Luria, M. Peleg, D. Perner, P. Russel;
Direct insight into atmospheric ozone formation by peroxy radical observation,
in: P.M. Borrell, P. Borrell, T. Cvitaš, W. Seiler (eds), *Proc. EUROTRAC Symp. '92*, SPB Academic Publ., The Hague 1993, pp. 186.

Wille, U., R.N. Schindler;
On the identification of oxiranes formed in the reactions of NO_3 radicals with alkenes,
Ber. Bunsenges. Phys. Chem. **97** (1993) 1447–1453.

Zellner, R.;
VOC oxidation under high NO_x conditions: Mechanisms and ozone formation potentials,
in: P.M. Borrell, P. Borrell, T. Cvitaš, W. Seiler (eds), *Proc. EUROTRAC Symp. '92*, SPB Academic Publ., The Hague 1993, pp. 339–345.

Zetzsch, C., M. Elend, R. Knispel, R. Koch, M. Siese;
Hydroxyl and aromatics: Fate of the adducts in the presence of NO_x and O_2,
in: P.M. Borrell, P. Borrell, T. Cvitaš, W. Seiler (eds), *Proc. EUROTRAC Symp. '92*, SPB Academic Publ., The Hague 1993, pp. 372–376.

1994

Bächmann, K., B. Göttlicher, I. Haag, M. Hannina, W. Hensel;
Sample stacking for charged phenol derivatives in capillary zone electrophoresis,
Fresenius J. Anal. Chem. **350** (1994) 368–371.

Barnes, I., K.H. Becker, A. Bierbach, B. Klotz, E. Wiesen;
Atmospheric chemistry of unsaturated dicarbonyl compounds,
in: P.M. Borrell, P. Borrell, T. Cvitaš, W. Seiler (eds), *Proc. EUROTRAC
Symp. '94*, SPB Academic Publ., The Hague 1994, pp. 158–162.

Barry, J., D.J. Scollard, J.J. Treacy, H.W. Sidebottom, G. Le Bras, G. Poulet, S. Téton,
A. Chichinin, C.E. Canosa-Mas, D.J. Kinnison, R.P. Wayne, O.J. Nielsen;
Kinetic data for the reaction of hydroxyl radicals with 1,1,1-trichloroacetaldehyde at
298+ 2 K
Chem. Phys. Lett. **221** (1994) 353.

Becker, K.H.;
The atmospheric oxidation of aromatic hydrocarbons and its impact on photo-oxidant
chemistry,
in: P.M. Borrell, P. Borrell, T. Cvitaš, W. Seiler (eds), *Proc. EUROTRAC
Symp. '94*, SPB Academic Publ., The Hague 1994, pp. 67–74.

Becker, K.H., E. Wiesen, I. Barnes, A. Bierbach, P. Biggs, C.E. Canosa-Mas,
P.S. Owen, R.P. Wayne, V. Daële, I.T. Lançar, G. Laverdet, G. Poulet,
G. Le Bras;
Etudes en laboratoire de processus d'oxydation atmosphérique de composés organiques
volatils: Application a la pollution photooxydante,
Final Report, Institut Français du Petrole, Paris 1994.

Becker, K.H., I. Barnes, L. Ruppert;
The OH radical-initiated initiated oxidation of isoprene,
in: G. Angeletti, G. Restelli (eds), *Physico-Chemical Behaviour of Atmospheric
Pollutants, Air Pollution Research Report* **50**, EC, Luxembourg 1994, pp. 54–60,
(ISBN92-826-7922-5).

Becker, K.H., K. Wirtz, M. Martin-Reviejo, M.M. Millán;
Photochemical ozone production in hydrocarbon-NO_x air systems,
in: G. Angeletti, G. Restelli (eds), *Physico-Chemical Behaviour of Atmospheric
Pollutants, Air Pollution Research Report* **50**, EC, Luxembourg 1994, pp. 163–168,
(ISBN92-826-7922-5).

Behmann, T., Weißenmayer M., J. P. Burrows;
Peroxy radicals in night-time oxidation chemistry,
in: G. Angeletti, G. Restelli (eds), *Physico-Chemical Behaviour of Atmospheric Pollutants, Air Pollution Research Report* **50**, EC, Luxembourg 1994, pp. 259–265, (ISBN92-826-7922-5).

Benter, Th., M. Liesner, R.N. Schindler, H. Skov, J. Hjorth, G. Restelli;
REMPI-MS and FTIR study of NO_2 and oxirane formation in the reactions of unsaturated hydrocarbons with NO_3 radicals,
J. Phys. Chem. **98** (1994) 10492–10496.

Benter, Th., M. Liesner, V. Sauerland, R.N. Schindler;
Mass spectrometric in-situ determination of NO_2 in gas mixtures by resonance enhanced multiphoton ionisation,
Fresenius J. Anal. Chem. **343** (1994)

Bierbach, A., I. Barnes, K.H. Becker, B. Klotz, E. Wiesen;
OH-radical initiated degradation of aromatic hydrocarbons,
in: G. Angeletti, G. Restelli (eds), *Physico-Chemical Behaviour of Atmospheric Pollutants, Air Pollution Research Report* **50**, EC, Luxembourg 1994, pp. 129–136, (ISBN92-826-7922-5).

Bierbach, A., I. Barnes, K.H. Becker, E. Wiesen;
Atmospheric chemistry of unsaturated carbonyls: butenedial, 4-oxo-2-pentenal, 3-hexene-2,5-dione, maleic anhydride, 3*H*-furan-2-one and 5-methyl-3*H*-furan-2-one,
Environ. Sci. Technol. **28** (1994) 715–729.

Biggs, P., C.E. Canosa-Mas, J.-M. Fracheboud, D.E. Shallcross, R.P. Wayne;
The reactions of alkyl, alkoxy and alkylperoxy radicals with NO_3,
in: P.M. Borrell, P. Borrell, T. Cvitaš, W. Seiler (eds), *Proc. EUROTRAC Symp. '94*, SPB Academic Publ., The Hague 1994, pp. 144–148.

Biggs, P., C.E. Canosa-Mas, J.-M. Fracheboud, D.E. Shallcross, R.P. Wayne;
Investigation into the kinetics and mechanism of the reaction of NO_3 with CH_3 and CH_3O at 298 K between 0.6 and 8.5 Torr,
J. Chem. Soc., Faraday Trans. **90** (1994) 1197–1204.

Biggs, P., C.E. Canosa-Mas, J.-M. Fracheboud, D.E. Shallcross, R.P. Wayne;
Investigation into the kinetics and mechanism of the reaction of NO_3 with CH_3O_2 at 298 K 2.5 Torr: a potential source of OH in the night-time atmosphere,
J. Chem. Soc., Faraday Trans. **90** (1994) 1205–1210.

Biggs, P., C.E. Canosa-Mas, J.-M. Fracheboud, D.E. Shallcross, R.P. Wayne;
An investigation of the kinetics of the reaction between CH_3O_2, CH_3O and CH_3 and NO_3,
in: G. Angeletti, G. Restelli (eds), *Physico-Chemical Behaviour of Atmospheric Pollutants, Air Pollution Research Report* 50, EC, Luxembourg 1994, pp. 77–83, (ISBN92-826-7922-5).

Biggs, P., C.E. Canosa-Mas, K.J. Hansen, P.S. Owen, R.P. Wayne;
The thermal decomposition of peroxyacetylnitrate in the presence of NO_3,
in: P.M. Borrell, P. Borrell, T. Cvitaš, W. Seiler (eds), *Proc. EUROTRAC Symp. '94*, SPB Academic Publ., The Hague 1994, pp. 139–143.

Catoire, V. E. Villenave, M.T. Rayez, R. Lesclaux;
Kinetic studies of the CH_2ClO_2 and CH_2BrO_2 radical reactions; reactions of the CH_2ClO radical,
in: G. Angeletti, G. Restelli (eds), *Physico-Chemical Behaviour of Atmospheric Pollutants, Air Pollution Research Report* 50, EC, Luxembourg 1994, (ISBN92-826-7922-5).

Catoire, V., R. Lesclaux, P.D. Lightfoot, M.T. Rayez;
A kinetic study of the reactions of CH_2ClO_2 with itself and with HO_2, and a theoretical study of the reactions of CH_2ClO between 251 and 600 K,
J. Phys. Chem. 98 (1994) 2889–2898.

Curley, M., J. Treacy, H. Sidebottom;
Kinetic and mechanistic studies of the ozonolysis of alkenes under atmospheric conditions,
in: P.M. Borrell, P. Borrell, T. Cvitaš, W. Seiler (eds), *Proc. EUROTRAC Symp. '94*, SPB Academic Publ., The Hague 1994, pp. 124–127.

Daële, V., A. Mellouki, S. Téton, G. Laverdet, G. Poulet, G. Le Bras;
Laboratory studies of NO_3 and OH reactions involved in the oxidation mechanisms of VOCs,
in: P.M. Borrell, P. Borrell, T. Cvitaš, W. Seiler (eds), *Proc. EUROTRAC Symp. '94*, SPB Academic Publ., The Hague 1994, pp. 134–138.

Eberhard, J., C. Müller, D.W. Stocker, J.A. Kerr;
Mechanism for the OH-radical initiated photo-oxidation of alkanes: alkoxy radical reactions,
in: G. Angeletti, G. Restelli (eds), *Physico-Chemical Behaviour of Atmospheric Pollutants, Air Pollution Research Report* 50, EC, Luxembourg 1994, pp. 169–174, (ISBN92-826-7922-5).

Elmaimouni, L., C. Bourbon, C. Fittschen, J. P. Sawerysyn, P. Devolder;
Reaction rates of the C_2H_5O (ethoxy) radical with O_2 and NO,
in: P.M. Borrell, P. Borrell, T. Cvitaš, W. Seiler (eds), *Proc. EUROTRAC Symp. '94*, SPB Academic Publ., The Hague 1994, pp. 153–157.

Fenter, F.F., B. Noziere, F. Caralp, R. Lesclaux;
Study of the kinetics and equilibrium of the benzyl radical association reaction with molecular oxygen,
Int. J. Chem. Kinet. **26** (1994) 171–190.

Granby, K., O. Hertel, O. J. Nielsen, T. Nielsen, A. H. Egeløv;
Organic sulphur compounds: Atmospheric chemistry and occurence,
in: P.M. Borrell, P. Borrell, T. Cvitaš, W. Seiler (eds), *Proc. EUROTRAC Symp. '94*, SPB Academic Publ., The Hague 1994, pp. 163–166.

Hayman, G.;
The oxidation of biogenic hydrocarbons,
in: P.M. Borrell, P. Borrell, T. Cvitaš, W. Seiler (eds), *Proc. EUROTRAC Symp. '94*, SPB Academic Publ., The Hague 1994, pp. 75–82.

Helleis, F., J.N. Crowley, G.K. Moortgat;
The reaction between CD_3O_2 and NO_2: A discharge-flow study at 298 K,
in: G. Angeletti, G. Restelli (eds), *Physico-Chemical Behaviour of Atmospheric Pollutants, Air Pollution Research Report* **50**, EC, Luxembourg 1994, pp. 90–96, (ISBN92-826-7922-5).

Hillmann, R., K. Bächmann;
On-line supercritical fluid derivatization and extraction/capillary gas chromatography of polar compounds,
J. High Res. Chromatogr. **17** (1994) 350–352.

Horie, O., P. Neeb, G.K. Moortgat;
Formation of formic acid and organic peroxides in the ozonolysis of ethene with added water vapour,
Geophys. Res. Lett. **21** (1994) 1523–1526.

Horie, O., P. Neeb, G.K. Moortgat;
Ozonolysis of *trans*- and *cis*-2-butenes in low parts-per-million concentration ranges,
Int. J. Chem. Kinet. **26** (1994) 1075–1094.

Kirschfeld, A., S. Muthusamy, W. Sander;
Dimesityldioxirane – a dioxirane stable in the solid state,
Angew. Chem., Int. Ed. **33** (1994) 2212.

Koch, R., R. Knispel, M. Siese, C. Zetzsch;
Absolute rate constants and products of secondary steps in the atmospheric degradation of aromatics,
in: G. Angeletti, G. Restelli (eds), *Physico-Chemical Behaviour of Atmospheric Pollutants, Air Pollution Research Report* **50**, EC, Luxembourg 1994, pp. 143–149, (ISBN92-826-7922-5).

Kukui, A., T. Jungkamp, R.N. Schindler;
Determination of the product branching ratio in the reaction of NO_3 with OCl
at 300 K,
Ber. Bunsenges. Phys. Chem. **98** (1994)

Kukui, A.S., T.P.W. Jungkamp, R.N. Schindler;
Determination of the rate constant and of product branching ratios in the reaction of
CH_3O_2 with OCl between 233 and 300 K,
Ber.Bunsenges. Phys. Chem. **98** (1994) 1298–1302.

Langer, S., E. Ljungström;
Rates of reaction between the nitrate radical and some aliphatic ethers,
Int. J. Chem. Kinet. **26** (1994) 367.

Langer, S., E. Ljungström;
Reaction of the nitrate radical with some potential automotive fuel additives.
A kinetic and mechanistic study,
J. Phys. Chem. **98** (1994) 5906.

Langer, S., E. Ljungström, J. Sehested, O.J. Nielsen;
UV absorption spectra, kinetics and mechanism for alkyl and alkyl peroxy radicals
originating from *t*-butyl alcohol,
Chem. Phys. Lett. **226** (1994) 165.

Locke, G.P., J.J. Treacy, H.W. Sidebottom, C.J. Percival, R.P. Wayne;
Kinetics and mechanism for the oxidation of bromomethyl radicals,
in: G. Angeletti, G. Restelli (eds), *Physico-Chemical Behaviour of Atmospheric
Pollutants, Air Pollution Research Report* **50**, EC, Luxembourg 1994,
pp. 405–410, (ISBN92-826-7922-5).

Mayer-Figge, A., F. Kirchner, H.G. Libuda, F. Zabel, K.H. Becker;
Thermal and photochemical stability of peroxynitrates,
in: P.M. Borrell, P. Borrell, T. Cvitaš, W. Seiler (eds), *Proc. EUROTRAC
Symp. '94*, SPB Academic Publ., The Hague 1994, pp. 175–179.

Mellouki, A., S. Téton, G. Laverdet, A. Quilgers, G. Le Bras;
Kinetic study of OH reactions with H_2O_2, C_3H_8 and CH_4 using the pulsed photolysis
laser induced fluorescence method,
J. Chim. Phys. **5** (1994) 473.

Mögelberg, T.E., O.J. Nielsen, J. Sehested, T.J. Wallington, M.D. Hurley,
W.F. Schneider;
Atmospheric chemistry of HCF-134a. Kinetic and mechanistic study of the
CF_3CFHO_2 + NO_2 reaction,
Chem. Phys. Lett. **225** (1994) 375.

Mögelberg, T.E., O.J. Nielsen, J. Sehested, T.J. Wallington, M.D. Hurley;
Atmospheric chemistry of CF₃COOH. Kinetics of the reaction with OH radicals,
Chem. Phys. Lett. **226** (1994) 171.

Moortgat, G.K.;
Radical reactions of NO₃ with HO₂ and RO₂. A review,
in: G. Angeletti, G. Restelli (eds), *Physico-Chemical Behaviour of Atmospheric Pollutants, Air Pollution Research Report* **50**, EC, Luxembourg 1994, pp. 66–76, (ISBN92-826-7922-5).

Neavyn, R., H. Sidebottom, J. Treacy;
Reactions of hydroxyl radicals with polyfunctional group oxygen-containing organic compounds,
in: P.M. Borrell, P. Borrell, T. Cvitaš, W. Seiler (eds), *Proc. EUROTRAC Symp. '94*, SPB Academic Publ., The Hague 1994, pp. 105–109.

Neeb, P., O. Horie, S. Limbach, G. K. Moortgat;
The formation of formic acid and organic peroxides in the ozonolysis of ethene under atmospheric conditions,
in: P.M. Borrell, P. Borrell, T. Cvitaš, W. Seiler (eds), *Proc. EUROTRAC Symp. '94*, SPB Academic Publ., The Hague 1994, pp. 128–133.

Nielsen, O.J., E. Gamborg, J. Sehested, T.J. Wallington, M.D. Hurley;
Atmospheric chemistry of HFC-143a: Spectrokinetic investigation of the CF₃CH₂O₂ radical, its reactions with NO and NO₂ and fate of CF₃CHO,
J. Phys. Chem. **98** (1994) 9518.

Noziere, B., R. Lesclaux, M.D. Hurley, M.A. Dearth, T.J. Wallington;
A kinetic and mechanistic study of the self-reaction and reaction with HO₂ of the benzylperoxy radical,
J. Phys. Chem. **98** (1994) 2864–2873.

Olzmann, M., Th. Benter, M. Liesner, R.N. Schindler;
On the pressure dependence of the NO₂ product yield in the reaction of NO₃ radicals with selected alkenes,
Atmos. Environ. **28** (1994) 2677–2683.

Peeters, J., A. Ectors, W. Boullart;
Reactions of OH with isoprene, 1-butene and isobutene: H-abstraction and OH-addition branching fractions,
in: G. Angeletti, G. Restelli (eds), *Physico-Chemical Behaviour of Atmospheric Pollutants, Air Pollution Research Report* **50**, EC, Luxembourg 1994, pp. 61–65, (ISBN92-826-7922-5).

Peeters, J., V. Pultau;
Reactions of hydrocarbon- and hydrochlorofluorocarbon-derived peroxy radicals with nitric oxide,
in: G. Angeletti, G. Restelli (eds), *Physico-Chemical Behaviour of Atmospheric Pollutants, Air Pollution Research Report* **50**, EC, Luxembourg 1994, pp. 372–378, (ISBN92-826-7922-5).

Peeters, J., W. Boullart, J. Van Hoeymissen;
Site-specific partial rate constants for OH addition to alkenes and dienes,
in: P.M. Borrell, P. Borrell, T. Cvitaš, W. Seiler (eds), *Proc. EUROTRAC Symp. '94*, SPB Academic Publ., The Hague 1994, pp. 110–114.

Sehested, J., Th. Ellermann, O.J. Nielsen;
A spectrokinetic study of CH_2I and CH_2IO_2 radicals,
Int. J. Chem. Kinet. **26** (1994) 259.

Sehested, J., Th. Ellermann, O.J. Nielsen, T.J. Wallington;
Spectrokinetic study of SF_5 and SF_5O_2 radicals and the reaction of SF_5O_2 with NO,
Int. J. Chem. Kinet. **26** (1994) 615.

Semadeni, M., D.W. Stocker, J.A. Kerr;
The temperature dependence of the OH-radical reaction of some aromatic compounds under simulated tropospheric conditions,
in: G. Angeletti, G. Restelli (eds), *Physico-Chemical Behaviour of Atmospheric Pollutants, Air Pollution Research Report* **50**, EC, Luxembourg 1994, pp. 150–156, (ISBN92-826-7922-5).

Seuwen, R., P. Warneck;
The oxidation of toluene in the gas phase induced by the reactions with hydroxyl radicals and chlorine atoms,
in: G. Angeletti, G. Restelli (eds), *Physico-Chemical Behaviour of Atmospheric Pollutants, Air Pollution Research Report* **50**, EC, Luxembourg 1994, pp. 137–142, (ISBN92-826-7922-5).

Siese, M., R. Koch, C. Zetzsch;
Cycling of OH in the reaction systems toluene/O_2/NO and acetylene/O_2,
in: P.M. Borrell, P. Borrell, T. Cvitaš, W. Seiler (eds), *Proc. EUROTRAC Symp. '94*, SPB Academic Publ., The Hague 1994, pp. 115–119.

Skov, H., Th. Benter, R.N. Schindler, J. Hjorth, G. Restelli;
Epoxide formation in the reactions of nitrate radical with 2,3-2-dimethyl-2-butene, *cis*- and *trans*-2-butene and isoprene,
Atmos. Environ. **28**A (1994) 1583.

Thomas, W., F. Zabel, K.H. Becker, E. Fink;
A mechanistic study on the ozonolysis of ethene,
in: G. Angeletti, G. Restelli (eds), *Physico-Chemical Behaviour of Atmospheric Pollutants, Air Pollution Research Report* 50, EC, Luxembourg 1994,
pp. 207–212, (ISBN92-826-7922-5).

Wallington, T.J., J. Sehested, O.J. Nielsen;
Atmospheric chemistry of $CF_3C(O)O_2$ radicals: Kinetics of their reaction with NO_2 and kinetics of the thermal decomposition of the product $CF_3C(O)O_2NO_2$,
Chem. Phys. Lett. 226 (1994) 563.

Wallington, T.J., M.D. Hurley, J.C. Ball, Th. Ellermann, O.J. Nielsen, J. Sehested;
Atmospheric chemistry of HFC-152: UV spectrum of CH_2FCHHO_2 radicals, kinetics of the reaction $CH_2FCFHO_2 + NO \rightarrow CH_2FCFHO + NO_2$ and fate of the alkoxy radical CH_2FCHFO,
J. Phys. Chem. 98 (1994) 5435.

Wallington, T.J., M.D. Hurley, O.J. Nielsen, J. Sehested;
Atmospheric chemistry of CF_3CO_x radicals: Fate of CF_3CO radicals,
the UV absorption spectrum of $CF_3C(O)O_2$ radicals, and kinetics of the reaction
$CF_3C(O)O_2 + NO \rightarrow CF_3C(O)O + NO_2$,
J. Phys. Chem. 98 (1994) 5686

Wallington, T.J., M.D. Hurley, W.F. Schneider, J. Sehested, O.J. Nielsen;
Mechanistic study of the gas phase reaction of CH_2FO_2 radicals with HO_2,
Chem. Phys. Lett. 218 (1994) 34.

Wallington, T.J., Th. Ellermann, O.J. Nielsen;
Pulse radiolysis study of CF_3CCl_2 and $CF_3CCl_2O_2$ radicals in the gas phase
at 295 K,
Res. Chem. Intermed. 20 (1994) 265.

Wallington, T.J., Th. Ellermann, O.J. Nielsen, J. Sehested;
Atmospheric chemistry of FCO_x radicals: UV spectra and self reaction kinetics of FCO and $FC(O)O_2$ and kinetics of some reactions of FCO_x with O_2, O_3, and NO
at 296 K,
J. Phys. Chem. 98 (1994) 2346

Warneck, P.;
Tropospheric oxidation chemistry,
in: P.M. Borrell, P. Borrell, T. Cvitaš, W. Seiler (eds), *Proc. EUROTRAC Symp. '94*, SPB Academic Publ., The Hague 1994, pp. 37–46.

Weißenmayer, M., J.P. Burrows, M. Schupp;
Peroxy radical measurements in the boundary layer above the Atlantic Ocean,
in: G. Angeletti, G. Restelli (eds), *Physico-Chemical Behaviour of Atmospheric Pollutants, Air Pollution Research Report* 50, EC, Luxembourg 1994,
pp. 575–582, (ISBN92-826-7922-5).

Wierlacher, S., W. Sander, C. Marquardt, E. Kraka, D. Cremer;
Propinal O-oxide,
Chem. Phys. Lett. **222** (1994) 319.

Zabel, F.;
Mechanistic studies of ozone reactions with alkenes,
in: G. Angeletti, G. Restelli (eds), *Physico-Chemical Behaviour of Atmospheric Pollutants, Air Pollution Research Report* **50**, EC, Luxembourg 1994,
pp. 197–206, (ISBN92-826-7922-5).

Zahn, C., O. Horie, G.K. Moortgat;
Gas-phase ozonolysis of butadiene and isoprene under atmospheric conditions,
in: G. Angeletti, G. Restelli (eds), *Physico-Chemical Behaviour of Atmospheric Pollutants, Air Pollution Research Report* **50**, EC, Luxembourg 1994, pp. 66–76.

Zellner, R., A. Hoffmann, W. Malms, V. Mörs;
Time resolved studies of NO$_2$ and OH formation in the integrated oxidation chain of benzene,
in: P.M. Borrell, P. Borrell, T. Cvitaš, W. Seiler (eds), *Proc. EUROTRAC Symp. '94*, SPB Academic Publ., The Hague 1994, pp. 171–174.

Zetzsch, C.;
Atmospheric oxidation processes of aromatics studied within LACTOZ,
in: G. Angeletti, G. Restelli (eds), *Physico-Chemical Behaviour of Atmospheric Pollutants, Air Pollution Research Report* **50**, EC, Luxembourg 1994,
pp. 118–128, (ISBN92-826-7922-5).

1995 + in press

Becker, K.H., I. Barnes;
Atmospheric chemistry relevant to urban pollution,
in: *NATO ASI Series, Monitoring and Control Strategies for Urban Pollution*, Springer Verlag, Heidelberg 1995, in press.

Behmann, T., M. Weißenmayer, J.P. Burrows;
Measurements of peroxy radicals in a coastal environment,
Atmos. Environ. (1995) in press.

Bierbach, A., I. Barnes, K.H. Becker;
Product and kinetic study of the OH-initiated gas-phase oxidation of furan, 2-methylfuran and furanaldehydes at ~300 K,
Atmos. Environ. A (1995) in press.

Biggs, P., C.E. Canosa-Mas, J.-M. Fracheboud, D.E. Schallcross, R.P. Wayne;
Rate constants for the reactions of C_2H_5, C_2H_5O and $C_2H_5O_2$ at 298 K and 2.2 Torr,
J. Chem. Soc., Faraday Trans. **91** (1995) in press.

Biggs, P., C.E. Canosa-Mas, R.P. Wayne;
The interaction of NO with organic peroxy radicals: a non-photolytic source of OH radicals,
in: K.H. Becker (ed), *Tropospheric Oxidation Mechanisms; Air Pollution Research Report*, EC, Brussels 1995, in press.

Brockmann, K.J., S. Gäb, W.V. Turner, S. Wolff, S. Mönninghoff;
Formation of alkyl and hydroxyalkyl hydroperoxides by the ozonolysis of alkenes,
in: K.H. Becker (ed), *Tropospheric Oxidation Mechanisms; Air Pollution Research Report*, EC, Brussels 1995, in press.

Cabañas-Galan, B., G. Marston, R.P. Wayne;
Arrhenius parameters for the reaction of the nitrate radical with 1,1-dichloroethane and (*E*)-1,2-dichloroethane,
J. Chem. Soc., Faraday Trans. **91** (1995) in press.

Coquart, B., A. Jenouvrier, M.F. Merienne;
The NO_2 absorption spectrum. II: Absorption cross-sections at low temperatures in the 400–500 nm region,
J. Atmos. Chem. (1995) in press.

Daële, V., G. Laverdet, G. Le Bras, G. Poulet;
Kinetics of the reactions $CH_3O + NO$, $CH_3O + NO_3$ and $CH_3O_2 + NO_3$,
J. Phys. Chem. **99** (1995) 1470.

Eberhard, J., C. Müller, D.W. Stocker, J.A. Kerr;
Isomerization of alkoxy radicals under atmospheric conditions,
Environ. Sci. Technol. **29** (1995) 232.

Finkbeiner, M., P. Neeb, O. Horie, G.K. Moortgat;
A simple method of calibration of the formic acid monomer concentration in the gas phase,
Fresenius J. Anal. Chem. (1995) in press.

Gäb, S., W.V. Turner, S. Wolff, K.H. Becker, L. Ruppert, K.J. Brockmann;
Formation of alkyl and hydroxyalkyl hydroperoxides on ozonolysis in water and in air,
Atmos. Environ. (1995) in press.

Jenkin, M.E., G.D. Hayman;
Kinetics of reactions of primary, secondary, tertiary b-hydroxy peroxy radicals: application to isoprene degradation,
J. Chem. Soc., Faraday Trans. (1995) in press.

Klotz, B., A. Bierbach, I. Barnes, K.H. Becker;
Kinetic and mechanistic study of the atmospheric chemistry of muconaldehydes,
Environ. Sci. Technol. (1995) in press.

Klotz, B., I. Barnes, A. Bierbach, K.H. Becker;
Atmospheric chemistry of unsaturated dicarbonyl compounds,
in: K.H. Becker (ed), *Tropospheric Oxidation Mechanisms; Air Pollution Research Report*, EC, Brussels 1995, in press.

Koch, R., M. Siese, C. Zetzsch;
Addition and regeneration of OH in the presence of O_2 and NO: Results on naphthalene, isoprene and acetylene,
in: K.H. Becker (ed), *Tropospheric Oxidation Mechanisms; Air Pollution Research Report*, EC, Brussels 1995, in press.

Malicet, J., D. Daumont, J. Charbonnier, C. Parisse, A. Chakir, J. Brion;
Ozone U.V. spectroscopy. II. Absorption cross sections and temperature dependence,
J. Atmos. Chem. (1995) in press.

Mellouki, A., S. Téton, G. Le Bras;
Kinetics of the OH reactions with a series of ethers,
Int. J. Chem. Kinet. (1995) in press.

Merienne, M.F., A. Jenouvrier, B. Coquart;
The NO_2 absorption spectrum I: Absorption cross sections at ambient temperature in the 300–500 nm region,
J. Atmos. Chem. (1995) in press.

Møgelberg, T.E., J. Platz, O.J. Nielsen, J. Sehested, T.J. Wallington;
Atmospheric Chemistry of HFC-236fa : Spectrokinetic investigation of the $CF_3CHO_2(\bullet)CF_3$ radical, its reaction with NO and the fate of the $CF_3CHO(\bullet)CF_3$ radical,
J. Phys. Chem. (1995) in press.

Møgelberg, T.E., O.J. Nielsen, J. Sehested, T.J. Wallington, M.D. Hurley;
Atmospheric Chemistry of HFC-272ca: Spectrokinetic investigation of the $CH_3CF_2CH_2O_2$ radical, its reactions with NO and NO_2, and the fate of the $CH_3CF_2CH_2O$ radical,
J. Phys. Chem. (1995) in press.

Peeters, J., J. Van Hoeymissen;
Tropospheric model for testing chemical mechanisms,
in: K.H. Becker (ed), *Tropospheric Oxidation Mechanisms; Air Pollution Research Report*, EC, Brussels 1995, in press.

Platz, J., O.J. Nielsen, J. Sehested, T.J. Wallington;
Atmospheric chemistry of 1,1,1-trichloroethane: UV spectra and self-reaction kinetics of CCl_3CH_2 and $CCl_3CH_2O_2$ radicals, kinetics of the reactions of the $CCl_3CH_2O_2$ radical with NO and NO_2, and the fate of the alkoxy radical CCl_3CH_2O,
J. Phys. Chem. (1995) in press.

Raber, W.H., G.K. Moortgat;
Photooxidation of selected carbonyl compounds in air: methylethylketone, methylvinylketone, methacrolein and methylglyoxal,
Advan. Phys. Chem. (1995) in press.

Ruppert, L., I. Barnes, K.H. Becker;
Tropospheric reactions of isoprene and oxidation products: Kinetic and mechanistic studies,
in: K.H. Becker (ed), *Tropospheric Oxidation Mechanisms; Air Pollution Research Report*, EC, Brussels 1995, in press.

Semadeni, M., D.W. Stocker, J.A. Kerr;
The temperature dependences of the OH radical reactions with some aromatic compounds under simulated tropospheric conditions,
Int. J. Chem. Kinet. **27** (1995) in press.

Siese, M., C. Zetzsch;
Addition of OH to acetylene and consecutive reactions of the adduct with O_2,
Z. Phys. Chem. **188** (1995) 75–89.

Thomas, W., F. Zabel, K.H. Becker, E.H. Fink;
A mechanistic study on the ozonolysis of ethene,
in: K.H. Becker (ed), *Tropospheric Oxidation Mechanisms; Air Pollution Research Report*, EC, Brussels 1995, in press.

Villenave, E., R. Lesclaux;
The UV absorption spectra of CH_2Br and CH_2BrO_2 and the reaction kinetics of CH_2BrO_2 with itself and with HO_2 at 298 K,
Chem. Phys. Lett. (1995) in press.

Wallington, T.J., W.F. Schneider, T.E. Møgelberg, O. J. Nielsen, J. Sehested;
Atmospheric Chemistry of FCO_X Radicals: Kinetic and Mechanistic Study of the $FC(O)O_2 + NO_2$ Reaction,
Int. J. Chem. Kinet. (1995) in press.

Weißenmayer M., J.P. Burrows, R. Gall, D. Hastie, A. Ladstätter-Weißenmayer, M. Luria, M. Peleg, P. Russel, D. Perner;
Observation of peroxy radicals during ozone formation in polluted tropospheric air,
Nature (1995) in press.

Wiesen, E., I. Barnes, K.H. Becker;
  Study of the OH-initiated degradation of the aromatic photooxidation product
  3-hexene-3,4-diol-2,5-dione,
  Environ. Sci. Technol. (1995) in press.

Wiesen, E., I. Barnes, K.H. Becker;
  Atmospheric degradation mechanism of alkylated benzenes,
  in: K.H. Becker (ed), *Tropospheric Oxidation Mechanisms; Air Pollution Research
  Report*, EC, Brussels 1995, in press.

Wirtz, K. K.H. Becker, M. Martin-Reviejo, M.M. Millán;
  Computer simulations of recent experimental results on the oxidation of aromatic
  hydrocarbons,
  in: K.H. Becker (ed), *Tropospheric Oxidation Mechanisms; Air Pollution Research
  Report*, EC, Brussels 1995, in press.

Zabel, F.;
  Unimolecular decomposition of peroxynitrates,
  Z. phys. Chem. **118** (1995) 119–142.

Theses

M. Sc. / Diploma

Aird, R.W.S.;
  Environmentally important reactions,
  University of Oxford, 1990.

Ashbourn, S.F.M.;
  Degradation of halogenated compounds in the troposphere,
  University of Oxford,1994.

Bagley, J.A.;
  Reactions of atmospheric importance,
  University of Oxford, 1989.

Berges, M.;
  Product quantum yields for the photolysis of pyruvic acid under atmospheric conditions,
  (in German),
  Johannes Gutenberg University of Mainz, 1991.

Van den Berghe, K.;
    Kinetic study of the reaction of the CF_3O_2 radical with nitric oxide,
    University of Leuven, 1991.

Bienfait, C.;
    Mise au point de codes de calcul sur les transitions rovibroniques dans les molécules
    triatomiques: application à l'ozone,
    University of Reims, 1995.

Blackburn, M.C.;
    Degradation of atmospherically important halogen compounds,
    University of Oxford, 1995.

Bourbon, C.;
    Mesure de la constante de vitesse de la réaction OH + méthylglyoxal → produits en tube
    à écoulement,
    University of Lille, 1993.

Cedroli, A.;
    Chimica della troposfera: Reattivita del radicale nitrato con i metilareni in fase gassosa,
    University of Milan, 1995.

Chambers, R.M.;
    Atmospheric iodine chemistry,
    Oxford University, 1991.

Chiodini, G.;
    Metologie analitiche per lo studio del comportamento di microinquinanti in atmosfera e
    in acque fluviali,
    University of Milan, 1990.

Cook, D.J.;
    Kinetics of radical reactions of atmospheric importance,
    University of Oxford, 1990.

Crotty, G.;
    Reactivity of halogenated compounds in the troposphere,
    University of Oxford 1995.

Dessent, C.E.H.;
    Reactions of the hydroxyl radical of tropospheric importance,
    Oxford University, 1991.

Fauvet, S.;
    Etude cinétique des réactions d'intérêt atmosphérique fasisant intervenir des radicaux
    peroxyles halogénés par la technique de photolyse modulée,
    University of Reims, 1994.

Ganne, J.P.;
Cinétique des réactions élémentaires faisant intervenir les radicaux peroxyles. Etude du
radical $C_2H_6O_2$: mesures expérimentales par la technique de photolyse modulée et
modélisation des résultats,
University of Reims, 1994.

Gamborg, E.;
HFC-143a,
University of Roskilde, 1994.

Geiger, H.;
Kinetische Untersuchung der OH-Radikalkette im System C_2H_4 + NO + Luft,
University of Wuppertal, 1991.

Haahr, N.;
Reactions of HCFC's and HFC's and some of the degradation products with the nitrate
radical,
University of Odense, 1991.

Hansen, K.J.;
Flow tube studies of the nitrate radical,
University of Oxford, 1993.

Harwood, M.H.;
Kinetics of radical reactions of atmospheric importance,
Oxford University, 1991.

Hauptmann, J.;
Trennung und Nachweis organischer Produkte,
TH Darmstadt, 1994.

Hirschberger, K.;
Konstruktion, Bau und Anwendung eines NO_2-Detektors,
Fachhochschule Weisbaden 1988

Joseph, D.M.T.;
Stopped-flow studies of gas-phase radical reactions,
University of Oxford, 1989.

King M.D.;
Reactions of peroxy radicals and related compounds in the troposphere,
University of Oxford, 1995.

King, A.D.;
Slow reactions of the nitrate radical,
Oxford University, 1991.

Kirchner, U.;
  UV-spektroskopische Untersuchungen an ClO und BrO und ihren
  Selbstreaktionsprodukten,
  University of Kiel, 1992.

Klotz, B.;
  Untersuchung über das luftchemische Verhalten von 2,4-hexadiendialen:
  Zwischenprodukte der Aromatenoxidation,
  University of Wuppertal, 1993.

Kutsenogii, P.;
  Photooxidation of isoprene and other unsaturated hydrocarbons by OH radicals in the
  absence of NO_x,
  Russian Academy of Sciences, Novosibirsk 1990.

Limbach, St.;
  Die quantitative Bestimmung organischer Peroxyde mit HPLC,
  Johannes Gutenberg University of Mainz, 1990.

Little, M.R.;
  Reactions of the nitrate radical in the gas phase,
  University of Oxford, 1989.

Mayer-Figge, A.;
  Untersuchungen am Propionylperoxinitrat : Untersuchung des thermischen Zerfalls von
  PPN sowie des Verhältnisses der Geschwindigkeitskonstanten der Reaktion von
  Propionylperoxiradikalen mit NO_2 und NO,
  University of Wuppertal, 1993.

Meller, R.;
  Absolute Absorptionsquerschnitte von Formaldehyd in der Gasphase,
  Johannes Gutenberg University of Mainz, 1990.

Moldanovà, J.;
  Oxidant formation in the troposphere; A modelling study of the role of photochemistry,
  transport and source strength of individual hydrocarbons,
  University of Göteborg, 1994.

Mönninghoff, S.;
  Bildung von Hydroperoxiden in der Ozonolyse von Ethen in Gegenwart von
  Wasserdampf,
  University of Wuppertal, 1995.

Mörs, V.;
  Zeitaufgelöste Untersuchungen der OH- und NO_2-Bildung in der Oxidation von
  Difluorchlormethan (H-FCKW 22), Chloroform und 1,1,1,2 Tetrafluorethan
  (FKW 134a),
  University of Hannover, 1991.

Müller, P.;
Entwickelung und Aufbau eines Probenaufgabesystems für die
gaschromatographische Analyse niedermolekularer Kohlenwasserstoffe,
Fachhochschule Wiesbaden, 1993.

Nasna, A.;
L'ozono, il perossido d'idrogeno e l'ossigeno in chimica ambiente. Reattivita
concomposti di rilevanza ambientale,
University of Milan, 1992.

Paratici, V.;
Un modello di chimica dell'atmosfera in arie densamente popolate: La razione del
nitrato radicale con metilsenzeni in fase gassosa,
University of Milan, 1989.

Percival, C.;
Kinetics of reactions of the OH radical of atmospheric importance,
University of Oxford, 1992.

Pleijel, K.;
Modelling of photochemical oxidants and atmospheric mercury,
University of Göteborg, 1994.

Poppel, M. Van ;
Kinetic Study of Reactions of some Haloalkylperoxy Radicals with NO,
University of Leuven, 1993.

Rathmann, T.;
Modellrechnungen zum erweiterten Chapman-Mechanismus unter Verwendung des
Sofwarepaketes LARKIN,
University of Kiel, 1989.

Roesdahl, H.;
Formaldehyde in the atmosphere: Chemical reactions, kinetics and formation of
products,
University of Odense, 1990.

Rothwell, K.M.;
Flow studies of radical reactions,
University of Oxford, 1990.

Ruppert, L.;
FTIR-Untersuchung der Selbstreaktion von Alkylperoxyradikalen,
University of Wuppertal, 1991.

Sauer, F.;
   Die Bestimmung von organischen Peroxiden und H_2O_2 in der Atmosphäre,
   Johannes Gutenberg University of Mainz, 1993.

Sauerland, V.;
   Aufbau und orientierende Experimente mit einem Photoionisations-
   Flugzeitmassenspektrometer,
   University of Kiel, 1991.

Schmidt, M.,
   Aufbau eines Strömungssystems zur absorptionsspektroskopischen Untersuchungen
   halogenhaltiger Radikale im Sichtbaren und nahen UV mit Hilfe eines Zweizeilen-
   Diodenarrays,
   University of Kiel, 1993.

Schütz, P.;
   Entwicklung einer analytischen Methode zur Bestimmung von
   1-Hydroxyakylhydroperoxiden und Wasserstoffperoxid mittels HPLC,
   TH Darmstadt, 1994.

Schwöpe, D.;
   Messung der Ozonphotolysenrate,
   Institut für Umweltphysik, University of Bremen 1995.

Simon, F.G.;
   Spektrum und Kinetik der Selbstreaktion des Methylperoxy Radikals CH_3O_2 und seiner
   Reaktion mit ClO,
   Johannes Guttenberg University of Mainz, 1989.

Sørensen, S.;
   Laboratory studies of tropospheric degradation mechanisms for HCHC's
   and HFC's,
   University of Odense, 1992.

Stewart, D.;
   Kinetics of gas-phase Reactions involving inorganic radicals,
   University of Oxford, 1992.

Thomas, W.;
   Die Bildung von molekularem Wasserstoff in atmosphärischen Reaktionen,
   University of Wuppertal, 1989.

Thomassen, K.;
   A simplified model of global tropospheric chemistry. Inclusion of detailed
   CH_4 chemistry and of PAN,
   University of Leuven, 1993.

Veronesi, G.;
Chimica della troposfera: Le reazioni del radicale nitrato con i metilareni,
University of Milan, 1992.

Vogt, R.;
Aufbau eines Strömungssystems für absorptionsspektroskopische
Untersuchungen im Sichtbaren und nahen UV. Eine Untersuchung des
Rekombinationsproduktes Cl_2O_2,
University of Kiel, 1988.

Weißenmayer, M.;
Der Chemische Verstärker, eine Methode zum Nachweis von HO_2-Radikalen in der
Außenluft,
Johannes Gutenberg-University of Mainz 1989.

Wiesen, E.;
OH-initiierte Oxidation von 1,4-Dimethylbenzol (p-Xylol) unter atmosphärischen
Bedingungen,
University of Wuppertal, 1990.

Wille, U.;
Massenspektrometrische Bestimmung absoluter Geschwindigkeitskonstanten in einem
Strömungssystem,
University of Kiel, 1988.

Wilson, M.R.;
Time-resolved studies of slow radical reactions,
University of Oxford, 1990.

Winstanley, J.;
Laboratory study of tropospheric pollutants,
University of Oxford, 1992.

Wolff, S.;
Untersuchungen der Ozonreaktion ausgewählter Olefine: Vergleich der
Hydroperoxidbildung in der wässrigen Phase und in der Gasphase,
University of Wuppertal, 1994.

Ph. D.

Balestra-Garcia, C.;
Etudes cinétiques de réactions de dégradation atmosphérique par le radical OH de
composés organo-halogénés et d´intermédiaires aldéhydiques,
Orléans Université, 1993.

Becker, E.;
Gaskinetische Untersuchungen zu den Reaktionen von NO_3-Radikalen mit F, OF, Cl,
ClO, H, OH und HO_2,
University of Kiel, 1992.

Benter, Th.;
REMPI-massenspektrometrische Untersuchungen zur Produktbildung in den
Reaktionen von NO_3 Radikalen mit ungesättigten Kohlenwasserstoffen,
University of Kiel, 1993.

Bierbach, A;
Produktuntersuchungen und Kinetik der OH-initiierten Gasphasenoxidation
aromatischer Kohlenwasserstoffe sowie ausgewählter carbonylischer Folgeprodukte,
University of Wuppertal, 1994.

Boyd, A. A.;
Kinetic studies of NO_3 reactions of atmospheric relevance,
University of Oxford, 1993.

Bridier, I.;
Etude cinétique de radicaux peroxyle d'intérét atmosphérique,
University of Bordeaux, 1991.

Brockmann, K.J.;
Untersuchungen zur Chemie der atmosphärischen Peroxide,
University of Wuppertal, 1992.

Catoire, V.;
Cinétique et mécanisme de réactions des radicaux méthylperoxyles chlorés et
éthylperoxyles d'intérêt en chimie de l'atmosphère et de la combustion,
University of Bordeaux I, 1994.

Connell, R.K.;
Tropospheric degradation of halogenated compounds,
University of Oxford, 1995.

Daële, V;
Etudes cinétiques de réactions de radicaux libres d´intérêt atmosphérique, NO_3, CH_3O,
CH_3O_2, Cl, ClO,
Orléans Université, 1993.

Eberhard, J.E.;
Kinetics and mechanisms of the OH radical initiated oxidation of volatile organic
compounds under simulated tropospheric conditions,
E. T. H. Zurich, 1994.

El Hag Ahmed, M. A.;
Kinetic Studies on the Reaction of Ozone with Alkenes and Haloakenes,
University of Dublin 1994.

Fracheboud, J.-M.;
Peroxy radical reactions in the troposphere,
University of Oxford, 1995.

Heard, A.C.;
Gas-phase reactions of halogen species of atmospheric importance,
Oxford University, 1991.

Heard, A.C. (née Brown);
Reactions of halogen species of atmospheric importance,
University of Oxford, 1991.

Heimann, G.;
Oxidation mechanisms for C_4-C_6 hydrocarbons in the atmosphere,
Johannes Gutenberg University of Mainz, 1991.

Hoffmann, A.;
Entwicklung und Anwendung von Absorptions- und Fluoreszenzmethoden zum
Nachweis von OH-Radikalen und NO_2 mit Lasern,
Universität Göttingen, 1991.

Jenkin, M.E.;
Kinetic and spectroscopic studies of peroxy radical reactions related totropospheric
photo-oxidation chemistry,
University of East Anglia, Norwich 1991.

Kinnison, D.J.A.;
Tropospheric chemistry of halogenated organic compounds,
University of Oxford, 1994.

Kirchner, F.;
Kinetische Untersuchungen an Peroxynitraten und Peroxy-Radikalen,
University of Wuppertal, 1994.

Knispel, R.;
Reaktionen von OH-Radikalen mit Aromaten und Folgereaktionen entstandener
OH-Addukte von Aromaten,
University of Hannover, 1993.

Koch, R.;
Kinetische Untersuchung der Folgereaktionen der OH-Addukte von Aromaten mit NO,
NO_2, und O_2 mit simultaner Auswertung von Kurvenscharen,
University of Hannover, 1992.

Ladstätter-Weißenmayer,A.;
Spektroskopische Untersuchungen von phtochemischen Prozessen in der Atmosphäre
unter besonderer Berücksichtigung von Stickstoffverbindungen
Johannes Gutenberg-University of Mainz 1993.

Lançar, I.T.;
Etudes cinétiques de réactions atmosphériques des radicaux BrO et NO$_3$,
Orléans Université, 1993.

Langer, S.;
Tropospheric chemistry of some automotive fuel additives,
University of Göteborg, 1995.

Monks, P.S.;
Kinetics of radical reactions of tropospheric importance,
University of Oxford, 1992.

Noziere, B.;
Cinétique et thermochimie de réactions des radicaux benzyle, allyle, et des radicaux
peroxyles correspondants,
University of Bordeaux I, 1994.

Owen, P.S.;
Radical reactions of troposheric importance,
University of Oxford, 1994.

Parisse, C.;
Contribution à l'etude des problèmes atmosphériques: Spectre d'absorption UV visible
de l'ozone,
University of Reims, 1995.

Parr, A.D.;
Radical reactions of atmospheric importance,
University of Oxford, 1991.

Percival, C.J.;
Laboratory studies of peroxy radical reactions,
University of Oxford, 1995.

Polesello, S.;
Metologie analitiche applicate a problemi ambientali: Distuzione termica di clorofenoli,
comportamento del radicale nitrato in faso gassosa, microinquinanti in corpi idrici,
University of Milan, 1989.

Polzer, J.;
Methodenentwicklung zur Untersuchung atmosphärenrelevanter Spurenstoffe mittels GC/MS,
TH Darmstadt, 1994.

Raber, W.;
Zur Photooxidation einiger atmosphärischer Spurengase in Luft:
Die Carbonylverbindungen Methylethylketon, Methylvinylketon, Methacrolein und Methylglyoxal,
Johannes Gutenberg University of Mainz, 1992.

Sehested, J.;
Atmospheric chemistry of hydrofluorocarbons and hydrochlorofluorocarbons,
University of Copenhagen, 1994.

Semadeni, M.;
Hydroxyl radical reactions with volatile organic compounds under simulated tropospheric conditions; tropospheric lifetimes,
E. T. H. Zurich, 1994.

Seuwen, R.;
Zur Oxidation von Toluol in der Gasphase eingeleitet durch Reaktionen mit Hydroxyl-Radikalen und Chlor-Atomen
Johannes Gutenberg University of Mainz, 1993.

Shallcross, D.;
Laboratory studies of radical reactions,
University of Oxford, 1995.

Skov, H.;
The interaction between nitrogen oxides and the naturally emitted isoprene (and structurally related compounds) in the night-time troposphere,
University of Odense, 1991.

Smith, S.J.;
Atmospheric chemistry: laboratory studies of kinetics of important reactions,
University of Oxford, 1989.

Teton, S.;
Etudes cinétiques de réactions de dégradation atmosphérique par le radical OH de composés organiques volatils (organo-halogénés, éthers et alcohols),
Orléans Université, 1995.

Träubel, M.;
Zur Thermo- und Photochemie von 1,2,3- und 1,2,4-Trioxolanen,
Technical University Braunschweig, 1994.

Vertommen, J.;
   A kinetic-mechanistic investigation of the reactions of CF_3 with CF_3, F and O_2; and of CF_3O_2, i-$C_3H_7O_2$, t- and s-$C_4H_9O_2$ with NO,
   University of Leuven, 1992.

Vogt, R.;
   Gaskinetische und photochemische Untersuchungen am HOCl. Aufbau eines Strömungsexperimentes mit einem Massenspektrometer als Detektor,
   University of Kiel, 1992.

Wängberg, I.;
   An experimental study of some nitrate radical reactions of importance to atmospheric chemistry,
   University of Göteborg, 1993.

Weßenmayer, M.;
   Peroxyradikalmessungen in der bodennahen Grenzschicht über dem Atlantik,
   Johannes Gutenberg-University of Mainz 1994.

Wiesen, E.;
   OH-initiierte Oxidation von p-Xylol und Produktuntersuchungen an ausgewählten Carbonylverbindungen,
   University of Wuppertal, 1995.

Wille, U.;
   Massenspektrometrische Untersuchungen zu Reaktionen des NO_3-Radikals mit ausgewählten ungesättigten Kohlenwasserstoffen,
   University of Kiel, 1993.

Wirtz, K.;
   Kinetische Untersuchung von Bildung und Reaktionen organischer Nitrate bei der Oxidation von reaktiven Kohlenwasserstoffen durch OH in Gegenwart von NO_x,
   University of Wuppertal, 1991.

Zahn, C.;
   Ozonolyse von Butadien und Isopren,
   Johannes Gutenberg University of Mainz, 1992.

Zhu, T.;
   The reactions of NO_3 with olefins and the atmospheric chemistry of their products,
   University of Wuppertal, 1990.

Subject Index

A

absorption cross-sections
  carbonyl compounds 59, 60
  formaldehyde 59
  NO_2 59, 158
  organic nitrates 62
  ozone 57, 158
  PAN 62
adducts
  aromatic OH 247
alkanes 243
alkenes 29, 167, 203, 243
alkoxy radical 54, 149
  production from oxidation of hexyl
    nitrites and hexane 131
alkylperoxy radical 225
aniline 247
Arrhenius parameters
  nitrate radical reactions 236
  ozone with cycloalkenes 220
  reaction of OH radical with aromatic
    compounds 130, 249
  reaction of OH radical with ethers
    130

B

benzene 102, 244, 247
benzyl radical
  reaction with NO 102
  reaction with NO_2 102
  reaction with O_2 102
biogenic compounds 68
branching ratios
  nitrate radical formation 49
  OH radical reaction with alkanes and
    alkenes 82
  OH radical with aromatics 249
  oxidation of alkanes 243

oxidation of alkenes 244
peroxy radical reactions 36
self-reactions of alkylperoxy radicals
  227
toluene 104
xylene 104
butane
  photo-oxidation 76

C

carbonyl compounds 59, 165
carbonyl O oxides 201
CH_3O_2 (methyl peroxy radical) reactions
  210
Cl atom
  reaction with alkyl nitrates 215
  reaction with alkyl nitrites 215
  reaction with nitroalkanes 215
continuous-flow reactor 233
Criegee radicals 201, 210

E

ethene 83
experimental techniques
  chemical amplification 94
  continuous-flow reactor 233
  cycling experiment 253
  detection of 1-hydroxy-
    hydroperoxides 75
  determination of hydroperoxides 74
  discharge flow method 186
  enrichment of hydroperoxides 75
  GC/MS determination of
    hydroperoxides 76
  HPLC flourescence 107
  laser flash photolysis 186, 233
  laser-induced fluorescence 186
  matrix isolation 203